D0154553

A Mind Always in Motion

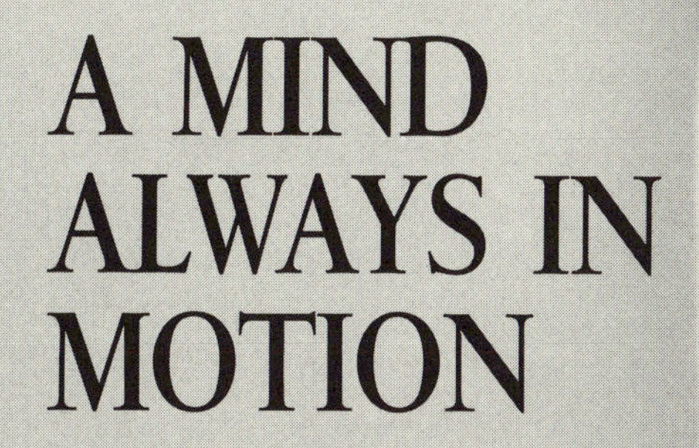

A MIND ALWAYS IN MOTION

THE AUTOBIOGRAPHY OF

EMILIO SEGRÈ

UNIVERSITY OF CALIFORNIA PRESS

Berkeley Los Angeles Oxford

University of California Press
Berkeley and Los Angeles, California

University of California Press, Ltd.
Oxford, England

Library of Congress Cataloging-in-Publication Data
Segrè, Emilio.
A mind always in motion : the autobiography of Emilio Segrè / Emilio Segrè.
p. cm.
Includes bibliographical references and index.
ISBN 0-520-07627-3 (alk. paper)
1. Segrè, Emilio. 2. Physicists—United States—Biography.
I. Title.
QC16.S35A3. 1992
530'.092—dc20 92-10722
[B] CIP

Printed in the United States of America
9 8 7 6 5 4 3 2 1

The paper used in this publication meets the minimum requirements of American National Standard for Information Sciences—Permanence of Paper for Printed Library Materials, ANSI Z39.48-1984. ∞

Contents

	List of Illustrations	vii
	Acknowledgments	ix
	Preface	xi
1.	Chromosomes: Family and Childhood (1905–1917)	1
2.	Discovering the World: Rome and High School (1917–1922)	24
3.	The Education of a Physicist (1922–1928)	37
4.	Scientific Springtime (1928–1936)	57
5.	On My Own: Professor at Palermo (1936–1938)	104
6.	In the New World: Refugee at Berkeley (1938–1943)	131
7.	Los Alamos: The Fateful Mesa (1943–1946)	179
8.	Returns: Science and Struggle, Berkeley and Italy (1946–1950)	207
9.	Ripening Crops (1950–1954)	228
10.	Triumphs and Tragedies (1954–1982)	249
	A Few Words from Rosa	297
	Notes	301
	Index	319

Illustrations

(following p. 130)

1. Emilio Segrè at age three, 1908
2. Notes of an experiment by Emilio, age 7
3. Segrè family on vacation, 1917

4 and 5. Giuseppe Segrè and Amelia Treves Segrè, 1937

6. Emilio in army uniform, circa 1930
7. International Physics Conference, Como, 1927
8. Enrico Fermi, circa 1928
9. Franco Rasetti, Enrico Fermi, and Emilio, 1931
10. Edoardo Amaldi, Rasetti, and Emilio, 1931
11. "The Group of Rome," 1934
12. Elfriede Segrè, circa 1937
13. Copenhagen Physics Conference, 1937
14. M. Stanley Livingston and Ernest O. Lawrence with the 27-inch cyclotron, 1932
15. Dinner at International House, Berkeley, circa 1939
16. Site of early fission experiments: forester's cabin, Los Alamos, 1943
17. Group leaders at Los Alamos, 1943
18. Segrè group at Los Alamos, 1943
19. Niels Bohr, Los Alamos, 1945
20. First atomic explosion, Alamogordo, 1945
21. Fermi and Segrè at Los Alamos, Armistice Day, 1945
22. Segrè family in Berkeley, 1946

23. Emilio, 1952
24. Antiproton research group and apparatus, 1955
25. Periodic table showing Emilio's discoveries
26. Notes on the antiproton experiment, 1955
27. Antiproton Emulsion Star, 1955
28. News of the day at the Lawrence Lab, October 24, 1959
29. Emilio receiving the Nobel Prize, 1959
30. Emilio's Nobel medal
31. Cover of *Time* Magazine, January 2, 1961
32. Glenn Seaborg and Emilio present plutonium sample to the Smithsonian, 1966
33. Segrè family, 1982
34. Rosa and Emilio in Switzerland, 1987
35. Emilio, 1980
36. Emilio doing his chinning exercise, 1981

Acknowledgments

When Emilio, who instructed me on so many things, said that he would leave the publication of his autobiography to me as a "consolazione," his advice was: "Just deliver the latest set of diskettes to the publishers, and they will do the rest." Well, it wasn't that easy, especially for a perfectionist like me, but I must say that the staff I worked with at the University of California Press were much more than just efficient and friendly. None of them had known Emilio personally, and they only got acquainted with him through his manuscript. When, one after another, the editors told me, "Once I started reading, I couldn't put it down," I knew the book was in good hands. Thus, to UC Press Director Jim Clark, to editors Eileen McWilliam, Erika Büky, and Peter Dreyer, and to designer Barbara Jellow, I convey my most heartfelt thanks for their interest and dedication in making Emilio's last book what it is now. To Professor Robert W. Seidel for his conscientious double-checking of the physics, my most profound gratitude. Last but not least, my deepest thanks to Professor Eugene Commins and his wife Ulla for their help and friendship (the saying "A friend in need is a friend indeed" comes to mind whenever I think of them), and to Professor Owen Chamberlain, whose spontaneous reply to my question, "How would you describe Emilio?" became the title of this book.

Rosa M. Segrè
Summer 1992

Preface

I have written this autobiography because I thought it might interest a public curious about the science-dominated period in which I lived. Many other physicists, my contemporaries, have done the same, among them Luis Alvarez, Freeman Dyson, Walter Elsasser, Richard Feynman, Otto Frisch, Werner Heisenberg, Sir Rudolf Peierls, and Bruno Rossi.

Each of them writes from his own point of view and according to his personality. This emerges clearly, for instance, in descriptions of the Los Alamos period; in comparing them, one recognizes the main facts, but the differences of interpretation and the importance assigned to those facts by the authors stand out starkly, as do judgments on persons and events. These differences are interesting and should not be suppressed.

Thus, for example, in reading Peierls's autobiography,[1] which occasionally refers to persons and circumstances I also discuss in this book, I found him to be a much more likeable and gentle person than I am. I am reminded of a remark made to me in the 1960s by the then governor of California, Edmund Gerald "Pat" Brown. He had invited me to a small intimate dinner at his official residence, and during our conversation he questioned me on some points pertaining to the University of California, of which he was an ex officio Regent. I answered as well as I could; he listened, thanked me, and then, laughing, added: "When you speak to a politician you should say those things like this"—and he repeated what I had said in the diplomatic form I should have used. The episode stuck in my mind, but I do not think I have profited from it.

Because the growth of science in this century is such an imposing phenomenon, I believe there is justification for this modest work. It is not just a manifestation of vanity, of gratitude to some, of disaffection with others, or a way of venting my spleen, as one might uncharitably surmise. It is rather a narrative of the life of one of the many scientists who have contributed to the phenomenon. Its appeal may be similar to that offered, in a different context, by the memoirs of one of Napoleon's generals or one of Lincoln's ambassadors.

I have tried to tell the unvarnished truth (as I see it) and to report events the way I believe they occurred, as well as what I felt and thought at the time. I do not like to speak ill of others, and even less of myself, but I have not sought to display manners and tact I never had, and I have tried to treat myself no better than anyone else. I believe it will be clear by the end of the book that scientists are only human.

I thank my friends Renzo de Felice, J. L. Heilbron, Anthony Walsby, Dr. Edgardo Macorini, and the late Avv. Goffredo Roccas, who helped me greatly in various ways. My wife Rosa Mines Segrè has valiantly and patiently helped me in writing this book, through her interest in it, her encouragement, and her criticism. Above all, she kept the old curmudgeon alive.

Chapter One

Chromosomes: Family and Childhood (1905–1917)

Smell of Skunk

... mi dimandò: "Chi fur li maggior tui?"

(... he asked me: "Who were your forebears?")
Dante, *Inferno* 10.42 (trans. Laurence Binyon)

A visitor to the Villa d'Este at Tivoli, near Rome, in May 1914 would have been able to see only its gardens and the reception halls on the first floor of the main building. The rest of the palace was closed to the public. Although it was worth seeing, the owner wanted to save on maintenance and did not want to spend money on custodians for these rooms. They contained frescoes by Taddeo and Federico Zuccari, of the same kind as those in the halls on the first floor. Although not masterpieces, these paintings were valuable. Thus the second floor could be used only by someone who, for some special reason, had privileged access to it.[1]

The visitor might one morning have found there a boy of about nine, wearing shorts. He liked reading; in particular he loved *La scienza per tutti* (Science for everybody), a popular magazine published by Sonzogno in Milan. Its illustrations in color and the many diagrams of machines and apparatus had a special fascination for the boy, whom the visitor might have found intent on reading an account of the working of the automobile engine, with its four phases. The subject was not easy, but in about an hour of concentrated attention he had succeeded in mastering it and, happy to have done so, inscribed it in his memory for life. He then passed on to another article in the magazine, which described the liquefaction of helium by the Dutch physicist Heike Kamerlingh Onnes, who had recently received the Nobel Prize and later discovered superconductivity.[2]

After a while, tired of reading, the boy descended to the garden to play with the gardener's son, his great pal, with whom he had secretly established a small vegetable garden and a tree house hidden by the box hedges that separated the garden paths descending along the slope of the hill. Many fountains and waterworks, as well as peculiar pieces of architecture, embellished the centuries-old park, and the boys had explored every corner of it.

In this peaceful atmosphere, nobody knew that we were on the brink of the first world war—perhaps least of all the owner of the Villa d'Este, Archduke Francis Ferdinand of Austria and Este, whose days were numbered and whose assassination at Sarajevo on June 28, 1914, was to launch the great tragedy that engulfed the world.

In the peace of that morning, Mother called aloud to the little boy from a terrace in the center of the palace, which faced the garden: "Pippi! It is time to go home for lunch." And Pippi, feeling hungry, joined her promptly.

I, Pippi, was born in Tivoli, on January 30, 1905. My father reported my arrival to the civil authorities later than prescribed by law, and to avoid complications, I was registered as having been born on February 1, which became my official birthday. I was called Emilio Gino; according to my mother, the first name reminded her of her great friends Emilia Treves and Emilia Pusterla, about whom more later. The second name honored my uncle Gino, my father's younger brother. However, as a child everybody called me Pippi, a nickname I coined as soon as I started speaking.

My birthplace was a house in a quarter then called Villini Arnaldi. My father, Giuseppe Abramo Segrè, was born in Bozzolo, near Mantua, in northern Italy, on February 2, 1859; my mother Amelia Susanna Treves was born in Florence on July 27, 1867. I was the youngest of three brothers. The eldest, Angelo Marco, was born in 1891 (died in 1969), and the second, Marco Claudio, was born in 1893 (died in 1983); thus my brothers were fourteen and twelve years older than I.

The Segrè Family had lived in Bozzolo for centuries. I believe they originally came from Spain, possibly at the time of the expulsion of the Jews in 1492. All my grandparents died old, but before my birth. The information I have about them comes from what I heard at home or found in documents. My paternal grandfather, Angelo Miracolo, was named Miracolo because his mother was fifty-four at the time of his birth. He was, I believe, a shopkeeper in Bozzolo. I have heard only of his physical strength, health, and gymnastic prowess. When he was about eighty, he retired with his wife to the city of Ancona, where one of his children, Claudio, worked. My grandfather Angelo seems to have liked to scare his family by showing off his acrobatics on the roofs of neighboring houses. He also commented to his children, all very successful in life: "If it were not for the worries you give me, I would live forever." He, his wife, and their son Claudio are all buried in Ancona.

His wife, Egle Cases, was an outstanding woman, both for her brains and for her character. Her children, daughters-in-law, and grandchildren always spoke of her to me with deep affection and high respect. My mother told me repeatedly how much she enjoyed having her mother-in-law in her home at Tivoli, and how she felt closer to her than to her own mother. I have found a subtle echo of this in letters written by my grandmother Egle to her future daughter-in-law, then engaged to my father. The letters, although short, show uncommon understanding and warmth. Egle had received an education well above what was customary for girls of her time. Her home was a center of attraction for the intellectual life, provincial, but not negligible, offered then by a small Italian town.

Angelo and Egle Segrè had four children. The oldest, and the only girl, by name Bice, married a Riccardo Rimini and had four sons. One of them, Enrico, became a professor of organic chemistry at the University of Pavia and is remembered for the Angeli-Rimini reaction of aldehydes. He was the father of Riccardo and Bindo Rimini, my contemporaries and close friends, who will often appear in this story.

The three sons of Angelo and Egle were Claudio (1853–1927), my father Giuseppe (1859–1944), and Gino (1864–1942). They were very close to each other, but not to their sister Bice. Possibly she was a difficult person, as is suggested by the many stories I heard about her. I had only a slight acquaintance with her.

My uncles Claudio and Gino attended the University of Pavia thanks to Collegio Ghislieri fellowships.[3] This college had been created by Pope Pius V at the time of the Counter-Reformation to help talented young men who could not afford an education. With the unification of Italy in the nineteenth century, the laws of earlier times discriminating on grounds of religion had been voided, and Uncle Claudio may have been the first Jew to be admitted to the college.

Claudio obtained his highest degree as an engineer from the University of Turin in 1876; subsequently he went to the Ecole des mines in Paris, where he studied geology. He remained fond of France for the rest of his life, but occasionally (and very loudly) criticized French chauvinism. He still resented unfriendly treatment he had received in the Mont d'Or region during geological field trips. On his return to Italy, in 1881, he went to work for the Ferrovie meridionali, a railroad company, and devoted himself to the application of geology to railroad construction. His personal experience led him to introduce more scientific procedures in the railroad industry, and in 1905 he succeeded in establishing an "experimental institute for the Ferrovie meridionali" in Ancona, which passed to the Italian government on the subsequent nationalization of the railroads. The institute's purpose was to help the railroads by putting the operations of planning, purchasing, building, and testing on a sound technical basis. This simple idea is now commonplace, but in my uncle's time it was new and unusual. Only his enthusiasm and persistence succeeded in carrying it through and in developing an excellent technological laboratory for its implementation.

Uncle Claudio was of short stature and in his later years somewhat deaf. He spoke very loudly and, when angry, tended to use profane language, not permitted at the time in polite company. In 1905 he moved, with his institute, from Ancona to Rome. He settled in a rented apartment in a then-modern building on the corner between Piazza della Chiesa nuova and Corso Vittorio. Uncle Claudio, for his times, was widely traveled, often representing the Italian railroads at international conferences. He told me that at one of these he had met Jules-Henri Poincaré, who, confusing him with the mathematician Corrado Segre (no accent and no relation), was most cordial, changing, however, almost to rudeness when he became aware of his mistake.[4] Claudio had visited all of Europe and Egypt. He never went to America. He used to tell me stories about the czarist Russia of 1908 reminiscent of what I experienced in the Soviet Union in 1956. For instance, he visited the Kremlin and the guide, in return for a suitable tip, allowed him to sit on the throne of the czar, but no tip would buy him an opportunity to take a photo. My uncle also noted that Russia was then, as subsequently, the only country requiring an internal passport.

Since Claudio never married, his nephews—that is, I and my brothers—were the beneficiaries of his paternal instincts. He was a sort of second father to us, someone with whom one could speak about anything, who told us his ideas and was always ready to help us. My brother Angelo often made trouble with some untoward deed, a poorly planned trip, an illness in some strange place, and Uncle Claudio always went to his rescue. It is not by chance that three of his grandnephews bear his name.

When we moved to Rome in 1917, we rented an apartment in the same building as Uncle Claudio, one floor below him, and I was sent to sleep in his apartment. My bedroom's temperature, in winter, was 52° F, but with a good eiderdown it was quite comfortable. In the morning I often had interesting talks with my uncle while showering; furthermore his library, from the Larousse encyclopedia down, nourished my curiosity.

A comical, long-remembered incident occurred around that time. Uncle Claudio's cat, which had disappeared for a few days, possibly wandering the neighboring roofs, reappeared one night in a sorry state, mewing desperately. My uncle let him in and, seeing the condition he was in, said loudly: "Dummy, why do you go looking for adventures? Don't you know that you've been gelded?" Next day a neighbor, highly embarrassed, approached my uncle and said: "Commendatore, you will excuse me, and you can count on my discretion, but I must tell you that last night I inadvertently overheard you dressing down your nephew. I regret that I heard some secrets that are none of my business." My uncle was stunned, but he soon understood his neighbor's confusion and reassured him as to the state of my testicles.

Uncle Claudio had a housekeeper, Annetta, who came from Urbania, in the Marche, and had been brought up by my grandmother. She was practically a member of our family, always ready to help in case of trouble. She died in my uncle's home and was replaced by her niece.

My Uncle Gino was a professor of Roman law. He had followed the usual career of an Italian university professor, passing from less important schools to more famous ones: Camerino, Sassari, Parma, and, ultimately, Turin. In his profession he had an international reputation, as I can confirm from an incident that occurred to me in Berkeley almost twenty years after his death. One rainy night, I was in the Faculty Club at the University of California and chanced to hear a distinguished gentleman who was Regius professor of law at Cambridge University in England telephoning a Berkeley law professor who was a friend of mine to apologize for being late for dinner because the weather made it difficult to find transportation. I introduced myself and volunteered to take him to my friend's home. He thanked me and said: "Ah! So you are the famous Segrè!" I felt flattered and assented, but he looked at me in a strange way, and I understood at once what he was thinking. For him, the famous Segrè was my uncle, and the ages did not jibe, because my uncle would have been about 100 years old. The worthy gentleman was disappointed, as Poincaré had been with my uncle Claudio.

Gino Segrè was unassuming and even timid, a disposition by reason of which he remained a little less famous, less well paid, and less honored than he deserved and wished to be. He often wrote important legal opinions for famous lawyers, who praised him highly, but paid him little, while they used his work in their briefs and collected fat fees themselves. My father successfully employed his brother in some difficult legal cases and always chided him about his low fees. A high school in Turin is named after Gino Segrè.

One's first impression of him was of a somewhat short man, lean, with an aquiline nose, luminous, impressive blue eyes, a high forehead, and a small blond, singed mustache, with half a cigarette in his mouth. I say "half" because he cut his cigarettes in two before smoking them and thus always burned his mustache. He had a strong constitution and liked gymnastics. In his familiar conversation you could hear the traces of his native dialect. Of course, in lecturing he used standard Italian, but when there was something he did not like he reverted to dialect to say, "Pias mia" (I don't like it).

When he came to Rome from Turin in the course of his official duties, his arrival was a joyous occasion for his two brothers, his sister-in-law, and his nephews. He had a room of his own in our house, with a bookcase containing the Corpus juris civilis and other ponderous Roman law texts in Latin. He could not live without them.

My uncle was frequently summoned to Rome, because both his colleagues and the government had very early discovered that when there was difficult work to do, requiring steadfast application, fairness, and acumen, and especially if it was not paid, Gino Segrè was the person to call for. His presence guaranteed success. Thus, for instance, although it was known that he was cool to the Fascist regime, the government entrusted him with heavy burdens in reforming the Italian Civil Code. He would explain to me, a young boy, the rationale for many sections of the Civil Code—for instance, the section on inheritance—and why it was fair to write the law in a certain way rather than in another.

As a boy, I lacked any special interest in the law or in history, and most of the dead classicism we learned in high school seemed a boring waste of time to me. Not so walks in the Roman Forum with my uncle. His detailed explanations of family relations in ancient Rome, their Latin names and legal implications, were sometimes a little ponderous, but he truly enlivened Roman ruins, inscriptions, and statues with the deep knowledge and familiarity of a person who had really mastered their history. To go with him to the Forum was like taking a walk with a learned ancient Roman bent on introducing me to his great city. "So and so did this, so and so did that," Uncle Gino would explain, pointing out their monuments or inscriptions. "They were related to each other thus, and these were their motives and interests." I have since had the fortune of meeting other great minds, but the first outstanding scientific personality I encountered was my uncle Gino. From him I had my first impressions of what it means to work with one's head, with absolute honesty, patience, stamina, precision, and devotion to the subject matter. These qualities, together with imagination and analytical ability, are among the requirements for any scientific enterprise.

Although he knew German well and was culturally close to his German colleagues and German science, he always kept a watchful, balanced detachment from German culture. He disliked narrow Italian nationalism as much as he did the subservient admiration for Germany that prevailed in Italy between 1910 and 1940. In politics he was a liberal, and a laicist, essentially in the traditions of the Risorgimento; he had a clear premonition of the future of Fascism, and of its consequences for Italy.

My uncles Claudio and Gino were well known in the Italian intelligentsia. Both were members of the highly selective Accademia nazionale dei Lincei, the Italian national academy, a fact especially remarkable in the case of my uncle Claudio, who had no academic connections.

My father, Giuseppe, never went to college. After finishing high school, he left Bozzolo and moved to the town of Urbania, in the Marche region in Central Italy. There he became an assistant to Count

Mattei, the administrator of the properties of the historical family Albani. My father was then about eighteen. At first he learned papermaking and ceramic techniques and, above all, acquired business experience. Later he worked for the famous Ginori ceramics works and for other manufacturers in Civita Castellana, not far from Rome.

In the 1880s, Tivoli's famous waterfalls, which had been painted by artists for centuries, were becoming important sources of power, on a very different scale from the small medieval mills of earlier years. The Società per le forze idrauliche ad usi industriali ed agricoli, a corporation devoted to the utilization of the waters of the river Aniene for power and irrigation, hired the young Giuseppe Segrè as an assistant to its general manager, and when the latter died, my father replaced him. He moved to Tivoli and devoted the corporation primarily to papermaking and the generation of hydroelectric power. The second was limited in scope, and that side of the business was ultimately sold to other companies, but I nonetheless remember spending many hours as a child in the generating plant, where the foreman tried to explain the workings of the generators and transformers to me. Unfortunately, I could never understand him. The foreman had mastered his trade well, but he had had only the most rudimentary formal education and could not communicate his ideas. In this he perhaps resembled, albeit in a modest sense, those great nineteenth-century physicists who ignored formal mathematics.

Papermaking became and remained my father's principal concern for the rest of his life. Slowly he increased his share in the mill. He first leased it through a partnership, in which he was the general partner. Later he transformed the partnership into a corporation, the Società cartiere tiburtine (SCT), of which he was a minority shareholder. In the course of time, he bought out the other shareholders, ultimately becoming the sole stockholder. He devoted years to disentangling relations between the concessionaries of the water rights of the river Aniene. Some of these rights went back to the Middle Ages and formed an extremely complicated legal and technical complex. In 1909 my father and others, with great patience and skill succeeded in persuading all the interested parties to come to an agreement that clarified and settled the situation in modern terms. My father was proud of this achievement, by which I expect he gained some water rights for his company.

My father also took upon himself an unpaid minor burden: the administration of the historic Villa d'Este, whose owner, Archduke Francis Ferdinand, never visited it. The archduke made it a condition that the property should not cost him anything. My father felt a lively responsibility for the preservation of the buildings and fountains, but could raise only very little money for this purpose, mostly from entrance fees. Nevertheless he accomplished the task, with the help of some willing artists, who contributed their work, and of a small but devoted and hard-working band of gardeners and artisans.

My mother's family, the Treves, were from Vercelli, in Piedmont, where my grandfather Marco was born in 1814. In search of better treatment of Jews, he migrated to Florence, where he studied architecture and married. After losing his wife and an infant son, he worked for a time in Paris under Eugène Viollet-le-Duc, Napoleon III's architect. The widower then married my grandmother Elisa Orvieto, who was from a Florentine family, and in 1857 the two settled in Florence. My grandfather's most important architectural work was the Florence synagogue, built by him and the architects Mariano Falcini and Vincenzo Micheli around 1882.[5]

My mother Amelia, born on July 27, 1867, was the youngest and favorite daughter of her austere father. Her formal schooling was limited in its subject matter but not superficial. She learned English well enough to read English novels and other books. In our home at Tivoli there was a good library, mostly literary, in Italian, French, and English.

The Treveses of my grandfather's generation were practicing Jews, and even the next generation would go to the synagogue at least on major holidays. The Segrès on the other hand did not observe any rites, and my mother, after her marriage, abandoned any formal religious practice. She told me only a minimum of biblical stories and the central Jewish prayer, with its monotheistic credo. However, if I was in Florence during some religious festivity, I participated in it with my uncle and cousins. I once was there for Yom Kippur when I was about twelve. On that solemn holiday, one is supposed to fast for twenty-four hours. We all went to the synagogue in the morning. Toward noon I left and, feeling hungry, went to a restaurant to eat. Lo and behold, I found my uncle Guido there! I thought he would pay for my lunch, but he was not pleased to see me and did not appreciate God's joke.

For the purposes of this account, my mother's most important siblings were Jacopo (1860–1912) and Guido (1864–1964).[6] Jacopo evidently had exceptional mathematical talent—at least this is what the famous mathematician Vito Volterra, who had been his friend and schoolmate, told me more than once. However, Jacopo died of syphilis he contracted on a trip around the world undertaken with the poet Angiolo Orvieto, his friend and cousin. He left a modest fortune to his nephews but, in drawing up his will, forgot me. My uncle Claudio was displeased by this omission and told me that he would leave me a special legacy as compensation, which in due course he did. He had not, however, reckoned with his executor, my father. Saying that Uncle Gino's three daughters, Egle, Bice, and Fausta, needed this inheritance more than I did, Father gave them my uncle's whole estate, including my special legacy. All this was done with my agreement, although, knowing my father, I believe that he would have done the same even without it.

My mother's other brother, Guido, studied law. After modest beginnings, he was soon involved in important real estate transactions on behalf of rich relatives. In a major building crisis in 1885, he negotiated default settlements between German capitalists and Italian contractors. He thus started a career as a financial expert that was to take him very far. Eventually, he became president of La Fondiaria, a major insurance company. He died a few days short of his hundredth birthday, and when over ninety was still actively presiding over the board of directors of the Fondiaria.

Guido Treves's wife, Emilia Finzi, who was a very close friend of my mother's, and greatly influenced her, died in 1922 of encephalitis. Guido and Emilia had four children, Silvia (Levi Vidale), Marcella, Marco, and Giuliana (Artom); these cousins were very close to me during my early years.

My father must have met Guido Treves on business. Father, who was then twenty-nine, was looking for a suitable bride, and Guido introduced him to his sister Amelia. They must have been favorably impressed with each other, because soon they were engaged.

My father traveled a great deal on business; his letters mention Pesaro, Pisa, Milan, Naples, Florence, and frequent commuting between Rome and Tivoli. Probably they deeply impressed my mother, who had lived in a protected and rather closed circle. The letters may also have frightened her slightly, because they intimated that the impending marriage would bring great changes in her habits and life-style. My father describes the comfortable and attractive home he was preparing for his bride (it even had running water in the kitchen!), the beauty of the location, and the view of the Roman Campagna.

My parents were married on July 7, 1889. Immediately after the wedding, my father brought his bride to the Villini Arnaldi quarter of Tivoli, where he leased a villa called Villino B Maria, which he eventually bought in 1920.

The house had three floors; for many years we occupied the first floor and some rooms on the top floor. Later, we took over the whole house. In my childhood we lived in the first floor. My father converted half of a large terrace into a bathroom, next to the master bedroom, with a zinc bathtub and primitive but adequate washing facilities. If bathrooms in Tivoli were not common, the city was among the first in the world to acquire electric light, with carbon filament bulbs that gave a reddish hue; at the time, they still used acetylene lamps in my uncle's house near Florence.

The Tivoli of my childhood was very different from the present-day city. Around 1915, Tivoli had a population of about sixteen thousand, confined in a town that had not changed very much since the Middle Ages. We lived in a new development, consisting of about a dozen houses on the slope of Mount Ripoli, outside of town, at the start of a country road that followed the hills at mid level among the olive groves, leading to villages and farms a few miles away. Between the houses there was a rustic park, well laid out with olives, plane trees, lilacs, and acacias. The three lowest houses, one of which was ours, overlooked the Viale Carciano, with a superb view of the Roman Campagna, at that time largely wild. The dome of St. Peter's in Rome was a mere bubble on the horizon. In between, the plain was crossed by the Aniene River, whose course was flagged by a prominent landmark: the large cylindrical mausoleum of the Plautii, dating from the early Roman empire. About a mile to the left of the river, one saw the dark cypress trees and ruins of the emperor Hadrian's villa. On the right side of the plain, there were two large hills, crowned by the Sabine villages of Montecelio and St. Angelo. Closer by, the eye rested on the silver gray of the olive trees covering the hill on which Tivoli itself was built. It was a stunning view, which changed vastly according to the hour, season, and weather conditions.

The Viale below our house was shaded by big elm trees, and in the fall the elm leaves formed a golden rug on the roadway. The wind made them swirl and sometimes heaped them up before blowing them away. I can still in my imagination smell the dry leaves and fresh rain. The stretch used by the people of Tivoli for the traditional promenade was still rustic, without sidewalks. Almost exactly below our house there was a small chapel with a Madonna, to whom the people of Tivoli addressed prayers. I remember crowds of shawl-wrapped women imploring the Madonna to keep Italy out of World War I. Soon thereafter the avenue was used for basic training of recruits destined for that war. Many of the soldiers had white handkerchiefs tied to their right arms to help them distinguish right from left.

At home the cooking was done on charcoal stoves. The food was very simple, but very good. Boiled meat, roasts, vegetable soups, peppers, and all greens in season: tomatoes, endive, eggplants, chicory, string beans, zucchini, squash, peas, and many other vegetables. The excellent bread, baked at home once a week, was dark; in addition we ate rice and a little homemade pasta. For the holidays, both Catholic and Jewish, there were other delicious dishes, often of Jewish origin, passed down through Uncle Claudio's housekeeper Annetta from my paternal grandmother. The coffee was always very weak because my mother liked it so: our home was notorious for it. All told, the food was not very varied, but wholesome and tasty. When I was about twenty and stayed at Tivoli either alone or with some friend to prepare for exams, the old caretaker, a maid trained by my mother, prepared the same fare for us: broth, boiled beef, roasted peppers, fruit. We loved it.

Many of the things we buy in shops nowadays were made to order: shoes by the shoemaker, my clothes by a seamstress, and so on. Needless to say, there were no automobiles; when we needed to, we rented a horse-drawn carriage.

Tivoli was linked to Rome by the railroad, which was used by Uncle Claudio, who visited us almost every weekend, and by a steam tramway, which was a little more rapid and convenient. The tram ran from Porta San Lorenzo in Rome and followed the Via Tiburtina, reaching the terminus in Tivoli, a five-minute walk from our house, in about an hour. The last two miles were quite steep, and the engine made loud noises.

Today (1987) the landscape has been devastated. The destruction is appalling: carelessness, speculative greed, and plain incompetence have destroyed most of the beauty of the place.

At the time of my birth, in 1905, the success of the paper mill was established and our family was, if not rich, more than comfortable. My parents had started traveling abroad and, among other trips, had been to England for the coronation of King Edward VII as guests of a London customer who bought cigarette paper from my father.

My first recollections go back to 1908: a red belt, certain striped socks, the Kodak camera of my brother Marco, a Japanese costume given to me by my uncle Jacopo Treves. Around that time my parents hired an Austrian nanny for me. She had a beard and had had an unlucky love affair with her brother-in-law, who belonged to an elite Austrian Alpine regiment. After lunch she repaired to her room on the upper floor of our house; she smoked strong cigars and sometimes drank cognac in her room. She loved me dearly, and her affection was fully reciprocated; she taught me many things, taking them from an illustrated encyclopedia, whose pictures, including those of tortures, occupied me for many hours. This Nanny, whom I called Tata, took me to the public gardens to play, and if anyone tried to kiss me, she would say severely, "Non si paciano i pampini" (One does not kiss children) with a strong Austrian accent. From her I learned German for the first time, but later I forgot it.

When I was about five years old, my mother taught me to read, and shortly afterward, my parents hired a young teacher for my private instruction. Signorina Maggini had just graduated from a teachers' training college, and I was her first pupil. She taught me with great enthusiasm and according to the latest educational theories she had learned at school. Besides reading and writing and the other usual subjects of the first grades (she was not too demanding on the Pythagorean table), she often took me hiking in the hills behind Tivoli. She would buy a one-penny tablet of Tobler chocolate, which had pictures I collected, and then we walked for a couple of hours in the hills. During those walks she taught me history, natural history, poetry, civics, and so on. I had an excellent memory and greatly enjoyed learning things such as the physiology of digestion, illustrated by the experiment of chewing a piece of bread until it became sweet through the action of the enzyme ptyalin on starch. I believe I was a rather extraordinary pupil, but since she had no experience and did not know what to expect of a child of six, she attributed everything to "family background." Later, with my own children, I often recalled the teachings of Signorina Maggini, with whom I remained friends until her death in 1971.

At the same time, my parents sent me to public school, mainly so that I would have the society of other children. It took me five minutes to walk from home to school, and I went alone from the very first. At the beginning of the town, the road crossed a pass often swept in winter by a bitterly cold wind. I would wrap myself in my hooded cloak and run as fast as I could past the critical spot.

Family strolls on the Carciano road were a firm habit; during the winter, my parents and I invariably used to take a walk there from about 2 to 4 P.M. On these strolls, when it was cold, as it usually was in winter, my mother wore a skunk fur and muff that preserved a slight skunk scent, which I liked and associated with her. My nose is thus imprinted on the skunk odor, and I still like it. Very often we were joined in our walks by Count Luigi Pusterla and his wife Emilia, my parents' closest friends in Tivoli. He was a handsome old gentleman with a white beard, a painter by profession. Count and Countess Pusterla lived in an eighteenth-century family palazzo in the center of Tivoli, but had very little money; the count worked as an agent of the Italian State Lottery. He jokingly called himself a "seller of nonsense on behalf of the state." In his palace he had beautiful old furniture, a good library, and a great number of rooms, which he had covered with frescoes illustrating Garibaldi's deeds. As was customary in the eighteenth century, all the rooms of the palace were in a row, allowing me to run from one end to the other. I regularly tripped at each threshold, and Pusterla took pity on me and ordered all the thresholds to be leveled so that I would not fall. When years later I saw the palace of the great poet Leopardi in Recanati, I was struck by its resemblance to the Pusterla palace, extending even to the books in the library. When I visited them, Count Pusterla would often make pencil drawings for me and help me to color them, to my great delight. He had a good classical education and spoke so much of the ancient Romans that I, seeing his white beard, asked him if he had lived in those days.

Pusterla was a liberal nourished in the ideas of the Risorgimento, the political movement that brought about the liberalization and unification of Italy. He was against any secular activity by the Catholic Church, admired Garibaldi and his movement, and had great faith in progress, education, and the future. He had introduced my parents to the idea, if not to the practice, of mountaineering and in general to love of the outdoors and of natural beauty. My parents and the Pusterlas shared a deep and devoted friendship; the Pusterlas often came for dinner at our house, and we saw each other almost daily. Luigi Pusterla was fortunate in dying shortly before the beginning of World War I; his widow survived him for many years and remained intimate with my mother.

Another member of the small world of Tivoli was Dr. Natale Allegri, an old-fashioned physician who knew more Latin than medicine, but treated his patients with great devotion and goodness of heart. He was a generous soul and secretly helped the poor, although he was poor

himself. He, too, was often a welcome dinner guest and always ate two eggs sunny-side up. He died during the influenza epidemic of 1917 and was universally mourned. Dr. Allegri cured our minor ailments. If there was something serious, we consulted Dr. Parrozzani, the chief of staff of the Tivoli Hospital, and a superior surgeon, who remained all his life in Tivoli because of an unhappy family situation. Dr. Allegri gave me a small tortoise, which I called Crocrò and tamed so that it would "run" to eat salad or cherries out of my hand. Crocrò used to hibernate underground in some flowerpot on our terrace.

Tivoli also had a national college and a *ginnasio liceo* (classical high school). Several of the professors in this high school were distinguished in their field. The botanist Lino Vaccari, an authority on Alpine plants, is still quoted in current literature. The teacher of Italian literature, Chiarini, had been a pupil of the important poet Giosuè Carducci and was well known as a critic. The musicologist Radiciotti (grandfather of the distinguished physicist Marcello Conversi, who was born at Tivoli in 1917) was an internationally recognized authority on Rossini. These people were not isolated from the rest of the world. They read and talked about current literature, and had relations with artists, some famous, who came to Tivoli. If, in the kitchen of the Pusterla palace, there was an inscription on the hearth "Vivitur exiguo melius" (One lives better with little), perhaps a consolation for the frugality, not totally voluntary, of the meals, in their parlor they had a beautiful concert piano, on which Liszt had played during his vacations, when he spent the summer at the Villa d'Este as a guest of Cardinal Hohenlohe. The poet Gabriele d'Annunzio, the painter Michetti, and the sculptor Costantino Barbella formed a trio of artists originating from the Abruzzi region who frequented Tivoli. The painters Ettore Roesler Franz,[7] Onorato Carlandi, and others from foreign countries often came to Tivoli, and my parents knew most of them. The fact that my father was in charge of the Villa d'Este helped us make contact, because the artists often worked there.

During the summer, my mother and some friends spent the mornings sewing in the shade on the central balcony of the villa, which had a magnificent view of its park and of the Campagna Romana. I too played in the park of the villa, together with the son of the gardener. Once my friend and I found out that a movie company had prepared a great scene involving fireworks for some film they were shooting. To signal when to let off the fireworks, they had hidden a Bengal light behind a tree, which we set off—needless to say at the wrong time, but innocently enough, I believe—with consequences that may easily be imagined. This was not my only adventure with fireworks. When I was about ten years old, I stuffed some sulfur, potassium chlorate, and charcoal into a bamboo cane and lit it. Luckily, I was not killed, but the tremendous explosion ended my playing with fireworks.

During the summer, I often went with Signorina Maggini to Acque Albule, a sulfur spa about eight miles from Tivoli, on the road to Rome. However, I learned to swim only when I was about twelve years old, at the seashore. At Acque Albule, I met and made friends with a young Australian priest, John Leyden, who gave me a most interesting book in English called *The Handy Boy*, which taught me how to build toys, airplane models, telegraphic apparatus, and so on. He also gave me a wonderful French book, *La Bannière bleue*, which introduced me to the exotic and fantastic world of the Mongols (my mother had started teaching me French shortly after my seventh birthday). After that I read and reread Marco Polo's *Il milione*. I spent pleasant hours with my friend on the slopes of Monte Ripoli; I believe my parents feared that he would try to convert me to Christianity, but, as far as I remember, the excellent young man never mentioned religion to me.

During the hottest part of the year, we went to the seashore at Viareggio or Forte dei Marmi, where I often found my Segrè or Treves cousins (but very seldom both together). Giuliano Bonfante, the son of a well-known professor of Roman law who was a friend and rival of my uncle Gino's, was my unpleasant playmate. We all lived in a pension in Viareggio, then still rustic and dominated by a famous and then-flourishing pine wood. Once in a while the composer Puccini would appear with his motorboat, and we knew that d'Annunzio had a villa nearby (until he was compelled to flee his creditors). We once went to visit Marconi's radio station at Coltano, and I still remember the noise of the sparks and the appearance of the complex electrical transmitter. Uncle Claudio visited us once in a while for short periods. Otherwise we followed the ordinary seashore routine: bathing, hikes, building canoes or castles of sand, and bellyfuls of grapes. Here my cousin Fausta taught me how to swim; with her I built many sand boats, which we used for imaginary travel.

In September we would move to Marignolle, the Treves villa, near Florence. It is a large building of medieval origin, with sizeable landholdings, then cultivated by sharecroppers. Olive oil, wine, vegetables, fruit, and some wheat were the main crops. There was also a handsome Italian garden in the grounds of the villa proper. The families Finzi and Treves, each with a large number of boys and girls of similar ages and more or less related to each other, occupied the villa. Once in a while, my parents would go for a vacation abroad and park me at the villa. When I was very young I suffered greatly being separated from my mother. My cousins Silvia and Marcella, who were about twelve years older than I was, tried unsuccessfully to take care of me. They were too prim and Victorian for a slightly wild child. Fortunately for me, I had never had an English governess like the Treves children. Despite their strict upbringing, my older cousins once organized a wonderful game. We smaller children, who were about twelve years old, had created a postal system for ourselves. It had its little letters, stamps, deliveries, and so on, and we enjoyed writing to each other. At a certain point, we started receiving mysterious communications commanding us to collect various objects and to bring them to preassigned places, with injunctions of strict secrecy. One night the messages, which we had scrupulously kept secret, called us to a fishpond that was in the territory of the villa. We all arrived there after overcoming several obstacles, such as masked enemies opposing our progress, and at the appointed place we found a feast, with fireworks, organized by the older girls, Silvia and Marcella.

I am unable to identify the earliest origins of my interest in physics. My first memories having some connection with physics have to do with tools and a camera belonging to my brothers. I called the camera a "*mappa sciafa*" (the correct term is *màcchina fotogràfica*), because I did not yet speak well. Later, at home, I must have heard talk about scientific or technical subjects, and as soon as I learned to read I got hold of books about science such as Tissandier's *Le ricreazioni scientifiche* (Scientific recreations)[8] and popular scientific magazines. I still have a notebook dated March 27, 1912, entitled "Physics," in which, in the handwriting of a seven-year-old boy, and with some misspellings, I describe the simple experiments I performed, possibly having read of them in Tissandier. My mother helped me in drawing the figures with which I illustrated the notebook. Colors such as those produced by the refraction of sunlight in a pitcher of water especially fascinated me.

A little later, my brother Marco, who was preparing for a chemistry examination at the university, bought chemicals at a local drugstore and repeated many of the experiments mentioned in his textbook at home. He allowed me to watch him, and I was completely enthralled by the color changes that I saw happening in his test tubes.

Uncle Claudio took me to visit his own institute, where for the first time I saw a real scientific laboratory, with all kinds of apparatus. My uncle also gave me an old physics text by A. Ganot, printed in 1863, in which, among other subjects, I found mention of "the recent experiments by Mr. Faraday."[9] This book became my constant companion, together with the history of France by Victor Duruy, which was given to me by my mother. A few years later, on November 21, 1916, seeing my interest in physics, Uncle Claudio gave me a 1913 French edition of Ganot's book.[10] On the flyleaf he wrote: "To my beloved nephew, with the wish that soon physics will serve the arts of Peace, Uncle Claudio." At the Pusterla house I also found a book by J. B. Dumas, from which I learned the composition of, and the difference between, sulphites, sulphides, and sulphates and similar facts of inorganic chemistry. I also admiringly read Faraday's *Chemical History of a Candle.*

I have a book, "For when I shall be grown up," in which one was supposed to write answers to questions about oneself. In it, I wrote as my wish for the future that I wanted to become a physico-chemist and die at thirty in the explosion of my laboratory.

All told, I had a happy childhood and was much cared for by my mother (less so by my father, who did not pay much attention to me). Needless to say, I also experienced my share of childhood scares and terrors. One day I read the vivid description of the famous plague in Alessandro Manzoni's classic novel *I promessi sposi.* As usual, I slept alone on the upper floor of our Tivoli house and, in the darkness, I was overwhelmed by fear of catching the plague. I was reluctant to get up and go to my parents on the lower floor to tell them of my fears, but finally I did. They calmed me down without making fun of me in any way.

From Tivoli I was brought once in a while to Rome. On these occasions we lunched at Uncle Claudio's, returning home in the evening. I remember the 1911 Exposition commemorating fifty years of Italian unity and its interesting regional ethnographic exhibits, as well as flying airplanes. On one of these trips, my parents took me to a puppet show, which delighted me, and for days I kept imitating Caliban. Without knowing it, I had seen a Podrecca production of Shakespeare's *The Tempest*, a classic still famous in puppet art, which revealed all the splendor of the original play.

When I was ten years old, I fell seriously ill with scarlet fever, complicated by nephritis. I was in mortal danger, and my mother took care of me day and night, with the help of Annetta. At the time there were no specific treatments for streptococcal infections. I was kept in bed for a couple of months on a milk diet. While I was ill my mother read me several books by Jules Verne, and Uncle Claudio gave me a mineral collection and a Brownie camera, which I learned to use, developing and printing my own pictures. An engineer who often worked for the paper mill gave me a Ruhmkorff coil, with which I performed many experiments as soon as I recovered. I also built myself a galvanometer, batteries, and other electrical apparatus.

After my bout with scarlet fever I remained susceptible to serious allergies, in which my skin peeled off and I had other symptoms closely resembling those of scarlet fever, except that I did not get nephritis (I checked by myself, testing my urine for albumin). This recurring "scarlet fever" came back almost every year until 1926, after which it disappeared. It was peculiar enough to have me reported in the medical literature. The last time I got it, I was skiing at Clavières and the doctors wanted to isolate me as contagious; later the professor at the medical school in Turin sent me to his former teacher, Frugoni, in Florence, and the latter's assistant, Giacomo Ancona, took my history. Many years later, Ancona and I met again in California as refugees and became very close friends. In 1948 I retrieved my medical history from Frugoni's archives; he had invited me to dinner in Rome, and when I reminded him that he had visited me in 1926, he found my papers in a couple of minutes!

I finished elementary school flunking Italian composition, but passed on a second try and entered the *ginnasio* at Tivoli. Of that school I remember an odd teacher of mathematics; he used to walk all alone on the Viale Carciano dressed in a morning coat, speaking to himself. This fellow told me that I did not understand any mathematics and gave me a flunking grade. I worried about it, but my parents sent me to a private tutor who was an excellent mathematics teacher. He gave me a few lessons and taught me the fundamental rule that a fraction does not change on multiplying numerator and denominator by the same number. Thus in a couple of hours he fixed my mathematics and told me that I did not need further coaching. As a bonus he taught me a little game in which each of two players alternately names a number between one and ten; the numbers are added together, and whoever succeeds in reaching 100 first wins. This teacher encouraged me and was of real help to me.

As my parents' youngest child by several years, I believe I was treated differently from my brothers. At the time of my birth, my father was forty-six, my mother thirty-seven. This must have influenced their attitude toward me. As a child, of course, I did not see this, but thinking it over now, it seems obvious to me.

In 1917, during the war, we moved from Tivoli to Rome, which produced a great change in our family's way of life. Business increasingly required my father's presence in Rome, and the day-to-day management of the paper mill could be entrusted to a technical director. Traveling back and forth between Tivoli and Rome was tiring, and my father suffered an angina attack. Scared, he consulted a noted doctor in Rome, who told him to set his affairs in order because he might die any moment. (My father lived over thirty years longer and, with a certain perverse glee, attended the funeral of the doctor he had consulted.) The physicians in Tivoli, Dr. Allegri and others, tried to minimize the importance of the episode, and they were clearly right. However, objectively, the center of my father's work had shifted from Tivoli to Rome.

Just then, a cardinal who lived on the floor below Uncle Claudio died; the apartment he left was large enough both for our family and for the business office of the paper mill. We had already spent a few months at a pension in Rome with a view to moving there permanently, and this finally decided us.

Both my Mother and I deeply regretted leaving Tivoli. For my part, I hated leaving my open air games and my old friends, as well as the space we had at home in Tivoli for my experiments. But the move to Rome, the end of World War I, and the new high school I now began to attend signaled the end of my childhood. I was twelve years old, and many things in me had started to change.

My mother had great difficulty in getting adjusted to life in Rome. All her friends were in Tivoli, where she had lived for almost thirty years. Life there suited her; she loved the freedom of the country. To cheer her up, my father kept the lease of our Tivoli house (later he bought it outright), and as a consolation we spent long periods in the spring and fall there. Perhaps it is not without significance, however, that there is a street in Tivoli named after Amelia Segrè, in part because of her tragic end as a martyr of the Nazis, in part because of the fond memories she left, but not one named after my father, Giuseppe Segrè, who did so much for the welfare of the city.

Chapter Two

Discovering the World: Rome and High School (1917–1922)

Scent of Florentine Wisteria

Che pensieri soavi,
Che speranze, che cori, o Silvia mia!
Quale allor ci apparia
La vita umana e il fato!

(What tender thoughts,
What hopes, what hearts, O Silvia mine!
How human life and fate
Seemed to us then!)
Giacomo Leopardi, "A Silvia" (trans. Arturo Vivante)

The move to Rome signaled the start of a second period of my life. I was no longer a child, and the change in residence happened to coincide with the onset of puberty. New feelings, new interests came to the fore. I started to see a wider world, to appreciate poetry, to recognize the beauty of intellectual constructions. It was a cloudy, not a happy period. I was confronted with new, seemingly dreadful problems, and I did not know how to cope with them or whom to turn to for help.

At Rome I was enrolled in the Ginnasio Mamiani, located in a palace next door to our home on Corso Vittorio. I hardly remember the teachers of the early classes. Soon I started taking some extra books in which I was interested with me to school in order to have something to read if classes became too boring. Usually the teachers let it pass, provided I did not disturb anybody. I got hold of a book on elementary geometry and amused myself in solving its problems, more or less as I would have solved crossword puzzles.

As usual, we spent the summer of 1918 in Tivoli; in the fall, the dreadful influenza epidemics of the previous year recurred; I, however, had already had the disease in 1917, without knowing what it was. We lingered in Tivoli, and in order not to waste too much time, I started translating Ovid's *Metamorphoses*, which was required reading for the coming school year, on my own. In that solitary fall at Tivoli, under the influence of this strange text, I felt poetic emotions for the first time. It was a period of deep upsets, certainly connected with puberty, and it left its marks on me.

All told, I remember my *ginnasio* years as rather boring. I did not learn much Latin or Greek. I was taught mathematics with a misplaced rigor, under the influence of the great mathematician Giuseppe Peano, but without adequate practical exercises. As my studies progressed, however, the teachers became of better quality. After five years I passed from *ginnasio* to *liceo*, where I spent three more years.

In the *liceo*, a Professor Rua tried to teach us, through sparse but appropriate remarks, what it meant to write well in Italian (and perhaps in any other language). He debunked the empty rhetoric of several of my schoolmates, derived from misguided imitation of writers such as the poet Giosuè Carducci (1835–1907) and d'Annunzio. He told us that if we succeeded in having one or two ideas and in explaining them clearly and concisely, we would write good essays and get high marks. Attempts to substitute even modest thoughts with empty words would be poor writing and earn low grades. Once I learned these simple rules, I started a small essay factory, not only for myself, but also for my cousins and friends. I produced them without effort and with good success. Professor Rua also said that had we been good writers, our styles would have resembled those of the great authors we studied. To me he assigned Leopardi.

The Latin and Greek teachers succeeded only in boring and disgusting me with their subjects. The professor of French, M. Grimod, a native Frenchman, was excellent, however, and seeing that unlike my schoolmates, I knew the language, he let me get acquainted with French literature without bothering to teach me all the subtleties of French grammar. An excellent anthology he had compiled helped me greatly to appreciate French writers.

Of history we learned only dates, without any sense.

Professor Monti in physics inculcated in me $F = ma$ until I really understood its meaning. He thought that one should teach few notions, but thoroughly. This was a very healthy attitude, which did not prevent him from explaining even some relativity. Einstein was fashionable in those days.

The mathematics professor drove me crazy with Dedekind cuts. I learned rigorous proofs of seemingly obvious things, useless at my level, and at a time when with a little effort, I could have learned calculus, which would have been invaluable to me. On my own I read sections of Enriques's *Collectanea,* an encyclopedia of elementary mathematics seen from a higher point of view, and some number theory. I regret the effort spent in those years, so important for learning, on non-Euclidean geometry, number theory, and other subjects, completely omitting applied mathematics. At home there were books on analytic geometry, algebra, and calculus that had been used as texts by my brothers at engineering school, but my brother Marco locked them up and forbade me to use them, on the pretext that they would "tire my head." In fact, he wanted to remain the only one at home to know "*Il càlcolo sublime,*" as he called simple infinitesimal calculus.

Being so much older than I was, my brothers Angelo and Marco virtually belonged to a different generation and were already at the university when I was barely learning to read. Angelo had been a difficult child, and Uncle Claudio frequently recounted his deeds. For instance, in a railroad compartment, he had insisted on being put on a baggage net above the passengers and, once there, used his vantage point to pee on those below him.

In 1911 Angelo volunteered for the army to satisfy his military obligations, but almost immediately came down with a serious case of pneumonia, from which he barely recovered. He was discharged from the army on health grounds and thereafter started a life of travels, strange adventures, and general disorientation that created serious problems for my parents, who did not know how to cope with him. Before joining the army, Angelo had started studying engineering, and he developed a passion for mathematics and physics, although he was not especially proficient in either. As a boy he had acquired a vast literary culture in Italian, French, German, English, and Spanish, which he mastered, as well as in classical Latin and Greek. The library he left at home when he went away was a great source of reading material for me.

Angelo did not like Tivoli and, especially in his early years, wanted to stay away from his parents and be independent. However, when he landed in trouble or fell ill, somebody, usually Uncle Claudio, had to come to his rescue. Ultimately, my parents unloaded him on Uncle Gino, who had a good opinion of the uncommon intelligence of his nephew and was not as close to him as my parents. Appreciating Angelo's talents, Uncle Gino downplayed his eccentricities and trained him with infinite patience in law and history. After this schooling by his uncle, Angelo studied with the noted philologist Gerolamo Vitelli, who introduced him to papyrology, and also with the historian Gaetano de Sanctis. Subsequently he went to Germany, where he worked with local historians and had an adventurous and somewhat nomadic life. In Germany he met Katja Schall (1899–1987), whom he married in 1936. About 1930 he won a university chair in economic history at Catania, in Sicily. He then started painting but refused to exhibit his work. More of him later.

Angelo disliked and despised his brother Marco, who reciprocated his feelings. The younger brother, although less intelligent, and something of a hypocrite, was often extolled as an example by our parents. They clearly favored him, because he gave them fewer problems. Marco on his side always tried to point out Angelo's weaknesses and to embarrass him. "There is no point in worrying about Angelo; he does not have the courage to get himself into serious trouble!" he would say.

As a young boy, Marco had performed scientific experiments, built some gadgets, and studied diligently, obtaining consistently good grades. Ultimately, he graduated in engineering. Compared with Angelo, always a problem child, Marco was a paragon of normalcy, and our parents tolerated some of his obvious faults, such as pompously preaching commonplace or ill-conceived trivialities, often seasoned with Latin quotations. One of his favorite subjects was "The Jews, a doomed race," referring to supposed Jewish physical and mental traits.

As a child I naturally admired and loved my older brothers. I still remember Angelo's charm in telling me a story based on Goethe's *Faust*. Furthermore, he had collections of old coins and of matches, which fascinated me. Marco showed me his tools and his camera, demonstrating their use to me. I have already mentioned the chemical experiments he performed in my presence.

After World War I, however, I had grown up, and my brothers appeared to me in a different light. Both had managed, with great prudence, to escape front-line service. Angelo ended up as an infantryman in a Tuscan garrison, but he was proficient as a cryptographer and succeeded in breaking a Greek code. He treated his military service honestly but rather cynically. Marco attended an officer-training school in Turin. My mother and I went to that city for some time to be near him. Afterward he joined a dirigible outfit, but he never went to the front. He gained valuable technical experience, and after the war he bragged about his heroism.

I then started recognizing some traits of his character that had escaped me when I was a child. Angelo was convinced that Marco had mightily contributed to estranging him from our parents, and I started to have some inklings that Marco might be trying something of the sort also with me. At home he managed to displace me from the room I shared with him, sending me to sleep on the upper floor in Uncle Claudio's home on the pretext that I snored and disturbed him. He was, in fact, trying to push me out of our parents' home.

Marco loved to pontificate on all occasions, even on subjects he was ignorant of. Much later he became famous among my physicist friends by lecturing Fermi on thermodynamics. Fermi, Amaldi or Rasetti, and I were present, and we all grinned at Marco's conceit; he certainly was the one who knew the least thermodynamics among us, and he did not realize how ridiculous the situation was.

In the spring of 1921, when I was sixteen, somebody persuaded me to join the Avanguardisti, a Fascist youth organization. At home opinions were divided: Uncle Claudio favored Fascism, which he saw as restoring order and national pride. Uncle Gino on the other hand, said that it would end badly, more or less as happened, and added, in dialect, "Pias mia" (I don't like it). Loud quarrels between the brothers followed, although they loved each other dearly. I believe my father listened to them without great feelings one way or the other; perhaps he tended to favor Fascism for reasons similar to Uncle Claudio's. In retrospect, I recognize that only Gino had sufficient historical and legal preparation to take the long view. However, all three brothers, by conviction or practical necessity, joined the Fascist Party. Their opinions, like those of most Italians, changed with time. I started having serious doubts about Fascism in the summer of 1924, after the murder of the Socialist leader Giacomo Matteotti.

In my first years in Rome, I also started wandering away from home. One day I discovered the Appian Way. I had gone for a walk alone and, I do not know how, found myself on an extraordinary road, flanked by cypress trees and Roman tombs and ruins, and with a Roman pavement. The surroundings were so beautiful and romantic that they reminded me of my beloved Tivoli. I did not know where I had landed, but the impression was enormous. When I returned home and told my parents, they explained to me that I had been on the Appian Way.

While I was at high school, I visited Florence several times, staying either at the beautiful Treves villa or at their home on Via Masaccio. The large three-storied building was surrounded by extensive grounds containing a splendid garden, with many small lemon trees in terra cotta pots, marble statues, a pond with red fish, tall bamboo thickets, hothouses, a green for playing *bocce,* and an orangery. The home, with much English furniture, was a display of my uncle Guido's taste and wealth. He smoked excellent Trabucos cigars, read English newspapers, and drank tea for breakfast. When he was about ninety years old, after World War II, he realized that it was impossible to keep his villa while all similar establishments in its neighborhood were being transformed into huge apartment houses. He then sold the estate without a word to his children, who later discovered the fait accompli.

I usually went to Florence around Easter. At the Treves villa and in its surroundings I could always smell the strong scent of wisteria. Even now I occasionally find a road or pass by a garden that by its smell reminds me of the Florentine spring and vividly evokes its mood. The poet Ugo Foscolo noted this olfactory peculiarity of Florence in his famous lyric *I sepolcri,* writing: "E le convalli / popolate di case e d'oliveti / mille di fiori al ciel mandano incensi . . ." ("while thy happy valleys dotted / with villas and olive groves send forth to heaven / the fragrance of a thousand flowers . . .")[1]

In Florence the family observed Jewish Passover rites, but what I was really interested in was sneaking away to Costa S. Giorgio to see my flame, J.H., a beautiful girl with whom I had fallen in love during the summer of 1919 at the seaside. It was a love unfortunately much too platonic; even a kiss would have been considered sinful. That did not make it less ardent, but both of us were too young, too innocent, and had been brought up too strictly.

The whole period from my fifteenth birthday on was dominated by repressed sexual desires. A puritanical upbringing, my natural bashfulness, lack of parental guidance, and some unhappy conversations with friends when I was about fifteen brought me to a difficult impasse. My brother Marco augmented my problems by giving me some unfortunate books by a Mr. Stahl, who advocated a Victorian credo of impossible chastity. Marco himself, however, while preaching to me, went to brothels. I found condoms in a drawer of his desk, but did not know what they were. I was afraid of and repelled by houses of prostitution, which were frequented by most young Italian men. Unlike some of my friends, I had not found a middle-aged matron willing to serve as a "*nave scuola*" (training ship). I had been infused with a deep sense of guilt about masturbation, which I had been told would have all kinds of dire consequences. I was thus confronted with an insoluble problem, since I could not marry at fifteen years of age and the few girls I met were more than chaste, at least in theory, and had been brought up similarly to myself.

I did not dare to speak to my parents about this. Of other adults, Uncle Claudio was the most understanding, but he was too old, born and brought up in a different century and possibly with wrong ideas. All he could give me was sympathy. However, he spoke openly and frankly and called a spade a spade.

After a few years, the problem became serious, as I suspect it may also have been for my brother Angelo. I was too shy to mention it to any of my contemporaries, even close friends. I am sure I was not the only one in this quandary; on the contrary, I believe it was common among my friends. Some turned to their priests, I do not know with what results; some were seriously hurt for a long time. Some young intellectuals formed chastity leagues.

My mother perhaps understood better than she showed, but she kept her counsel. However, she encouraged me in any sport I took up, and particularly in mountaineering and skiing. My father was afraid I might hurt myself, and for his own peace of mind would have preferred me not to try skiing. I started in 1921, when skiing was hardly known in central Italy. Before then, with my mother's encouragement, I had already made cycling trips, played tennis, and fenced. Attempts to teach me to dance failed miserably. I attended classes with young ladies of good family, who were mostly ugly and clumsy; a dancing teacher at a school in Florence gave lessons skimpily clad and was attractive, but I would have preferred her to transfer her teachings to bed.

As I have said, Papà did not pay much attention to me, but with the passing of time I lost my fear of him and started to recognize his uncommon qualities. Still in Tivoli, when I was about twelve years old, he had sent me as an apprentice to a cabinetmaker to improve my manual dexterity. My father justly thought that the use of the hands is not less important than that of the brain. Later, when I was perhaps fourteen years old, he made me work in a small laboratory in the paper mill, where I badly burned my fingers trying to dissolve some rosin in boiling alcohol, which caught fire. I was lucky that nothing worse happened.

To get to the paper mill I had secured a key to the lower gate of the Villa d'Este, which opened onto an old medieval dirt road. Starting from home, I first passed through the silent and solitary gardens of the Villa, until I reached the last, quiet lane leading to the lower gate, a pathway overgrown with emerald green moss that looked like a Persian silk rug. Beyond the gate, one found noise, stench, and mud. The contrast could not have been greater, and it perturbed me deeply.

Papà had also put me to work in an office preparing electric power bills for his customers; furthermore, he forced me to study German. He did all this with yells louder than necessary and insufficient patience, but he nonetheless had a positive educational impact. Papà did not speak to me frequently or at length, but I listened very carefully to his conversation during meals, from which I learned a great deal. He thus taught me how banks work, what a corporation is, and about stocks and bonds and other business subjects. I enjoyed his explanations thoroughly, and my esteem for him grew further. With the passing of time, he taught me much more on business subjects and I came to realize the penetration and fairness of his judgments. I also recognized his kindness, often hidden by his gruff manner, and his generosity to those who merited it. Later, as an adult, I treasured his advice, which deserved to be listened to all the more carefully inasmuch as it was given so sparingly.

My schoolmates in Rome were very different from those in Tivoli, but I made a few friends. In addition to my classmates, I also became friends with two boys who later became famous; they were in the same school but in different classes. One was Enzo Sereni, a future Israeli leader; the other was his brother Emilio, a future bigwig of Italian communism.[2] They headed a group that ardently discussed political subjects, and they invited me to some highbrow meetings. I disagreed with many of the ideas I heard; indeed, several of them seemed absurd to me, but I did not know how to defend my opinions with adequate rhetorical and dialectical skill. Ultimately, I lost patience with them and went my own way, but the experience taught me to distinguish between well-founded conclusions and those that prevail only through skilled advocacy. At the time I acquired a distaste for what I later called *"pappagalli parlanti"* (talking parrots).

With the move to Rome, I saw my second cousins Riccardo and Bindo Rimini much more often. They were the orphans of my cousin Enrico and about my own age. Enrico, who had been very close to my father, died in a railroad accident in the summer of 1917, leaving his family in difficulties. My father helped them materially in spite of the obstacles he encountered in the exaggerated pride of the widow, Ada, who did not want to accept financial help.

Ada and her children became intimate with us. I was especially close to Riccardo; his younger brother Bindo was an intractable young boy, and Ada ultimately enrolled him in a military college to give him some discipline. Riccardo was like a brother to me. He had a very sharp mind, with scientific inclinations, and was a hard worker, conscientious, and observant. In due course, all this made him into a superior physician. In 1938 the Italian racist laws forced him to emigrate, and he went to Montevideo in Uruguay, where he wrote some good papers on blood circulation, besides achieving great professional success. When life separated us, we kept up an active confidential correspondence, and when we met again after ten years of separation, we felt as if we had met only a few days earlier. Our correspondence lasted until his death in 1977 and contained our most intimate thoughts.

During my first years at high school I again studied German, which I had forgotten since learning it from my Tata Giuseppina. My parents had hired a young Swiss governess for me around 1915, and thanks to her I could use German fluently. I also digested a fair amount of German poetry, including Goethe's *Faust,* several of Schiller's plays, and Heine. My Treves cousins were almost bilingual in English thanks to a long succession of witch-governesses their parents had forced on them, and they decided to teach English to me during their visits to Tivoli. Marco Treves was an excellent teacher, somewhat pedantic, but patient, insistent, and effective. Later I also had some private coaching.

At school I followed the prescribed courses reasonably well, but without shining in them, so that usually I was third (in grades) in my classes. Besides what was taught at school, I studied some physics books, often in German or English, on my own. I still have Glazebrook's *Light*, Ball's *Elements of Astronomy*, Maxwell's *Theory of Heat*, and above all Reiche's *Die Quantentheorie*, which greatly impressed me.[3] Angelo had bought these books and left them at home. I cannot claim that I understood all I read. I labored over Maxwell, but could not fathom it. I had not yet learned that in order to study physics, one has to use paper and pencil and work through the calculations as one goes along. Usually I read these books at school during boring classes that I disdained.

My experience with modern languages shows that I was not refractory about learning languages, and that if in eight years of *ginnasio* and *liceo*, I did not learn Latin or Greek, not all the fault was mine. The methods used to teach these dead languages, and the teachers, were at fault. By the time I was nearing the end of high school, the teacher of Greek had completely disgusted me, and my grades were failing. At that time it was possible to obtain a high school diploma without a special examination provided one had grades above a certain minimum in each subject. In Greek I was falling below the minimum, and I knew that a Greek exam would have been disastrous for me. Luck had it that one day Giorgio Pasquali, a famous Greek scholar and a family friend, came to our house for lunch. I had known Pasquali since I was a small child, and I told him about my difficulties with Greek. "Who is your Greek teacher?" he asked. "What a strange coincidence!" he remarked when I gave him the man's name. "He is the secretary of the committee I am chairing in the Ministry of Education. I am here just for one of its meetings." He said no more, but my grades in Greek miraculously started to improve and by the end of the year, taking into account my progress, I had reached the minimum required to graduate from high school without an examination.

By 1919 I had become a young man, and I naturally started to ask some of the eternal human questions about the purpose of life, good and evil, the foundation of morals, and the essence of the soul. I read then some of those books I innocently thought to be the pillars of our culture. I chose them by hearsay or because they were available at home. Among them were Descartes's *Discours de la méthode*, Galileo's *Dialogo . . . sopra i due massimi sistemi del mondo* and *Il saggiatore*, Tolstoi's *War and Peace* and *Resurrection,* several novels by Victor Hugo, some of Plato's dialogues, and other classics. I had an iron stomach for any reading, and some of it stuck, but I do not remember much of what I read.

I never had a religious crisis. I read Renan's *Life of Christ* and Sabatier's life of Saint Francis with pleasure, as well as some short books on Buddhism and Judaism. My parents asked Dante Lattes, a noted Jewish scholar, to give me some lessons on Judaism, but they did not impress me, and neither did I read the Bible with veneration or celebrate a bar mitzvah. Whenever I tried to read the Old Testament, I had the impression of a very unhomogeneous text. Certain parts seemed to me great, sublime, and rich with moral teachings; other seemed barbarous and cruel. It seems to me that Adonai is very different in different parts of the Bible, and in some cases so churlish and vindictive as can be conceived only by the mind of a priest.

The only Jewish religious rites I attended when I was young were in the Florence synagogue. I found them interesting and, especially later, after I had grown up, even moving, but not for theological or religious reasons. Their compelling force came from the traditions they evoked, from family history and from feelings rooted in the subconscious. In 1984, on the occasion of my grandson's bar mitzvah in Israel, I saw an old cantor in the small Herzliya synagogue ecstatically embracing the Torah, as if it were a child. I understood him, but I did not share what I thought were his feelings.

I do have some religious feelings, which I rationally recognize as childish; nonetheless, they comfort me because they remind me of people I once loved and of old times. On a more intellectual level, I find myself close to Einstein's position as he described it in response to a letter from a schoolchild asking whether scientists prayed, and, if so, what they prayed for:

> Scientific research is based on the idea that everything that takes place is determined by laws of nature, and therefore this holds for the actions of people. For this reason, a research scientist will hardly be inclined to believe that events could be influenced by a prayer, i.e. by a wish addressed to a supernatural Being.
>
> However, it must be admitted that our actual knowledge of these laws is only imperfect and fragmentary, so that, actually the belief in the existence of basic all-embracing laws in Nature also rests on a sort of faith. All the same this faith has been largely justified thus far by the success of scientific research.
>
> But, on the other hand, everyone who is seriously involved in the

> pursuit of science becomes convinced that a spirit is manifest in the laws of the Universe—a spirit vastly superior to that of man, and one in the face of which we with our modest powers must feel humble. In this way the pursuit of science leads to a religious feeling of a special sort, which is indeed quite different from the religiosity of someone more naive.[4]

I have seen innumerable tragedies and horrors caused by religious and political fanaticism, enough to inspire in me a great dislike for "absolute truth," "fundamentalism," and similar attitudes. On the other hand I have met truly religious people, in whom religion inspired the noblest conduct.

Two such saintly men come to my mind. One was a Catholic priest, Don Nello del Raso, who, after World War II, moved by the horrors he saw among Italian children, built a home for them at Tivoli and directed it until his death. I met Don Nello after the war, and he instantly became my friend. His personal charm was extraordinary and felt by everyone who came in contact with him, whether it was a brutalized child, a rich landowner, or an agnostic scientist. I remember my visits to him as warm, enriching experiences.

The other was Professor Burton J. Moyer (1912–73), who was for a time my colleague. As a young man, he had wanted to be a Protestant missionary and trained for that calling, but the war made him into a superior physicist. He worked at the Radiation Laboratory in Berkeley and in due course played a major part in the discovery of the neutral pion. However, his real desire was to help his fellow men. Moyer became head of the physics department at Berkeley at the time of the worst student unrest, and he was one of a handful of people who managed to gain the confidence both of the administration and of the rebellious students. When things quieted down, he went to India to help in the setting up of a technical institute there. Some time after his return to Berkeley, he was called to the University of Oregon to revitalize its scientific departments, which he did with outstanding success, although he died before he could finish the job.[5] I felt deeply attracted to him in spite of our greatly different backgrounds.

Chapter Three

The Education of a Physicist (1922–1928)

Scent of Roman Hay and Alpine Snow

> . . . *chè non fa scienza,*
> *sanza lo ritenere, aver inteso.*
>
> (. . . for knowledge none can vaunt
> who retains not, though he have understood.)
> Dante, *Paradiso* 5.41–42 (trans. Laurence Binyon)

The end of high school materially changed my studies, which were still my primary occupation. No longer was I forced to study subjects in which I was not interested, and neither was I subject to pedagogues whom I often could not admire. At the university I found several teachers who were universally known as eminent in their fields, and who obviously dominated their subjects. What I learned was new to me; interesting and challenging, it stimulated me to think further about what I heard, which had only rarely happened in high school. I was much freer in organizing my time and chosing my friends. I did a fair amount of fencing at first, but later I became an enthusiastic mountaineer and skier. Love, girls, and sex also came to the forefront.

I finished high school in July 1922, matriculating in engineering by a process of elimination. The first two introductory years of study were common to engineering, physics, and mathematics. I knew, from a practical point of view, that my father would give me a job in the paper mill or find me work in some other industry. The idea of a career as a physicist seemed farfetched because it offered little hope of finding any employment.

I and my parents certainly favored a university career, but to try for this was very risky because at the time there were very few openings in Italy—in total about twenty chairs in physics, all of them occupied. Perhaps one position became available each year because of retirements and deaths. Furthermore, I had some misgivings about whether physics was up to date in Italy. I do not know from where I got that idea. Possibly in reading Reiche's book I had noted that hardly any Italians were mentioned, and even in the old Ganot, there were few Italians. In any case the choice of the preparatory biennial course postponed making a final decision, so I registered for it.

In my first year at university, I studied algebra under Professor Francesco Severi, geometry with Professor Guido Castelnuovo, and chemistry with Professor Nicola Parravano. I also took a drafting course. Severi gave excellent lectures, and I was pleased by the change of level from high school; here there was real intellectual stimulus and a challenge to understand; perhaps even to try to invent something new. Castelnuovo was a paragon of clarity, and in spite of his monotonous voice, which induced sleep, one learned new and interesting things. I soon came to suspect, however, that the chemist, Parravano, did not always know what he was talking about. At home I had found a treatise on physico-chemistry by Walther Nernst,[1] and comparing what I had learned from it with what Professor Parravano taught, I concluded that he had misunderstood several things, or at least that he understood them differently from me. The professors' assistants were more accessible than the great men, and one of Severi's explained Fourier series to us in a startling and profound way, which gave much food for thought.

In my second year, Severi taught us analysis; Senator O. M. Corbino, the head of the department, physics (in practice only dealing with electricity); Pittarelli, descriptive geometry; and Tullio Levi-Civita, rational mechanics. In my first two years at university, nobody taught us any thermal physics or optics, let alone more modern subjects.

Levi-Civita's course on rational mechanics was poorly attended, although the professor was famous and the lectures were good, even if slightly verbose.[2] Levi-Civita was very short and also short-sighted;

nevertheless, he strove to reach the top of the blackboard, putting his nose very close to it, raising his arm, and writing blind. In this position, he was once struck on the back of the head by a missile from the peashooter of some nasty student. Levi-Civita turned around and, with the most innocent expression, asked: "Have I written a wrong sign?" His candor and good faith were so obvious that nobody laughed, and no peashooter ever dared disturb him again. For many months we heard the simplifications that occur in mechanics if $F \times dP$ is a total differential without the professor ever explaining what a total differential was, and without us ever asking. Levi-Civita trusted our analytical competence, but unfortunately Severi had not mentioned total differentials.

No less important than the courses were my new fellow students. Among them Giovanni Ferro-Luzzi was the closest to me. We came from the same high school, but from different sections. Soon we started studying together. We liked to compete in solving problems and we quizzed each other on the theory.

I also discovered that there was a very great advantage in studying steadily during the year, avoiding a cramming period before the examinations. Good, paternal Professor Castelnuovo had warned us on this subject; I was surprised in discovering the pertinence of his advice. I found that I needed time for digesting many new ideas and that a steady diet nourished infinitely more than occasional feasts (and bouts of indigestion). I believe that the discipline of steady work helped me immensely then and later in life.

Ettore Majorana, who subsequently acquired a well-deserved reputation as a mathematical genius, was another of my fellow students.[3] Once, not having sufficiently prepared a lecture, Severi started a proof of a theorem the wrong way. Majorana immediately whispered that he would soon be in trouble, so we all anticipated what was to come. After a minute or two, Severi's face reddened, and it became obvious that he did not know how to proceed. Some voices then murmured: "Majorana predicted it." Severi did not know who Majorana was, but said haughtily, "Then let Mr. Majorana come forward." Ettore was pushed to the blackboard, where he erased what Severi had written and gave the correct proof. It is noteworthy that Severi neither complimented him in any way nor made any effort to become acquainted with him. On a different occasion, while I was waiting to be called to an oral examination, Majorana gave me a synthetic proof for the existence of Villarceau's circles on a torus. I did not fully understand it, but memorized it on the spot. As I entered the examination room, Professor Pittarelli asked me, as was his wont, whether I had prepared a special topic. "Yes, on Villarceau's circles," I said, and I proceeded immediately to repeat Majorana's words before I forgot them. The professor was impressed and congratulated me on such an elegant proof, which was new to him.

My friend Giovanni Ferro-Luzzi and I prepared for examinations on a bench in the marvelous Palatine gardens, near the house of the superintendent of the diggings at the Roman Forum, the humanist and archaeologist Giacomo Boni (1859–1925). It was an extraordinary location, quiet and evocative, with laurel thickets and Italian gardens. On the days of ancient Roman festivals, somebody put up rich festoons of flowers and fruits, hanging them in appropriate places.

In the fall, Ferro-Luzzi and I repaired to Tivoli for concentrated, undisturbed study, as I was later to do with Edoardo Amaldi. My parents' housekeeper there fed us excellently; after studying we went for hikes under the olive trees or on the Viale Carciano, as I had done since my childhood. If by chance it rained, the rain was followed by the clear, cool, scented air of the beginning of autumn.

Mountaineering became a serious avocation with me, and every summer I and some friend went to the Alps for rock or ice climbing; during the winter we practiced cross-country skiing, often in the Abruzzi. In the Dolomites I climbed the Vajolet Towers, and, all alone, the Cinque Dita, and many other mountains in the vicinity of Cortina d'Ampezzo, in the Pale di San Martino group, and elsewhere. Our climbs reached today's fourth class and would now be deemed easy, but in the 1920s they were thought fairly difficult. We always went without a guide, for sport as well as to save money.

My active mountaineering lasted until about 1930. Later physics absorbed my summers too, and still later, after my marriage, I discovered that while I still liked hiking, camping, and outdoor activities a lot, I had no more stomach for difficult climbs.

In my third year at university, I transferred to the Engineering School, where I found the courses much less interesting than in the preparatory biennium, except for one by Professor U. Bordoni, who taught us thermodynamics according to Clausius, emphasizing all its subtleties. The other professors taught ordinary engineering practice, at a low technical level and without imagination. To refresh myself, I attended a mathematics course on the theory of functions of a complex variable, given by Professor Ugo Amaldi, the father of my future friend Edoardo. The lectures were at 1 P.M., not exactly a pleasant hour in Rome, but the teacher presented the material in a fascinating way. The exposition resembled a soap opera, and at the end of each lecture I asked myself what the next would bring: new singularities? new power series developments?

At about this time, my fellow student and mountain-climbing companion Giovanni Enriques told me that he had heard from his father that there was at Rome a sort of genius, a certain Enrico Fermi, who had recently got his physics degree in Pisa. When I went to hear Fermi speak at a mathematics seminar,[4] I soon realized that the rumor was not exaggerated; at last here was somebody fully conversant with modern physics. However, I did not approach Fermi at that time. At other meetings of the same seminar, I heard E. Landau, whose talk confirmed my conviction that pure mathematics was not my cup of tea. Another time a professor from Bologna spoke for an unconscionably long time. For some reason, the light went out for a few minutes; when it came back on, the room was almost empty.

In my third or fourth year in engineering, why I do not remember, or perhaps never grasped, I grew a Charlie Chaplin moustache and started going to school wearing a bowler hat and kid gloves, carrying a cane. This lasted for several months and provoked a certain amount of mockery among my fellow students, to which I responded with haughty disdain.

In 1927 my mother had the happy idea of giving me a Fiat 509 automobile. I suspect she may have thought it was the simplest way of getting me a girlfriend. If she had any such idea, however, she never even hinted at it to me.

Cars were then still relatively uncommon in Rome, and the 509 made big changes in my life. It ranked me as affluent among my fellow students, and everybody wanted to use my car. It also greatly facilitated mountain trips. On the other hand, it did not have great success in procuring me a girlfriend. My friends and I took several girls of a good family background to the seashore or to places in the vicinity of Rome, but always within the limits of strict decorum and prevailing Italian rules. Often we went in a party of four: Giovanni Enriques and I with two girls, whom we may have liked even more than we let on. To give an idea of the difficulties we faced, once at the seashore at Castel Fusano, about twenty miles from Rome, we lost the ignition key. We were supposed to return before nightfall, and being late would have caused a scandal. By combining all our technical ingenuity, we succeeded in bypassing the ignition switch and starting the car. This shows how restricted we were; a simple accident would not have excused us for a few hours' delay in getting back.

In those days, the surroundings of Rome were of an unsurpassed beauty, now almost entirely vanished. In springtime, places such as the Pratoni di Nemi, a plateau at about 3,000 feet in the volcanic Alban Hills, overflowed with wild jonquils and violets; Pratica di Mare, almost unknown except to a few cognoscenti, resembled a Pacific island, with palms, tropical vegetation, and wild buffaloes wallowing in the mud; Veio's Etruscan ruins, covered with scented honeysuckle, were interspersed with green meadows, on which I would lie for hours talking to my girlfriend. It was hard to tell which was more exciting: her perfume or the scent of the hay and of the wildflowers. Even Ostia and Fregene, still sparsely populated, were true gems.

In the beginning of the 1920s, I had carefully toured Rome, often together with my cousin Fausta, who came visiting from Turin and stayed with us. She had rather romantic tastes, and with the help of the Touring Club Italiano guidebook, knew where to go and at what time of the day. Little wonder that two young people under such conditions should develop tender feelings; however as time went on, friendship prevailed over love, and endured solidly until Fausta's death in 1982.

At home, Marco, after attempting some deals on his own, in which he lost money to an impostor, had accepted a job in the paper mill, and he expected, in due course, to succeed my father. Father told me openly that there could be only one person in command of the paper mill, and that he would help me start a different business for myself or find a job elsewhere, according to my preference.

In the meantime my distaste for engineering studies steadily increased. Luck had it that in the spring of 1927, Giovanni Enriques introduced me to Corbino's newly arrived assistant, Franco Rasetti.[5] Rasetti, who was then about twenty-five, was a close friend of Fermi's; he had studied with Fermi at Pisa, had followed him to Florence, and had been hired by Corbino on Fermi's suggestion when Fermi was appointed professor of theoretical physics at Rome. In addition to being an excellent physicist, Rasetti was a skier, a mountain climber, an insect collector, and in general a person of the most diverse interests.

My car allowed us to go to places that were otherwise rather inaccessible, and at the end of May 1927, Enriques, Rasetti, and I went to Castel del Monte in the Abruzzi. From there, we followed a long ridge to the Gran Sasso, the highest mountain in the Apennines. It was a hike of a couple of days, requiring us to sleep in the open. These conditions helped create a fast friendship between us, and soon thereafter Rasetti introduced me to Fermi.

The two of them were looking for physics students to educate, and I was looking for teachers, so we suited each other. At the end of July, I went with Rasetti to the Pizzo d'Eta from the valley of the river Liri in central Italy, a lengthy and very beautiful excursion. The long, deep valleys of this isolated and rarely visited part of the country, with their well-preserved ash, oak, and maple woods, preserved the aspect of Italy before its deforestation in recent centuries. Near the top of the mountain, Rasetti started looking for some tiny insects of the genus *Bythinus,* which lived under the bark of trees. He sucked them into a small glass container he had brought with him and saved them for his famous collection of coleoptera. At the same time, speaking very loudly, he taught me the principles of statistical mechanics and Boltzmann's distribution. He asserted that except for Fermi and himself, there was no physics professor in the Italian universities conversant with such theories. On the Pizzo d'Eta, I also learned some calculus of variations and analytical mechanics. I greatly enjoyed these strange lectures and on my return home made notes on them.

A few weeks later I went to Ostia with Fermi, Rasetti, and other young men and women. Fermi started talking physics and asked me what I knew. He challenged me to calculate the vibrations of a heavy rope dangling vertically, which I did to his satisfaction. Thus started our friendship.

In the summer of 1927, always in my car, we went to the Val d'Herens, in the Alps. I stopped on the Riviera near Genoa, to say hello to Renata J., a great and unhappy love of mine, who was spending her vacations there. Our company consisted of Giovanni Enriques, Rasetti, Ferro-Luzzi, Piero Franchetti (Enriques's cousin) and myself. We settled in a very primitive refuge and from there started several rather difficult climbs: Dents des Bouquetins, Dent d'Herens, and others. I had with me a guidebook and carefully studied the itineraries. It was a sensible thing to do, but I was rather pedantic, and my friends teased me because I would say things like: "Here according to the book we should find the plaque marking the spot where X lost his life." However, recognizing landmarks and critical spots on the climb helped us achieve our goal. Once we were caught in an electrical storm. The sight of the sparks coming out of our ice axes and of our hair standing on end was truly spectacular, and scary.

That summer I experienced for the first time a strange, almost pathological, peculiarity of Rasetti's. Whenever he saw a chance of ditching his companions, whether because of darkness or fog or any other reason, he took it. Later he rejoiced in having done so as though it had been a funny joke. A psychologist could have a field day with such behavior.

From the Val d'Herens, we passed to Val Tournanche and from there on August 14, 1927, we climbed the Matterhorn. The weather foiled a first attempt, but the next day, taking advantage of a clear spell, we bounced back from Breuil, slept at the Luigi Amedeo di Savoia Hut and, after a very cold and windy climb, reached the top. We descended from the Swiss side, sleeping at the Solvay Hut. That year the ice and snow conditions on the mountain were difficult, and ours had not been an easy enterprise. During our stay at Breuil, we met the duca degli Abruzzi, a cousin of the king of Italy and a noted explorer, as well as the famous writer and mountain climber Guido Rey. The latter invited us for tea at his villa and gave me some pictures. The Breuil too has been disfigured by much new construction and automobile traffic since World War II. In our day, the easiest access was by a mule trail from Val Tournanche.

After the Matterhorn climb, I met my uncle Claudio in Aosta, where he happened to be attending a geological meeting. I accompanied him on a field trip and he showed me the landscape through the eyes of a geologist. It was a revelation and a fascinating lesson; I do not know why my uncle had never taken me on a field trip before. Unfortunately, it was also the last possible chance. At Aosta, while we were together, Uncle Claudio suffered a small stroke; it did not seem serious at first, and he recovered fast, but after his return to Rome, more strokes, of increasing gravity followed, and on March 18, 1928, he died. He was the first person dear to me that I saw dead, and the sight affected me deeply. From what I have written about him, it should be clear what a loss it was for me.

From Aosta I went to Como, where an International Physics Conference was held that September to commemorate the hundredth anniversary of the death of Alessandro Volta in 1827.[6] Needless to say, I was not invited. I was not even a physics student, but I thought, with some impudence, that by tailing Rasetti, who in turn was tailing Fermi, I might be able to attend some of the lectures and see what was going on. Indeed, this came to pass. Besides going to the lectures, I collected several free publications that were given to those attending the conference, and in particular an excellent series of articles on modern physics, written by K. K. Darrow of the Bell Telephone Company. I read them carefully during the conference.

Many great physicists were present; among them Max Planck, Ernest Rutherford, Niels Bohr, Robert Millikan, Wolfgang Pauli, and Werner Heisenberg. Einstein was not there because he did not want to enter Fascist Italy. Corbino, Quirino Majorana (Ettore's uncle), Fermi, and a few others represented Italy, and it was easy to see that Fermi was the only Italian who counted in the eyes of the foreign participants. The organizers offered lavish receptions and excursions on a lordly scale. Corbino privately commented that Italy should have exhibited more physics and less hospitality and that it should not deceive itself that sponsoring a conference was a substitute for scientific achievement. From Como, the participants in the conference went to Rome, where Mussolini received them.

What I had seen at the conference tipped the scales in my decision to switch from engineering to physics. I had been brooding over the idea since the spring, when my meetings with Rasetti and Fermi had convinced me that a unique opportunity for seriously learning and practicing physics had arrived. The young professors, who were only slightly older than I was, treated me not as a mere freshman but as a future colleague. They were prepared to let me have immediate access to the laboratory of the physics department, to show me what they were doing, and to give me an opportunity of helping them in their work. What more could have I wished for?

On my return to Rome, I started going regularly to the lab in the Physics Institute at Via Panisperna 89a, where I found my mentors dressed in none-too-clean gray smocks, so much so that my olfactory memory can still evoke the characteristic slight smell of those garments. They explained to me the purpose and the techniques of the experiments they were performing and the results they expected.[7]

At the time, having read much physics privately, I was acquainted with classical physics at an intermediate level. My mathematical preparation was pretty good and derived mainly from the excellent university courses I had attended. I also knew some chemistry and had worked in a good analytical laboratory at the Engineering School. My practical experience derived from amateurish experiments and apparatus; I had built a crystal radio receiver and played with a Ruhmkorff coil. All told, I knew how to use my hands and simple tools.

Fermi started tutoring me privately almost from the outset. At about 6.30 P.M. he would call me to his office and there, mostly with Rasetti also in attendance, he would explain to me whatever came to mind, or whatever I proposed. I listened, when I did not understand I asked questions, and then at home I wrote down what I had learned, with the help of sheets of paper that Fermi had filled with formulae as he progressed in this cross between a lecture and an informal conversation. I did this work mostly the day after the lecture, and I included some problem proposed by Fermi or, more frequently, of my own invention. Here are some samples of the subjects treated, according to my extant notes: light diffraction from a slit; mean free path in gases; fluctuations; classical theory of light resonance; molecular rotatory power; diffusion vacuum pumps; X-ray diffraction according to Max von Laue and to W. H. Bragg. This was possibly the most influential "class" teaching I ever had. A little later Edoardo Amaldi joined the "class."

Rasetti showed us how to perform spectroscopic experiments using interferometers, spectrographs, and simple techniques. Fermi thought that one could teach theory, but that the only way of learning experimentation was a laboratory apprenticeship. Furthermore, it was his rule that beginners should study theory and experiment equally; specialization would come only later. He used the same teaching method again and again, particularly in Chicago.[8] The subjects too, even after many years, were often the same as in Rome, almost stereotyped. In fact, they were the methods and results that seemed important to Fermi. I was able to retrace them to an extent even in notebooks of his own studies at Pisa or earlier. Some of these notebooks, which are astounding for the unerring choice of materials, are deposited in Chicago among Fermi's papers. I saw them only after his death.

Fermi did not give himself airs; he was very simple in his manner, courteous, and easily accessible. These outward appearances nonetheless concealed great reserve. Outside of physics, he was much more inclined to listen than to speak and refrained from private confidences.

Although we spent many hours together every day, and ostensibly on a footing of equality, I do not believe there was the same degree of intimacy with him as with the other members of our group. Perhaps his manifest scientific superiority contributed to this situation, but its main cause was Fermi's disposition and the care with which he set reason over feelings. With the passing of time and given the special circumstances that obtained later during the war, Fermi's reserve increased rather than decreased. This was contrary to what one might have expected; relative differences in age and scientific standing tended to diminish with time. Although affable, Fermi always inspired a certain awe, and perhaps more so in those who knew him best than in those who had only a cursory acquaintance.

Under Fermi's ministrations, we rapidly gained an incredible enthusiasm for science. We loved physics with an intensity comparable to that of physical human love; we thought and talked only about physics. Rapid progress followed, which further enhanced our passion. We spent all our available time at the Physics Institute—that is, according to the holy Italian schedule, from 8 A.M. to 1 P.M. and from 3 P.M. to 8 P.M., Monday to Friday and on Saturday mornings. We never went to work after dinner and very seldom on Sunday. The institute closed on a regular schedule, and none of us had a key to the door.

Saturday was a very interesting day because we frequently devoted Saturday mornings to planning future work. On Sunday we usually went on hikes with friends of both sexes, including many nonphysicists, but the physicists often formed a separate group after a while and started talking shop. Under the influence of our common life, we developed some strange tics, such as the habit of speaking with an intonation or cadence characteristic of Fermi. It was not a pose, only mimicry.[9]

Everything was going well for me, but I was still registered at the Engineering School and not in the physics department. Any mention of the fact that I was inclined to switch met with distinct coldness at home. Uncle Claudio, who was already ill, wanted me to get my engineering degree first; my parents said that as an engineer I would easily be able to find a job, but that as a physicist, if I did not obtain a university chair, I would have to make a living as a high school teacher, a dim prospect.

Rasetti and Fermi had rapidly convinced me that they were supermen (not a totally erroneous opinion, merely exaggerated), and that whoever associated with them would become one too. They were positive they were the only up-to-date physicists in Italy (with the possible exception of their students) and the only ones that counted scientifically.

In the beginning I tried to keep up with both engineering and physics, but I soon realized that this was not possible. Studying physics with the intensity and dedication I desired, and that was necessary if I was to meet Fermi's expectations, did not leave time or energy for anything else. In the meantime I had been introduced to Corbino, who in spite of his benevolent cordiality inspired me with considerable awe too.[10] This feeling paradoxically manifested itself in my often acting as if I did not see him, or in my being curt with him. I always regretted not having been more open with him, as I desired and as he deserved.

Finally, I decided I had to burn my bridges and transfer to physics. Unexpectedly, I met with a bureaucratic difficulty. Because I had not taken a required laboratory course in practical physics, I stood to lose a year. It was then that I really saw Corbino in action for the first time. He telephoned a relative of his who was registrar at the Engineering School and told him: "Segrè wants to register in physics, but in the certificate you have given him there is no mention of the exam in practical physics. How come? He passed it in my institute last year and got a grade of 30/30." The registrar apologized and corrected the "error," and I did not waste the year. The following year, however, Corbino assigned me to teach that course, and now, sixty years later, it dawns on me that perhaps it was not just by chance. I repeatedly saw Corbino circumventing obnoxious regulations or managing some well-planned scheme, but never without a clear justification from a higher point of view. He was a master at overcoming obstacles standing in the way of raising the level of physics in Italy. Ultimately, he succeeded brilliantly.

A few months after I transferred to physics, I spoke to Ettore Majorana about doing so, encouraging him to follow my example. I told him that the Engineering School was not for him, just as it was not for me, and that the present situation in the physics department offered a unique opportunity. Ettore listened and then decided to come to see in person. At the institute, he found Fermi calculating the function central to the Thomas-Fermi statistical method for calculating atomic properties,[11] cranking a small Brunsviga adding machine by hand. With its help, in about a week of work, he had obtained a numerical table of the function. Majorana informed himself in detail about the mathematical problem and went home without further comment. At home, he transformed Fermi's nonlinear equation into a Riccati equation and solved it numerically using his brain as calculating machine. After a few days he returned to the physics department and asked Fermi to show him his numerical results. He compared them with his own and verified that they agreed. "Surprisingly, Fermi has made no errors," he said. After this experience, he too converted to physics, but being mathematically vastly superior to all of us, and in some respects even to Fermi, he did not come regularly to our instructional sessions, although he participated in many of our conversations and discussions. He never tried experimental work.

In the early days of our informal Rome group, Giovanni Gentile, Jr., son of the philosopher, powerful senator, minister, and Fascist bigwig of the same name, often came to the institute. He had recently graduated from the Scuola normale in Pisa and, possibly because of their common Sicilian roots, became very close to Majorana and wrote a few papers with him. Gentile remained our friend and occasionally visited the institute, but never became a regular member of our group.

Already in the fall of 1927 we had a new recruit, Edoardo Amaldi.[12] "In the present state of rapid change now prevailing in physics all over Europe, and with Fermi's appointment at Rome, an exceptional period has opened up for young people who have already shown sufficient ability and are willing to make an exceptional effort in theoretical and experimental study," Corbino had announced in one of his lectures to the second-year students. Amaldi was the only one who answered the appeal and joined our group. He was the youngest, and seemed even younger than he was because of his rosy complexion, so he was often called "the little boy" (*il fanciulletto*).

Jokingly, we assumed nicknames that originated from a ribald poem popular at the Scuola normale at Pisa, where it had been transmitted orally for a long period. Fermi and Rasetti knew by heart long excerpts from this poem and occasionally quoted it. Another origin of the nicknames was a parody of the offices of the Vatican Curia. Corbino was the Heavenly Father; Fermi, the Pope; Rasetti, the Cardinal Vicar; I, the Prefect of the Libraries, because I was interested in the library and knew the physics literature; however, I was also the Basilisk, because I was supposed to spit fire when mad; Majorana's extremely critical attitude earned him the title of Grand Inquisitor.

Amaldi had the most common sense and was also by nature and upbringing warmer and more humane than the rest of us. Majorana, with his profound skepticism and pessimism and his ironic bent, was not always easily accessible. Rasetti was pure brains, with childish forms of selfishness fostered by his mother, but in spite of his peculiarities, he was easily approachable. Of Fermi I have already spoken. All in all, they were excellent friends, loyal, generous, and honest. There were, however, differences in the degree of intimacy prevailing among us. Amaldi, Rasetti, and I could speak of anything with one another, including girls, love, politics, and career; with Fermi or Majorana, there was more reserve.

In that period, or shortly before it, I fell desperately, but unfortunately platonically, in love with Renata J. In the famous 509, we went to the most romantic spots in the vicinity of Rome, where she deployed much detachment. A high school friend whom I still saw occasionally told me stories aimed at inflaming my jealousy and sufferings. Many years later I learned astounding things about that period from Renata; had I known them around 1927, they might have changed my life. Both Renata and I were victims of our upbringing in a world where there was no pill, and where girls married as virgins (at least most of them). I still see myself trying to adjust the fringes of an interferometer while thinking of and sighing for her.

I was registered as a physics student only for one year. I registered in the fall of 1927 and obtained my doctorate on July 14, 1928. Because I had already studied engineering for four years, Corbino and Fermi shortened my formal study period in physics as much as possible. The most important subject I studied in class that year was theoretical physics with Fermi. His course followed his book *Introduzione alla fisica atomica*, which he had recently written. The book was not yet published, but Fermi gave me a set of proofs. It is an elementary book, still treating the atom in terms of the semiclassical Bohr-Sommerfeld approach. Only in the last chapter is there brief mention of the recent novelty "quantum mechanics." I studied the book thoroughly and soon supplemented it with information from Erwin Schrödinger's memoirs, which Fermi had explained to us privately as they appeared.

I was required to pass also an examination in "Higher Physics" as taught by Professor Antonino Lo Surdo. He was Fermi's avowed enemy, and his feelings were reciprocated. He had opposed Fermi's call to the Rome chair, stating that such a call was a personal slight to him. Scientifically, the man was badly out of date. He knew Drude's optics and J. J. Thomson's gas discharge book, but he was about thirty years behind his time, both in his teaching and in his anemic research. In class he showed beautiful experiments, but his lectures did not convey anything of vital import. Since he could no longer fight Fermi, he took it out on me. As the Italian saying goes, "He beats the donkey, being unable to hit the master."

As a third required course, I attended the lectures on mathematical physics given by Vito Volterra. I should add that the subsequent year I again attended his course, because he changed subject every year and from him one learned interesting notions of classical mathematical physics. Volterra's lectures were well organized and the subject matter was skillfully chosen (as I realized later), but his delivery, in a thin and slightly high-pitched or nasal voice, tended to put me to sleep. There were no textbooks, and one had to take notes; I therefore asked Amaldi to write for both of us, since he wrote faster than I, and also to wake me up if I fell asleep. Volterra used to close his eyes while lecturing and somebody said that this was because, being kindhearted, he did not want to see the students' sufferings. Except for these superficial shortcomings, the lectures were profitable. One learned the mysteries of the Laplacian, Green's functions, Poisson brackets, and similar topics. It seemed sometimes that Volterra did not want to reveal the physics underlying the equations and the analogies between different theories. After taking a course on elasticity and one on analytical mechanics, I passed the exam with the highest grades. I remember that to show my proficiency I mentioned the analogy between some elasticity coefficients and the capacity coefficients of electrostatics. I should not have done it. Volterra cut me short, remarking that he had not mentioned this in his lectures, almost as if he disliked my revealing a secret.

Volterra always treated me very kindly. I visited him in his residence in a palazzo at the center of Rome. He received me in his magnificent library and gave me some reprints of his papers on the applications of mathematics to biological population problems; later he helped me to secure a Rockefeller fellowship. He was, however, far removed from current physics, which did not seem to interest him. Corbino, in one of his sharp remarks, once said that mathematical physics as practiced in Italy was the "theoretical physics of 1830."

University rules required preliminary discussion within the department of one's doctoral dissertation prior to the final public examination. I presented a modest piece of experimental work on anomalous dispersion in mercury and in lithium vapor.[13] It was certainly no great shakes, but it was above the average then prevailing at Rome. When I went for the discussion, I found a committee composed of Fermi, Rasetti, and Lo Surdo. Lo Surdo immediately swamped me with questions on all possible types of interferometers, which I did not know, although, of course, I had mastered the ins and outs of the Jamin interferometer I had used. He then passed to optical features of crystalline quartz, although my own apparatus had an optics of amorphous fused quartz. I correctly told him that my work was not affected by the properties of crystalline quartz. Lo Surdo concluded, however, that I did not know what I was doing, that I was superficial and ignorant. This behavior was peculiar and unprecedented. I was furious, and after the examination I waited until Lo Surdo had left the building and then went to his office, where I found just what I had expected. Piled on his desk lay a couple of big German treatises, open at the pages dealing with interferometers. The good man had prepared for the occasion in order to make a fool of me.

A few weeks later, on July 9th, the formal discussion of my thesis took place, with eleven professors in attendance, among them Severi, Levi-Civita, Volterra, Fermi, and the young Beniamino Segre, subsequently a well-known mathematician. Corbino was absent. I discussed as a subordinate subject, given to me by Volterra, some properties of partial differential equations of the second order, which are much more profound than I realized at the time. Finally, I was excused and left the room, but I overheard the following exchange:

SEVERAL VOICES:	"He did quite well. Let us give him the maximum."
SEVERI OR LEVI-CIVITA:	"Yes, for sure, but I propose also cum laude."
ALL TOGETHER:	"Very well, then let it be 110 cum laude."
LO SURDO:	"No, I object."
SEVERI:	"But we are ten to one in favor."
LO SURDO:	"Yes, but the regulations require unanimity."
SEVERI:	"The regulations are idiotic! Too bad!"

I must say that the exchange I heard compensated me for the laurels I missed. Fifty years later, Beniamino Segre, by then a famous president of the Accademia dei Lincei, still chuckled when reminding me of the scene.

After completing my doctorate, I had to satisfy the military obligations prescribed by law, which I did at a recently instituted army officers' training school in Spoleto, an ancient town in Umbria, now best known for its summer festivals. Not long before I went there, I met Lo Surdo in the street one day. He stopped me, inquired what I was doing, and commented: "So; you will go to serve in the army.

Well, you will forget everything. Ordinarily it is a fatal interruption in one's scientific career. After your discharge it would be better for you to look for another profession." I thanked him for the friendly advice and touched wood (Italian style one touches something else). In our love for the gentleman, we had given him a reputation of casting the evil eye, and we had ample corroboration of this—for example, of an apparatus blowing up with catastrophical consequences as soon as he looked at it.

Lo Surdo occupied a wing of the physics building on the same floor as us; he also had an assistant, usually an insignificant fellow. He treated him in a way no scientist, even a beginner, would have tolerated. Lo Surdo's research is insignificant, with the exception of a method for observing the splitting of spectral lines when the source is in an electric field. The phenomenon is the electric analog of the magnetic Zeeman effect and is usually called the Stark effect. It is likely that Lo Surdo had observed it before Stark, but certainly he did not understand what he saw, possibly confusing it with a Doppler effect. When, a couple of months later, Stark announced his discovery, Lo Surdo was deeply disappointed at having missed the boat and tried to establish his priority on shaky grounds. Corbino had favored Lo Surdo's call to Rome hoping he would help to raise the level of the place. His expectations came to nothing, and several years later Corbino bet on Fermi with the same aim. He was disgusted by the jealousy shown by Lo Surdo and on further provocation gave him a memorable lesson.[14]

On Corbino's death in 1937, Lo Surdo was appointed director of the Physics Institute in preference to Fermi. I do not know how he wangled the appointment. After the promulgation of the infamous racial laws, he showed unusual and unnecessary anti-Semitic zeal. For instance, he locked the venerable Professor Castelnuovo, his colleague for many years, out of the library of the physics department. He earned Edoardo Amaldi's gratitude, however, by helping him to return from military service in Libya during World War II.

At the end of the war, Lo Surdo was dismissed from the Accademia dei Lincei and experienced some retribution for his fascist zeal. When I returned to Italy for the first time in 1947, he asked to see me. The undignified and servile manner in which he greeted me did not improve my opinion of him. I told him coolly that I knew how he had behaved during the war, that I had a good memory, and that he did not need to pay his respects to me. He died about a year later. Let him rest in peace. He had been generous to me in letting me use his spectroscopes at the beginning of my career.

Chapter Four

Scientific Springtime (1928–1936)

Smell of Amsterdam's Canals

Ci lasciaron talune una fragranza
così tenace che per una intera
notte avemmo nel cuor la primavera
e tanto auliva la solinga stanza
che foresta d'april non piú dolce era.

(Some of them left us a fragrance
so lingering that we had spring
in our hearts for a whole night;
and the lonely room was so perfumed
that no forest in April was sweeter.)
Gabriele D'Annunzio, "Le mani," from *Poema paradisiaco*
(trans. Louise George Clubb)

The *laurea* that entitled me to call myself Dr. Segrè completed my formal scholastic career, but my study of physics was to be a lifelong occupation. In fact, most of the physics that was to form the subject of my later work did not exist when I was at university; not even in an embryonic state. The neutron and artificial radioactivity were far in the future, not to mention particle physics.

Having finished university, I was thinking of the next step, which I hoped would be an assistantship at Rome. I did not have to worry about earning money, because my father could support me without any sacrifice and would do so gladly, but I wanted a salary as an acknowledgment of my work, although not at the price of taking a job that might endanger my scientific prospects.

For the present, fulfilling my military obligations would give me time for reflection. At the officers' training school in Spoleto to which I was sent, I entered a world utterly new to me. My new comrades came mostly from the Italian bourgeoisie, from every region of Italy; most of them were lawyers, literary men, small businessmen or landowners, with very few engineers or technicians.

Discipline was strict and unreasonable. One of the main occupations was changing one's uniform at high speed many times a day. There was enough food, but it was of poor quality. We were taught mainly by noncommissioned officers, who were happy to display their authority and zeal, using methods reminiscent of the Catholic catechism, occasionally peppered with cruelty. I was assigned to the artillery section, and a major gave us theoretical gunnery courses, whose content went back to about 1890.

One of my comrades whom I vividly remember was a Marquis Lignola, who belonged to an aristocratic Neapolitan family. He considered the Bourbons, ousted from Naples in 1860, to be his legitimate rulers and regarded the House of Savoy as usurpers. Lignola was small and somewhat clumsy. The other students provoked him into expressing his extreme opinions and then scurrilously made fun of him. The victim turned his eyes to heaven and "offered his sufferings to Jesus," showing courage, extreme firmness in his opinions (the pope had erred in dealing with the usurper), and strength of character. Soon his tormentors started to respect him and left him in peace.

The strictly disciplined military life and lack of freedom were boring, but restful, because one did not have to think of anything or make any decisions; the sergeants prescribed all our activities. I had taken with me several books: Courant and Hilbert's *Methoden der mathematischen Physik,* Oscar Wilde's *The Picture of Dorian Gray,* and similar esoteric reading. I used to read them during a compulsory siesta period after lunch, lying on my field bed. The books, in foreign languages to boot, allowed me to put on airs with my comrades, but also spiritually transported me far from my military surroundings. A few weeks after the beginning of instruction, we had a free Sunday, and a comrade and I used the short furlough by going to Gubbio and other places near Spoleto. I still remember the exhilarating impression of having regained our freedom, albeit only for a short time. We manifested it by carrying our heavy army swords on our shoulders like hoes. It was a childish gesture, which could have brought unpleasant punishment if detected. A little later I was called to headquarters and given three or four days' furlough for Yom Kippur; I did not know I was entitled to this leave and was pleasantly surprised. Several of my comrades proposed converting to Judaism if that produced leaves of absence. I went to Rome by train, put myself in mufti, rejoicing in contact with the soft flannel of my elegant trousers, and, well groomed, went to court Renata J.

Quite unexpectedly one day, Fermi and his wife Laura showed up in Spoleto in their small yellow Peugeot car. The visit greatly raised my spirits because it gave me an opportunity to resume contact with physics and because it demonstrated that my Physics Institute friends remembered me.

At the officers' training school I had my dose of small adventures, such as falling off my horse right in front of an inspecting general. I immediately got to my feet, unharmed, and grabbed the horse by its bit. The good general kindly commented to me that anybody could be thrown by a horse, and that he had appreciated the way in which I had done the correct thing by preventing the animal from running away! I noted that whenever there was an inspection by some visiting bigwig, I was always called upon to aim the guns. My superiors had soon discovered that I was to be trusted to make the required simple calculations correctly, without sign errors, which occasionally pointed the guns backward.

I graduated from the officers' training school on January 11, 1929, and resumed work at the laboratory until the following July 1, when I was commissioned a second lieutenant in the anti-aircraft artillery and stationed at Forte Braschi, very near Rome, as I had requested. On this assignment, I did not have much to do. I slept at home, going to the barracks early in the morning with some book on physics to study. However, in my military service, I learned many other things besides physics. My captain taught me a card game called *scopone*, which I greatly enjoyed, and also revealed to me novel attitudes to life. I had been brought up with the idea that I should work, that everything had to be taken seriously, and that I was expected to excel, or at least to do well. From my captain I learned that zeal was a grave fault; that many problems took care of themselves provided they were left alone; and that when one received an order, contrary to what we had been repeatedly told, prompt execution was imprudent, and that it was advisable to await its countermanding. Furthermore, contact with the soldiers gave me a chance to get to know contemporaries of very different education and from diverse social and economic conditions. I could also see firsthand the differences between soldiers from the various Italian regions. The unhomogeneousness of Italy's population was such that soldiers often did not have a common language, used as they were to speaking dialect. We had orders, moreover, to see to it that soldiers from the rice-eating north and those from the pasta-eating south did not throw away their rations on alternate days, when the food unusual for them was served.

As a commissioned officer, I bought myself secondhand a gorgeous blue cape, which, although not a very efficient protection against cold, was certainly elegant. When I finished my service I sold it to Edoardo Amaldi, then also an officer.

During my service in Rome, I managed to go once in a while to the laboratory and keep in touch, but I did not have time for experimental work. One day I was urgently called from the barracks at Forte Braschi to the Physics Institute at Via Panisperna. For some reason, all the scientists were away, and the factotum of the institute, who did not know English, was faced with an obviously important Indian visitor with whom he could not communicate. I rushed down and found that the visitor was none less than Sir Chandrasekhara Venkata Raman (1888–1970), who in 1930 received the Nobel Prize in physics for his work on the diffusion of light and the discovery of the Raman effect, on which Roman physicists had done important work. I did the honors as well I could, unexpectedly helped by being in dress uniform, with a blue sash and conspicuous gold epaulettes. Raman believed that I had dressed like this to honor him and thanked me; I did not disillusion him by revealing that the true reason was H.M. the Queen's birthday!

During my military service, I was once unjustly placed under arrest for something I had not done. By chance, I mentioned this to a friend of mine, who without my knowledge spoke about it to his father, a powerful general. With surprising speed, my punishment was commuted to a much lighter penalty, and the colonel who had condemned me without even talking to me must have found himself in serious trouble. I was discharged as a second lieutenant on February 15, 1930, and placed in the reserve.

In 1928 I had published my first physics paper jointly with Edoardo Amaldi, a short note in the *Rendiconti* of the Accademia dei Lincei, introduced by Corbino, summarizing my doctoral thesis.[1] The following year, Amaldi and I published a second paper, dealing with the Raman effect.[2] I did much early work with Amaldi, but because the rules then prevailing in university competitions penalized collaborative efforts, we often divided the work in a friendly fashion after we had finished it together.

Next I wrote a paper on anomalous dispersion in molecular band spectra. The ensemble of the absorption lines near the head of a band gives a peculiar variation of the refractive index, which I explained with Fermi's help.[3] In about the same period, Amaldi and I produced a couple of papers in the wake of Fermi's study on quantum electrodynamics. They contained results similar to those of a famous and often quoted paper by Eugene Wigner and Victor Weisskopf. Our work was done earlier, although less detailed.[4]

Fermi observed strict rules concerning the publication of his work and that of his students. He did not permit publication of completely insignificant results. Results of little importance appeared only in Italian. He allowed publication in the *Zeitschrift für Physik,* or as a letter to *Nature,* only of papers he considered important. This wise policy was motivated by his desire to establish an international reputation for our Rome group. He did not want any foreign reader ever to be disappointed in reading one of our papers; there should always be something interesting in them. He applied this rule strictly, and his judgment on the quality of an investigation rarely erred.

Furthermore, when Fermi developed a theory capable of many applications, such as his quantum theory of radiation, he explained the principles to us, gave some examples, and then left to us the satisfaction of finding further applications. Not a little of the work of this period started out like this.

"You live off Fermi's crumbs," my father, who knew no physics, but was a shrewd observer and knew men, once told me. This was quite true, and I did not forget it. Only later was I able to do something truly my own. My mother, who wanted to know whether my studying physics would lead anywhere, once asked Fermi what he thought of my ability. The ever-truthful Fermi answered, correctly and objectively, that it was too early to pass judgment and make predictions. I was not present at this conversation, but knowing both parties, I am sure that my mother continued worrying about the perfectly honest answer she received.

In the late 1920s and early 1930s, we worked intensely, but in a relaxed way. We used to read the most important journals, such as the *Zeitschrift für Physik, Nature,* and the *Proceedings of the Royal Society,* eclectically, hunting for experimental inspiration. My official job—a relic of times past—was that of "Conservator of the Tuning Fork," a sinecure that amounted to an assistantship. When students came looking for research subjects, even in theoretical physics, they were often referred to me, because I usually had a good supply of ideas. I would suggest a problem and explain it to the student in detail. However, when it came to technical details, I often had to send the student to Fermi, who instantly gave the key to the solution. The problems I set were usually in atomic physics or connected with it. I asked Fermi why he did not give out the problems himself; he answered that those he thought of were usually too difficult for students, and that those at their level did not interest him. When Fermi and I co-authored papers, I was often entrusted with the writing. I did not mind writing, and I was proud of the assignment. "It is clear that since you will never get the Nobel Prize for Physics, you are preparing yourself for the Literature Prize," my friends teased.

The study of theoretical physics at Rome progressed under full sail. Fermi was obviously not only a first-class theoretician but a superb teacher; one could not ask for more. He had contacts through conferences and visits with the principal theoreticians of his own age, as well as with Arnold Sommerfeld, Paul Ehrenfest, and some others of the previous generation. Soon postdoctoral fellows from abroad started to come to him, to learn and work under his inspiration. Among the first to arrive were Hans Bethe, Rudolf Peierls, and George Placzek, who acclimatized himself to Rome better than the others, learning Italian and striking up a solid personal friendship with Amaldi and myself. Others who followed included Edward Teller, Fritz London, Felix Bloch, D. R. Inglis, and Eugene Feenberg, who became Majorana's particular friend.

The situation in experimentation was different. One cannot learn the experimental art from books, and we felt the need to go see what happened elsewhere and to learn techniques on the spot where they were practiced. Franco Rasetti was the first to take off, in 1929, going to the California Institute of Technology at Pasadena, then dominated by R. A. Millikan. I do not know what influenced his choice. The Rockefeller Foundation granted him a fellowship.

The Rockefeller Foundation was a great benefactor of physics in that period, helping it through the general economic depression and, later, Hitler's persecutions. A shrewd and farsighted choice of Fellows was at the base of the Foundation's success. Looking today at a roster of Fellows of that period, one wonders at the sagacity of the selection, and the wonder grows when one considers that Fellows were appointed at an early age, often before the work that later distinguished most of them. The selection occurred on the basis of recommendations by two or three established professors whom the Foundation trusted, mostly in the applicant's country of origin. In Italy, it seems that these advisors were Volterra, Levi-Civita, and Corbino, a choice that in itself shows the Foundation's sagacity. All three were honest, experts on their subjects, and well informed about the local situation. They were not in the good graces of the Fascist government, but if this was resented in Italian high places, it did not matter to the Foundation.

Rasetti did very well scientifically and personally at Caltech. He accomplished important work on the Raman effect in gases and fostered the good reputation of the Rome group. He also visited Berkeley, from where he sent me a postcard. At the time he had the impression that Caltech was way ahead of Berkeley.

When he returned home, he spoke only of California's wonders: of Mount Whitney, which he had climbed in winter, of Pasadena's orange groves, of the wealth of American laboratories, of the attractiveness for him of the American way of life. He also proudly showed off a toothbrush he had bought in Hawaii. Needless to say, he would drive only an American car and bought a Ford.

Back in Rome, Rasetti continued to do fruitful work on the Raman effect. I too tried something on the subject without obtaining anything of importance. Amaldi and Placzek studied ammonia with better results. All this work on the Raman effect lasted into 1931.[5]

In November 1930, in the form of a very short letter, I sent *Nature* the first paper of any importance thought out and executed entirely by myself.[6] Atomic theory gives the laws according to which the electron jumps from one atomic energy level to another, emitting photons. Sometime these rules are violated and there are so-called forbidden transitions. My little discovery concerned certain (S-D) forbidden transitions in the spectra of the alkaline metals, and the paper shows that they are owing to electric quadrupole radiation, neglected in the usual first approximation calculations. The proof is obtained by observing the Zeeman effect of these lines. I examined absorption lines because emission lines are too weak. The experiment was very simple, although at the limit of the resolving power of the instruments available to me in Rome.

The best instrument I could use was a large Hilger prism spectrograph bought by Professor Lo Surdo, and located in his rooms. Lo Surdo very kindly and generously gave me permission to use this instrument. After a few days of work, I succeeded in seeing with my own eyes the potassium absorption lines delineated on a violet continuum produced by a hydrogen discharge. When I energized the magnet in which I had placed my absorption tube, the lines broadened and almost disappeared. With a little more work, I adjusted the instrument to obtain its maximum resolving power. I still remember my great elation in recognizing that the Zeeman pattern I was seeing was the one I expected for quadrupole radiation. This was my first small discovery, and it made a permanent impression on me. If my previous work could be called "Fermi's crumbs," this was my own. Furthermore, it had been obtained rapidly, which enhanced its impact.

My friends at the Physics Institute bestowed upon me the title of "Lord Quadrupole," and, more important, Fermi told me to publish the work in the *Zeitschrift für Physik*.[7] The self-confidence of young scientists is a delicate plant. Even Fermi, who looked so self-assured in later years, and had performed extraordinary feats when very young, was not sure of himself until he went to Holland at about the age of twenty-one.

What I had seen in my study of the Zeeman effect of the S-D combinations in potassium was conclusive, but did not reveal all the details one would have liked to know. Unfortunately, the instruments at my disposal in Rome could not give more; I had squeezed them to their limit. I was thinking about what to do next when the great Dutch physicist and chemist Peter Debye visited Rome. There was a reception in his honor at Enriques's house and I was invited. Debye inquired in a friendly way about what I was doing, and I told him about my quadrupole work, adding that I was at a dead end for lack of adequate instruments. To my surprise, Debye sternly answered that my complaints were mere excuses; only lazy people were stopped by so-called lack of means. At the time, I was hurt, but the lesson sank in and was highly beneficial then and later. Debye himself suggested that I try going to some foreign laboratory. Four laboratories seemed likely to have a diffraction grating (a device consisting of narrowly spaced slits used for measurement of wavelengths) adequate to my project: E. Back's in Tübingen, H. Cohnen's in Bonn, F. Paschen's in Berlin, and Pieter Zeeman's in Amsterdam. I wrote four letters explaining what I wanted and asking for hospitality. (My father was more than happy to pay my expenses, so I did not need financial help.) Back did not reply; Cohnen said that his grating was at the moment out of commission, because his institute was being rebuilt; Paschen told me that he liked my idea, and that he had just put one of his doctoral candidates to work on it. I was furious at this unexpected answer, but the project must have come to naught, because I never heard any further news of it. Zeeman, a Nobel Prize winner and the discoverer of the celebrated Zeeman effect,[8] told me to catch a train and come to Holland.

I did so without delay, and arrived in Amsterdam at the beginning of the summer of 1931. After finding lodgings at a pension, I introduced myself to Zeeman and told him my precise work plan. Personally, Zeeman was most courteous, benevolent, and affable. He was then sixty-six years old and had ceased active laboratory work. I had the impression that he was not conversant with modern theory, and in particular with quantum mechanics, which was then still a relatively new field. On the other hand, he was a superb experimenter and a master of optics. In any case, talking to him was most instructive. His way of considering an experiment was new and unexpected to me. He had a refreshing diffidence about theory, and while he did not underestimate its power, he knew that nature had more imagination than we did. He thus pushed for thoroughness in experiments, saying that something unexpected was likely to happen. He was right, even in my simple case, when everything seemed predictable.

Zeeman immediately told me that his diffraction grating was the greatest treasure of his laboratory, and that he could not entrust it to me alone, since I did not have any experience in its use. He suggested that I collaborate with Cornelius J. Bakker, a doctoral candidate of his, who was familiar with the grating and with other delicate instruments in the laboratory. I found the proposal reasonable and fair and accepted it at once. Fortunately, it turned out that Bakker was a very nice person and soon we struck up a close friendship that lasted until his untimely death. I still have his portrait in my study.[9]

I immediately started preparing an absorption tube containing potassium. For this purpose, I cut off a piece of potassium and replaced the rest of it in what I believed was the bottle from which it had come. I had overlooked the presence, next to it, of an open bottle of acid. Without looking, I put the residual piece of potassium in the wrong bottle. For about half a minute nothing happened; then a tremendous explosion shook the laboratory. Everybody ran to see what had happened, and I can hardly say how I felt, although I was, fortunately, bodily unhurt. I deeply admired the calm Dutch, who gave no sign of commotion and did not ask me to leave.

The work, which started so dramatically, continued smoothly and successfully. We rapidly obtained all the expected results, as well as several more that rounded off and completed the picture. Zeeman took a liking to me, and one day he asked me about my plans for the near future. I told him that I had applied for a Rockefeller fellowship, but that nothing seemed to be happening. Zeeman remarked to me that he knew somebody in the Paris office of the Foundation, but said no more. By strange coincidence, about a week later I received a letter announcing the grant of the fellowship. Although he never told me so, I suspect that Zeeman may have had a hand in it. When I left Amsterdam to return to Rome, Zeeman told me that whenever I wanted to come back to work in his laboratory, I would be welcome. I subsequently took advantage of this cordial invitation, returning to study different types of forbidden lines, always together with my original co-worker and friend, Bakker. Zeeman also gave me a picture of himself with a warm inscription. He was not far from retirement and possibly thought of me as one of his last disciples.

During my stay in Holland, I became acquainted with its cigars. Zeeman invited me to a couple of very formal and elegant dinners at his home, on the occasion of which he offered his guests exquisite cigars, which I liked. I noted the brand and kept buying it, although I later switched to Otto Stern's favorite brand, which was available in Germany and was equally good.

Many months after publication of our papers on quadrupole radiation, I received Back's response to my application to work in his laboratory; in reply I sent him a reprint of our work.

In the summer vacation of 1931, I went to England for the first time. Between the day Zeeman's lab closed for the summer vacation and a date I had with Amaldi and Rasetti for a hiking tour in Norway, there was time for a short visit to London. I boarded a ship and saw a young man, about my age, with skis. It occurred to me that he was possibly a student who had spent the winter in Germany at some university and was now returning home. Hazarding a wild guess, I asked him whether he had been studying with Sommerfeld. Incredibly, my surmise turned out to be right, and the young man was absolutely flabbergasted. He was very friendly, found me a suitable hotel in Russell Square, and offered to help me to orient myself in London. In exchange, he asked that I join him for my first English breakfast, because he wanted to see my reactions. Undeterred, I ate porridge and haddock and drank tea, all of which I liked, much to my new friend's surprise.

My first impression of England in 1931 was of a great imperial power at sunset. As I wrote home, one saw a thousand signs that the country had passed its zenith. My parents had witnessed the coronation of Edward VII in a very different, splendid period. While in England, I visited Cambridge, where I saw J. J. Thomson but did not talk to him.

Soon thereafter, I met Amaldi and Rasetti in Oslo. We had planned a long hiking trip on Norway's glaciers. We left Oslo by train and alighted at 2 A.M. at Finsoe in daylight. We started walking on the Hardanger Fjell until we reached the sea. Later we explored other fjords by boat and on foot.

On my return to Rome, I kept working on forbidden lines and found other interesting features of their Zeeman effect, revealing their origin, in cases where they could not be the result of quadrupole radiation. Furthermore, theory indicated that there should also be forbidden quadrupole lines in X-ray spectra. I made a systematic search of the literature to see if by chance somebody had observed them without understanding their origin. To my joy, I found that that was indeed the case.[10] All this work was noted, and I had the satisfaction of seeing myself quoted in a new edition of Sommerfeld's famous treatise, *Atombau und Spektrallinien.*[11] Sommerfeld was always ready to help young scientists and had excellent relations with the Rome group.

In line with our program of learning new experimental techniques abroad, Fermi suggested I spend my Rockefeller fellowship in Hamburg, where I would be able to study vacuum technique (one of our weaknesses) and molecular beams under Otto Stern.[12] He made the necessary arrangements with Stern, and at the end of 1931, I set out.

Fermi and other friends had described Hamburg's wretched climate and dark, wet winters to me, and I found that they had not exaggerated (they did not, however, know what came after—that is, the long, beautiful northern spring, unknown in Italy). I installed myself satisfactorily in a rented room in a private house. The Rockefeller fellowship stipend of $150 per month made me rich and, to satisfy the usual pride of spoiled children (*figli di papà*), I did not want any supplement from home. Soon after my arrival at Hamburg, following Fermi's advice, and as an act of courtesy, I visited the Italian consul. This gentleman opened my eyes to the German politics of the time, and even more so to Italian foreign policy, giving me a well-reasoned lecture on anti-Fascism. It seems that among Italian officials there were some who thought independently and had the courage of their own convictions.

Stern was then in the process of making important discoveries and was entirely submerged in his work. He used to arrive at the laboratory every morning at about 10 o'clock; at noon he had lunch with his assistants and guest workers, then returned to the lab, if necessary until late in the evening. The schedule depended considerably on the behavior of the instruments and the vagaries of the vacuum.

Stern suggested that I finish an experiment on the dynamics of space quantization and explained the motivation and theory behind it to me, as well as the details of the existing apparatus. This had been built by my predecessor, the American T. E. Phipps, whose fellowship had expired before he could obtain results. I asked Stern to teach me some physico-chemistry, and in several conversations he gave me interesting illuminations of thermodynamics.

After a while he left me to myself. I tried to learn techniques by watching Stern, as well as younger people such as F. Knauer, Otto Frisch, R. Schnurmann, and B. Josephy. Stern's institute was small in size and in number of scientists, but in spite of this there were not many exchanges between its workers. Schnurmann was the most open, and he introduced me a little to German life. Others had their girlfriends or other concerns, and as soon as the day's work was done, they left on their own private business. Frisch served as Stern's personal assistant at that time and was involved in two major experiments, a demonstration of de Broglie's waves with helium atoms, and the measurement of the magnetic moment of the proton.[13]

Stern taught me a way of experimenting that I had not seen before. He calculated everything possible about his apparatus, such as the shape and intensity of the molecular beams he expected to generate, and did not proceed until preliminary experiments were in complete quantitative agreement with his calculations. This modus operandi slowed down the preliminary work, but it shortened the total time by making it possible to avoid errors and was absolutely necessary for the extremely difficult experiments Stern was conducting. The method allowed him to localize sources of misbehavior in his apparatus and of failures, and to come to a firm decision as to whether there were new and unexpected results, which occurred repeatedly. It was a rigorous and most useful schooling, very different from Zeeman's, but just as valid. I learned much from both, more in the philosophy of experimentation than in technical details. Years later, I saw the totally different, much more pragmatic and empirical, approach taken by Ernest Lawrence. Tutte le strade portano a Roma.

There was also an active theoretical seminar at Hamburg. Pauli had been there until recently; William E. Gordon and later H. D. Jensen followed him. Hermann Minkowski, who later became a noted astronomer, was a lively member of the company. W. Lenz was older, and he seemed to me less interested in current problems.

As to my own work, having thoroughly studied the apparatus I had inherited, I concluded that to make it work, there needed to be a radical change in the method used for producing certain magnetic fields, although I had no idea how to achieve this. I had long admired James Clerk Maxwell's *Treatise on Electricity and Magnetism* (1873), however, and one day, while looking at an illustration in it of the magnetic field produced by a rectilinear current in a homogeneous magnetic field, I immediately saw the solution to my problem with molecular beams. The theoretical analysis had to be changed, but that seemed to me more feasible than following Stern and Phipps's original experimental plan.

Stern approved my idea as soon as he heard it, got me the few extra parts I needed, and told me to rebuild the apparatus in the way I proposed. As to theory, I knew whom to look to for help. I wrote a letter submitting my problem to Ettore Majorana in Rome, and soon received the answer I needed. By then it was spring, and over the Easter vacation I went to Rome, where I reported all I had seen and discussed at Hamburg. There was enough to keep my theoretical friends busy, and Majorana, Gian Carlo Wick, and Ugo Fano were soon struggling with three different problems strictly connected with the Hamburg experiments.

I longed to find a girlfriend in Germany with whom to escape from my priestly Italian life. My experiences with girls of "good family" in Italy had on the whole been negative, and I had reached the conclusion that if I wanted to lead a normal life, I had better look further afield. Once in Hamburg, I got busy in that direction, but it was not easy for a foreigner who spoke the language imperfectly, who did not know local usage and customs, and who was not able to judge people from subtle signs and gestures, which had different meanings in Germany and in Italy, to make the necessary contacts. I reproduce here some notes I made in 1932:

> "125, 130!"
>
> "Ein Moment! Langsam!"
>
> Stern's impatient voice, emerging with effort from the crevice left in his mouth by the large cigar he is smoking, recalls me to my duties. Galvanometers are insensitive to the impatience of their users, and one has to wait 156 seconds between readings, even when one is in a hurry.
>
> Reading the light spot of the reflecting galvanometer, Stern turns his back to me. I sit higher than he; in the semi-darkened room I can see only the light spot of the instrument and his shining skull. I take advantage of my position to look at the clock: 7:42 P.M. I must arrive at the Hotel Esplanade by 8; I am meeting a girl for a blind date. My upbringing and a certain sense of chivalry urge me not to be late, not even one minute, especially in this wintery weather—it is full February in Hamburg. In eight minutes at the latest I must get away; no easy matter, because when enthused with his measurements, the boss—"the Chef," as they say here—becomes intractable and does not recognize that there may be obligations more pressing than molecular beams. "If the

apparatus should suddenly fail?" I hardly dare to let the thought emerge to consciousness, much less surreptitiously to turn a stopcock and violently end the evening's work. In the meantime, I turn my pointer too fast and earn a second scolding. If Stern turned around, he would detect my impatience.

A few more points and the curve is completed. . . . Will it suffice for the evening? I know the scene too well, and I know what to expect. There is no point in deluding myself. After one measurement, another. This could go on until 11. This time, however, I will overcome my shyness. At 7.50, I have decided, I'll quit.

Stern turns around exactly at 7.48. "Enough for the evening!" Without another word, he stands and exits the lab. Telepathy? A patron saint? I switch off six circuit breakers and turn four or five stopcocks to secure the apparatus. I put on my overcoat, wash my hands (Stern's fine Cologne soap has an excellent fragrance, and I use it with pleasure, despite the freezing water), take a last look at the various dials, and rush away.

Outside, the freezing air blows on me and I want to run. I cross the space to the Esplanade trotting.

"White gloves, a black hat, a fur overcoat, rather tall." From this description I am supposed to recognize the Unknown. However, in this cold and at this hour, there will be few people. In fact, as soon as I arrive at the designated spot, I see a girl stepping off a streetcar. She approaches; it is she.

"Good evening, Fräulein F.?"

"Ja!"

"Segrè."

"Ah!"

A few perfunctory sentences. It is cold; we have to decide quickly where to go. She suggests the Alster Pavilion, and I accept with alacrity. I have already noted that she speaks clearly enough for me to understand her German easily. Thank God.

We hurry to the Alster Pavilion and take shelter in the coffee house. My glasses, as usual, immediately fog up; I barely have the time to notice that the place is nearly empty. All the better: we want privacy and a minimum of noise, not the crowded freedom of dancing floors. The place is elegant, but almost deserted, which makes me suspect its prices. I know from experience what dangers lurk in menus. The thought is unworthy, but I cannot hide it from myself. A waiter approaches to take our orders. She orders a cup of tea. I ask for chocolate, because tea gives me insomnia. The finance minister smiles. Tea and chocolate are economical, a good omen. German girls know the value of money in these days of depression. Many struggle hard to make a living—who can forget the faces in the streetcars, morning and evening, during the rush hour? Other girls have learned the rewards of frugality from the movies. In movie romances with happy endings, the heroine always drinks lemonade; champagne is reserved for those that end badly. Pleased by her frugality, I suggest she try the cake; she refuses and I insist, but to no avail.

In the meantime, we have started a conversation and I see her for the first time, or rather I look at her carefully, perhaps with snake's eyes, as we jokingly say of Rasetti. She is wearing a blue suit, of no particular distinction, as most middle-class girls here do. I study her face. She is definitely pretty, I decide, but a few seconds later I change my mind. This reversal occurs several times more, and I feel the need to find the reason for it. It is the extreme mobility of her face, its most outstanding characteristic. A smile, a movement of her mouth, a blink of her eyes, changes her expression profoundly.

Otherwise she has indifferent teeth (oh, memories of the Buddenbrooks), a somewhat large nose, vivid, short-sighted eyes, short black hair, parted in the center, combed with a certain coquetry. She looks almost Italian, I tell her. I suspect she might be Jewish, but I do not ask. I have learned to ask as little as possible and to wait for information to emerge by itself. Indiscretion is the most direct route to aversion.

The conversation begins with easy subjects: travel and books, and I understand immediately that I have to do with an intelligent person. We come to politics—usually it would be the inconclusive, disorganized, internal German politics, but this time we start with foreign politics. My pacifist remarks meet with cold and skeptical answers; in part out of conviction, in part out of contrariness, I heatedly insist. I receive a declaration of faith in National Socialism. It is not an ardent declaration, but I am glad I did not ask whether she was Jewish. (Later I was to learn that I would not have been the first to inquire.) Glad also because the same question would have been bounced back at me.

Politics is always dangerous ground, especially in foreign countries. I turn the conversation to a safer subject: travel. By now I know half Europe and it is not difficult to find places familiar to both of us. We talk of Norway. Italy is next, or rather the Italy she knows, which is the Lago di Garda region. I try to recall my memories of Gardone, Riva, Sirmione, but I soon perceive that her impressions and recollections are more vivid and precise than mine. She visited Gargnano two years earlier with an aunt and someone else. I believe Gargnano, but the company is not clear to me. Eyes do not shine at the recollection of landscapes only, nor aunts inspire such strong emotions. A motorcycle seems to have played an important part in this trip. But the events at Gargnano are no concern of mine, and it is getting late.

I propose to accompany her home. We start on foot. We pretend not to notice that streetcars are available. Instead, we walk the dark and nearly deserted streets. We say little to each other as we walk: I imagine how these streets would appear in the summer, with their flanking trees in full leaf.

We say goodnight at her door. I return by streetcar on an itinerary destined to become disagreeable to me from sheer repetition. Suddenly I note that in the long ride I have landed in an unfamiliar part of Hamburg. I am angry at my mistake. I think of the Rockefeller Foundation that pays me, of Fermi, of my work, and I discover I am murmuring to myself in a deep voice (a souvenir of Fermi), "What a life, my son, what a life."

• • •

Now we meet frequently; going to the movies, to the theater, to dances, or at other places where we can be together without too much talking and without boring ourselves. Slowly we start sharing each other's lives. Sometimes I recognize names that recur in her talk. I try to remember them and to connect different facts. Once in a while, I discover something, and involuntarily my face lights up. The period is interesting, though without event. It resembles the hour before sunrise, which seems to contain the future of the next day, although one is ignorant of what weather to expect. I realize that now we are seeding the subconscious, for the future, and that I must be delicate, agreeable, and avoid opening unbridgeable chasms. Not everything is amusing, but occasionally a small accident, a short exchange, an apparently meaningless fact, make up for the great sacrifice: the loss of sleep. I dimly feel that the tenuous tie binding us is becoming stronger. . . .

One evening we return home about midnight; it is not very cold, but windy and it is snowing heavily, unusual weather for here. No living soul is on the street; only we persist in walking. The snow slips off her ponyskin, but sticks to my overcoat and changes me into a snowman. I feel elated, enjoying the cold air, turning my face toward the snow. Suddenly I remember the huge white dogs, gigantic sheepdog pups, that, many years ago, accompanied me to the Sebastiani Hut in the Abruzzi and dug their den in the snow to sleep at night. I recall also the wolf dogs I saw in Norway, and I rejoice almost with animal pleasure. I take her arm and try to explain my thoughts and sensations to her. Comparisons with animals do not please her; she must have some metaphysical idea of the superiority of man in creation that I do not share. However, the charm of the juncture does not escape her, and I suspect she must share my mood. Inadvertently, I stroke the ponyskin; it has short hair, almost shaved, and I feel that, in this weather, it must be comfortable. Anyway we keep close to each other, so that we feel our bodily warmth. We continue walking in silence from the center of the city, always through new streets (in how many ways can one go from the center to her home?). We arrive, a greeting, a warm handshake, and then I run off. Only now do I become aware that snow consists of water, and that my overcoat has become unpleasantly waterlogged. The heat of a streetcar finishes melting the snow and, almost, the poetry attached to it. This evening something new was in the air, and maybe the first spring flowers were blooming under the snow.

• • •

One dance more. I have arrived at the "Boccaccio" on the run, directly from the lab, and hungry. The hall is still empty, and we choose a good table with the utmost care. I look for something to eat; she takes her usual tea, but it is early, the orchestra does not play, and our mood is not jubilant. Conversation languishes. I tell some of my usual stories, and in response I hear some more tales of Lake Garda. By now I know that a romantic interlude and something important in her life occurred there. By now it is also plain that the participants in that expedition were two and not three. The aunt has vanished into thin air. . . .

Now we should dance. I regret my poor dancing and hesitate to ask her. For the present, we are good enough friends to sit at a table without boring each other, and dancing with a partner like me might not be pleasant. I tell her all this frankly, and she replies, equally frankly, that she will tell me when she is ready to dance. I am reassured.

A band plays "Good night, sweetheart, all my prayers are with you.". . . Its effect is manifest: she asks me to dance. It is already late, and I was thinking of going home, but it now seems that a mood is waking up in the hall, and I find myself committed to a fox-trot on a shining, overcrowded floor, in a hot hall. Somehow the first rounds of the fox-trot wake me up. Suddenly I realize that I like dancing; I like it more than I ever liked it before. A lock of my partner's hair has brushed my face. Now some small gas-filled balloons are thrown into the room, starting a lively competition to catch them. Many of the balloons are sucked to the ceiling and stop at the openings of the air ducts, strange bubbles, lazy and slow. The narrow parquet floor is full of people, and every few steps we collide with somebody. I suggest we move to an emptier room, and we briefly debate the proposition, but the conversation is soon moot, because the dance takes me over entirely. I do not know how, but now something strange has entered me, from the surroundings, from the general cheerfulness, and above all from the living thing I am holding in my arms. She too, I easily see, is in a golden moment, and from now on we do not miss a dance.

Tacitly, we have come closer. I am aware that our relationship is changing in nature. We are no longer restrained by the cold and correct forms of social rules, but something warmer, livelier, more personal creeps over us as I keep turning on the shining floor. It is a waltz, and I am slightly dizzy. Spontaneously, but somewhat timidly, I lean my face against hers. Many here dance this way; it is not unusual. The air of the hall is slightly fogged now, and the waltz's tones and the sweetness of the moment lull me into a happy state of semi-intoxication. It is an enchanted moment.

But now the last dance has ended, and we must leave. We have not noticed that it has become very late. At 11 I thought of ending the evening because I was sleepy; it is now 2 A.M., and I think only of staying longer. As always, we walk home. On the way, however, my mood suddenly changes. I rejoice, but I do not dare say a word. The silence is heavy, but stupid words do not fit the moment, and I do not want to spoil my inner happiness with some trivial phrase. I feel a great urge to kiss her, but in spite of training I have not lacked recently, I am even shyer and more indecisive than usual. I think one minute more, and then I kiss her at length on her mouth. We continue our way in silence. I only say to her that I am very happy indeed, but that I do not want to speak. She understands, and we continue our way, arm in arm. We have arrived; I kiss her hand and rapidly go away. D'Annunzio's lines: "e tanto auliva la solinga stanza / che foresta d'april non più dolce era" ring in my ear.

• • •

Today I discover a whiff of spring in the Hamburg air. We are approaching Easter, but the season called *Vorfrühling* (early spring) here

> is only starting now. Halfway between winter and spring, it is an uncertain period when two seasons compete, each in turn overcoming the other. The trees are still bare, but small green spots sprout on the brown branches; they are almost invisible buds, but so many that looking from my window at the tree-lined street, I discover a new green tone. Across the street, the plants in the botanical garden too seem to rediscover the existence of chlorophyll. A stretch of blue and transparent sky, the lengthening of the day, an occasional balmy afternoon with a softness in the air portend the equinox. Then suddenly a snowfall and everything reverts to winter. The Alster freezes in spots, and the seagulls on the ice floes give a polar aspect to the center of Hamburg. Today I leave for Italy. Not willingly, I have too many ties to Hamburg now, but my absence will be short, and in my thoughts I anticipate my return. I am happy, with a deep, serene joy that brings a smile to my face and fits well with the spring.

Although my girlfriend lived at home with her parents and worked in an office and I was busy at Stern's institute, we were occasionally able to meet during the week, and we always tried to spend the weekends together. We did not have a car, but we did not miss one. Public transportation was efficient and took us wherever we wanted to go. Once we crossed the Lüneburger Heide—whose vegetation, chiefly juniper, called *Wacholder*, greatly resembled the piñones I was to see later in New Mexico—on foot.

It was natural that we should think of marrying, but there were tremendous obstacles, and as time went on, they became an insurmountable wall. She was a German nationalist and was becoming more and more of a Nazi. I knew enough of Nazi ideas to realize their wickedness, but we avoided the subject. We preferred to live from day to day, enjoying what we both knew was one of the richest periods of our lives. We also felt, at least subconsciously, that the hour of reckoning would come. In an album she gave me at the end of 1932, she copied Schiller's lines:

> Dreifach ist der Schritt der Zeit
> Zögernd kommt die Zukunft hergezogen,
> Pfeilschnell ist das Jetzt entflogen
> Ewig still steht die Vergangenheit.

(Threefold is the march of time:
While the future slow advances,
Like a dart the present glances,
Silent stands the past sublime.)[14]

During the summer we found some excuse for a vacation together in England. Waiting for her, I saw some more of London, and from there we went first to the south coast at Brighton. Soon we moved to the Lake District, near Windermere, romantic places full of literary reminders. We seriously thought of going on to Gretna Green and taking advantage of the local privileges allowing a fast and simple wedding. I did not know of the existence of the place and of its traditions, about which she informed me. Other counsel prevailed, however, and we did not proceed with the plan.

I then returned to Holland and my forbidden lines, to which I was attached, although I had dropped them during my stay with Stern. My mood at the time is reflected in a diary entry I made in September 1932:

> Now I am back in old Amsterdam, in the same room where I read *Die Katrin wird Soldat.*[15]. . . Autumn is starting in Vondelpark, and the first yellow leaves have fallen to the ground. The city . . . speaks to me of the past: Moses en Aaron Straat; Sarphatistraat; Wertheimpark; this . . . occidental Jerusalem has a fascination . . . and Spinoza has perhaps imparted a . . . piece of his soul to me. But another image is much more vivid . . . ; all that has happened since May fills . . . my heart. . . . I possess an indestructible and incalculable treasure, because I have now had my share, and a rich one, of the sweets of life. I have lived . . . what I desired in vain for years, and now it is fair that I submit, but it will not be without a struggle, to approaching fate.

Upon my return to Rome I continued asking myself what we should do. An intense correspondence followed, and as soon as I could, in January 1933, I returned to Hamburg. We spent more delicious days together, but the fatal January 30 on which Hitler came to power was approaching.

Europe now seemed to me to be destined for catastrophe. Naturally I did not acquire this foreboding in one day; it grew slowly. Already in 1929, during a trip to Germany with Angelo and Rasetti, I was dismayed by the fanatic enthusiasm with which a group of children on a boat on the Rhine were singing "Deutschland, Deutschland, über Alles." Later events, my life in Hamburg in the last years of the Weimar Republic, and conversations with colleagues convinced me of the deadly seriousness of the Nazi menace; they were a band of fanatics ready for anything. I could not anticipate what "anything" meant, but I included in it a major war—that is, a world war, or at least a European war.

Italy's position seemed to me ambiguous, and I did not know on which side of the fence she would come down. I realized that the Duce's big talk was mostly empty bombast. My profession fostered cosmopolitan attitudes and relations. Both Fermi's and Rasetti's horizons consistently extended beyond Italy, and they often considered emigration. It was fundamental for me that my wife share my mobility, because I might find emigration from Italy desirable or, God forbid, necessary. Any future wife must agree with me on this in advance. I sometimes touched on this issue with my physicist friends, and I spoke freely about it to Riccardo Rimini. At that stage, I was not thinking of the dangers of anti-Semitism, but rather of the situation as a whole.

At home my parents realized the obvious—that there was a German girl in the offing, and that I was in love with her. They wanted to avoid experiences like those they had had with Angelo, which were a cause of continuing worry to them. As early as May 27, 1930, my mother had written a brief note to me about choosing a wife, a most unusual precaution for her. It reflected her preferences or prejudices, but in the end it left the final say to me.

There are traces of the prevailing conflict between reason and feelings in many of the letters I exchanged with my girlfriend in 1933. She nicknamed me "t.d.," the doubter, because I was addicted to doubting. I was not the first in my family to deserve such a nickname. My uncle Gino was dubbed "Cacadubbi" by his brothers, but his doubts were legal subtleties, and, I should add, much appreciated by his peers.

In the end, with Riccardo's support, I faced up to the fact that my current relationship could not be conducive to happiness, and that marriage would be foolish. A sudden, final break would have hurt us too much, and I saw my German girlfriend once more in Hamburg in 1937 on my return from a physics conference in Copenhagen. Nonetheless, the end of the affair was dismal and left its marks.

Perhaps a little cynically, I decided to adopt the principle "The king is dead; long live the king!" The sooner I found someone else, the better. This was easier said than done, however. I did not expect too much from Italian girls. Indeed, perhaps I expected too little of them. My parents on the other hand were eager to see me married to an Italian Jewish girl and suggested I meet some young women who looked suitable to them. One was in Ferrara, where Riccardo was then working in a hospital. On the pretext of visiting him, I met the girl, but I barely remember her. After the war, I learned that she had been murdered by the Nazis.

Later it was the turn of a Neapolitan beauty, who had many advantages. We went for a walk together and I said something like: "Neither of us is a child, and we both know perfectly well why we are on this walk together. I will speak openly; because of my profession and other circumstances, it is possible, or even probable, that I shall end by emigrating. What do you think of it? Could you adapt to life outside of Italy?" The girl was somewhat taken aback and haltingly said that she could not live far from her mother and from Naples. That ended the conversation.

There was another Italian girl, whom I had loved for over ten years, and about whom I had thought seriously many times. The trouble was that I did not succeed in conquering her heart in spite of many efforts and great sufferings on my part, and she married somebody else. I have remained her constant friend.

The first university competition I entered was sponsored by the University of Ferrara and decided on October 31, 1932. There were many competitors older and more ignorant than I. Among my contemporaries, Bruno Rossi already had a reputation for his work on cosmic rays. The judges were Quirino Majorana, Alfredo Pochettino, Carlo Somigliana, Luigi Puccianti, and Fermi, who was the only one who understood contemporary physics. Fermi, preserving the secrecy demanded of the panel, never told me what happened, but I learned about it from other sources. It seems that the majority of the committee said something like: "We have the votes to do whatever we want, no matter what you may say, Professor Fermi. We are, however, considerate, and we shall permit you to choose one candidate. We shall choose the other two." And they selected O. Specchia and C. Valle. Fermi pondered the situation and chose Rossi. I was thus not one of the three winners. Fermi did not know it, but by not selecting me, he had conferred upon me an inestimable benefit. It was a true blessing in disguise, not the last of my life.

This did not prevent me from sulking for a while. Rossi and I were virtually the same age, and our papers, although in very different fields, might have been seen as of comparable quality. It was noted that Fermi had preferred a Florentine to a Roman, and to his pupil. I am sure that Fermi's vote reflected his deeply considered evaluation of merit, and today, so many years later, I think he was probably right. At the time, however, Fermi saw that I was angry and unhappy. With a rare show of solicitude and affection, he told me that I should not be angry, that there would be other competitions, that I was young, and that what counted above all was to do good physics.

He proposed that we do some research together, which quickly cured my sullenness. We investigated hyperfine atomic structures with a view to showing that they could be completely explained by the nuclear magnetic moment, and that there were no other nuclear forces at play. The paper we subsequently co-authored contains the standard Fermi-Segrè formula.[16] During this work we labored together for hours on end, and on a couple of occasions I fell asleep out of exhaustion while Fermi was talking to me.

In the summer of 1933 I visited America for the first time. After his 1929 experiences, Rasetti had stuffed our heads with fabulous descriptions of the United States—the promised land, according to him. He declared he could not drive Italian cars anymore and imported a Ford Model A, in which we traveled extensively. Rasetti's tales persuaded Fermi to go and see for himself, taking advantage of an invitation in

1930 from the University of Michigan at Ann Arbor, which held famous summer schools on theoretical physics. Invitations were extended to young European luminaries, as well as to about thirty American students and postdoctoral fellows, and Fermi's old friend G. E. Uhlenbeck, who was on the Michigan faculty, always attended. In Rome, Fermi had reformulated P. A. M. Dirac's quantum theory of radiation in a form much easier to understand than the original papers and had made many illuminating applications of the theory. In his course at Ann Arbor, he reported on this investigation with extraordinary success. He was reinvited many times later, and whenever he could, he accepted. Fermi loved the American atmosphere, and in particular Ann Arbor and its stimulating school. In fact, he became one of its mainstays.

In 1933, Fermi suggested that I accompany him, and I gladly accepted. I first stopped at the Long Island home of G. M. Giannini, a contemporary of mine who had also studied physics in Rome. Then I moved to Ann Arbor, where I shared a room in a filthy fraternity building that had been vacated for the summer.

Both Fermi and I were eager to improve our English pronunciation, and we asked some of the students to point out where we were weakest. It seems that our pronunciation of the letter *r* was particularly bad. I accordingly invented the exercise "Rear Admiral Byrd wrote a report concerning his travels in the southern part of the Earth," which we declaimed at least twelve times a day, if possible in the presence of somebody who could correct us. Fermi and I bought a secondhand car from D. R. Inglis, which we named "The Flying Turtle" because of its performance, and toured the state of Michigan in it, eating very well as the paying guests of local farmers. We thus discovered delicious rural American dishes, which are difficult to obtain in the cities because they require very fresh vegetables.

To justify my presence at Ann Arbor, I tried doing some experimental work, but the humid heat of the place prevented me from working efficiently and I did not get anywhere. In trying to fit a rubber tube onto a glass tube, I badly cut my left middle finger. I went to the hospital, and as soon as he saw the wound, before I could open my mouth, the doctor said: "You have been trying to fit a rubber tube to a glass one." Later I saw other victims of the same accident.

On my return to Italy, I felt that although I had spent a year learning about molecular beams, and had even performed a creditable experiment with them, I was more interested in forbidden lines, which had a special attraction for me. After the study of quadrupole radiation, it was clear to me that there were also other mechanisms producing forbidden transitions, among them the presence in a discharge of ions that created random electric fields. I devised a simple theory to explain this effect and verified it with Bakker's help by studying the Zeeman effect in a suitable case.[17]

Besides using the field produced by ions in the discharge, one could think of using an external electric field, under more controllable conditions. Bakker and a colleague named Kuhn performed this experiment, while G. C. Wick and I developed the theory. There were some reasons for thinking that the behavior of potassium and sodium would be different. Bakker had studied potassium. Amaldi and I experimented on sodium. During these experiments we observed high quantum states, corresponding to enormous orbits. I called them "swollen atoms"; today more scientifically, but less pictorially, they are called "Rydberg states."

We noted then that the foreign gases we had introduced into our absorption tubes to prevent distillation did not broaden the lines as much as we had feared, but rather, to our surprise, shifted them. We mentioned this unexpected phenomenon to Fermi, who thought about it a little and then said that it was probably because of the dielectric constant of the gas we had added to the alkali vapor. This effect had to be reckoned with, and he calculated it at once. However, for some gases the observed effect had the sign opposite to that expected. Surprised, we went back to Fermi with the puzzle. This time it took Fermi several days to come up with an additional cause of level shifts, and he wrote an important paper on the subject, which for the first time introduced the idea of what is now called a "pseudopotential." Subsequently I noted that swollen atoms should show a term in their Zeeman effect, quadratic in the applied field. In the usual theory, this term is justifiably neglected, but I experimentally showed its importance in suitable cases.[18]

The quadratic Zeeman effect, the shift of the lines, and the effect of the electric field on lines near the series limit have been extensively investigated and now form a subdivision of spectroscopy. Amaldi and I summarized our investigations in a review article that appeared in a festschrift for Zeeman's retirement.[19]

At about this time my father made an important decision that was to affect the future of everyone in our family. He was over seventy years old and was increasingly inclined to leave the management of the paper mill to Marco, by now his undisputed successor, because Angelo was busy with completely different matters and I was committed to physics. In his business, my father carefully observed developments, gave general directions, and as the sole stockholder had the final say on everything, but he left the day-to-day management to Marco, whom he tried to groom as his successor.

The paper mill had prospered substantially, but especially in later years, Father had chosen to invest profits in real estate rather than in enlarging and modernizing the mill. I do not know why; possibly there was a certain amount of tacit mistrust of Fascism and its economic policies. He also planned the disposition of his estate, which he wanted to divide into three parts of equal monetary value. He thought the mill should go to Marco, who worked there, and that Angelo and I should receive assets we could easily administer independently of Marco. In recent years I had been abroad for extended periods and had often hinted that I might emigrate. Furthermore, I had seen Nazism with my own eyes and had no illusions on the subject.

Up to 1933 I had not paid any attention to financial problems. Papà handled them with much greater ability than I could hope to muster, and I had no money of my own. I had occasionally intervened in personnel questions at the paper mill, in particular defending Bindo Rimini, who had been attacked by Marco. In this connection I once made a trip to Florence to investigate certain paper sales, and proved that Marco had blamed Bindo unjustly. Bindo remained deeply grateful to me for my help. I also introduced my friend Giovanni Ferro-Luzzi at the paper mill, hoping he might keep an eye on what was going on.

In years past, his foreign business had produced some assets outside of Italy, and my father had left them there, as was legally permitted at the time. About 1933, however, the Fascist government promulgated new laws demanding the repatriation of foreign money, with severe penalties for transgressors. These laws worried my father, and one day in the fall of that year he invited a prominent banker who was his friend and advisor to his office, along with Marco and myself, and asked for our opinions on the subject. The banker at once said something along the lines of: "You did not leave this nest egg overseas for times in which the export of assets was permitted. You left it there for times like the present. The new laws are the best proof of the importance and prudence of keeping such a reserve." He also pointed out that there was currently a flurry of capital exports on the part of industrialists, professionals, and people of means. Why should my father give up his own safety net? The argument convinced me at once. Our friend had hit the nail on the head.

Father said that at his age he wanted a quiet life. Marco signaled his own importance, saying what I expected of him: he was in the limelight and could easily be subjected to an investigation; he had a family to protect, and being such a prominent industrialist, he had to be prudent.

It was my turn to speak, and I said something like: "I am a physicist who often works abroad, and I might emigrate. I am not a person in the public eye. You may transfer the funds to my name, and I shall keep the money for the benefit of the whole family. I insist however in being the only one with access to the account." Papà agreed on the spot and told me to make the necessary provisions for disposal of the money in the event of my death. Angelo was not consulted; my mother was not present, but I am sure that my father had informed her.

To make the necessary arrangements, I needed to go to Switzerland without attracting attention, so over the Christmas vacation I went skiing near the Italian border with some friends, including the physicist

Giulio Racah. We encountered foul weather, and, what with storms and avalanches, I had several close shaves. Escaping with nothing worse than a broken ski, however, I was able to enter Switzerland without seeing any police or custom official. There I transferred the money to my name and deposited with a notary a letter, to be delivered only upon presentation of my death certificate, giving access to my account. In this way we created a secret fund that was to be providential five years later when Italy started its racial persecutions.

From Switzerland I returned to Italy by train and then went to the Val Gardena to join some physicist friends who were there for a Christmas skiing vacation. Several days had passed since my adventurous crossing of the Alps, and as a result of my falls on the frozen snow, my buttocks had acquired impressive green and black spots, of hues rarely seen. I could not deprive my friends of the fun of such a sight. Later, Fermi, sitting on a bed in the small hotel room, explained to us his new theory of beta decay, as yet unpublished.

Physics in the meantime was taking an important new turn for us. For some time Fermi, and we as a consequence, had been making long-range scientific plans. Fermi felt that the golden age of atomic physics was coming to an end, and that the future lay with nuclear physics. In a letter dated September 9, 1932, he wrote to me: "I have no program for next year's work: I do not even know whether I shall start fooling around with the Wilson Cloud Chamber again, or if I shall again become a theoretician. . . . The problem of equipping the Institute for nuclear work is certainly becoming ever more urgent if we do not want to fall into a state of intellectual slumber."

My personal reaction was that we had just learned spectroscopic techniques, with which we were reaping good results, and that we might persist in that field a little longer. I was, however, open to Fermi's arguments. Amaldi and Rasetti also had their points of view, and we had long, lively discussions on the subject. As was to be expected, Fermi's ideas prevailed, although everybody was left free to do what he liked best. Thus I continued to work experimentally on spectroscopy until we started our neutron work. However, we all increased our reading on nuclear subjects. As a bridge between spectroscopy and nuclear physics, Fermi and I actively investigated hyperfine structure, as already mentioned.

Even my work in Hamburg on the dynamics of space quantization had an unexpected nuclear ramification. In the last weeks of my experiments, Otto Frisch had helped me, and Stern suggested that he sign the paper with me, to which I consented. In spite of all our efforts, we found that the experimental results did not agree with the theoretical expectation, but the experiment had been difficult, for its time, and we were able to find experimental excuses for the discrepancy. A few months later, however, we received a letter from I. I. Rabi inquiring about some experimental details we had not published. I sent them to Rabi, and he answered that the reason for the apparent disagreement between the theory and our results was that in the theoretical calculation, we had neglected the effect of nuclear spin! Had we included it, attributing to potassium a spin of $3/2$, theory and experiment would have agreed. In other words, we had without knowing it measured the nuclear spin of potassium. Rabi most generously published all this in the *Physical Review.*[20]

By 1933 Fermi had started an intensive investigation of nuclear subjects. Amaldi organized a seminar to study Ernest Rutherford, James Chadwick, and C. D. Ellis's recent book *Radiations from Radioactive Substances.*[21] Soon Rasetti and Fermi started learning experimental nuclear techniques. Together they built a gamma-ray spectrograph using a bismuth crystal; then, following a model used by Lise Meitner, they designed a cloud chamber, which was built in a machine shop in Rome. All of us together built some Geiger-Müller counters that more or less functioned. Rasetti was the moving spirit in this preparatory work; he had also been working in Meitner's laboratory to learn some radiochemistry. In particular, he had learned how to prepare Po + Be neutron sources and had taught me that art. Fortunately, G. C. Trabacchi, director of the Istituto fisico della sanità pubblica, which was located in the same building as the Physics Institute, had a gram of radium for medical purposes and benevolently lent us a fraction of it, making our neutron work possible. We thus laid a respectable experimental foundation for nuclear studies.

To further enhance our readiness, Fermi used his clout as a member of the Accademia d'Italia to promote a small international nuclear physics conference, which was held in Rome in October 1931 and attended by about thirty well-chosen physicists. At the conference I had the privilege of cleaning the blackboard for Marie Curie. Regrettably, I did not do it to her satisfaction, and she told me so in no uncertain terms. The timing of the conference was unfortunate, because it was a few months before the discovery of the neutron, which opened a new era in nuclear physics.

By 1932 Fermi had already accepted an invitation to report on nuclear physics to a large international conference in Paris, and he also participated in the famous 1933 Solvay conference devoted to nuclear physics. Shortly thereafter, on his return at Rome, he invented the beta-ray theory—in his own opinion, his theoretical masterpiece. In it, developing Pauli's neutrino hypothesis, he formulated a quantitative theory of beta decay. This theory introduced the so-called weak interaction, which turned out to be a new "force of Nature," as Faraday would have said. After Fermi's death, the weak interaction revealed startling properties, such as the nonconservation of parity and, ultimately, deep relationships with electromagnetism.

Great events, however, were incubating in a different field. In February of 1934 we were stunned by the announcement of Irène and Frédéric Joliot-Curie's discovery of artificial radioactivity. By bombarding light elements with alpha particles, they had obtained new radioactive isotopes of common elements that decayed by positron emission. Fermi thought at once of the advantage of using neutrons as projectiles. Although the available neutron sources emitted many fewer neutrons than the alpha particle sources emitted alphas, the much superior efficiency of neutrons overcompensates this handicap. This is because the alphas are repelled by the nuclear charge and do not penetrate the nucleus. The neutrons on the other hand always end by penetrating a nucleus.

Thanks to the previous year's work, we had all the tools ready for testing these ideas. Rasetti, who had contributed so much to it, was in Morocco, where the king was decorating him with some order. Fermi recalled him by telegram, but he answered that he did not want to be disturbed. Fermi proceeded alone and, using a Rn + Be source, tried to form new radioactive isotopes in all elements, in order of increasing atomic number. He first succeeded with fluorine ($Z = 9$).

The next step was to try to activate all the elements, and to study all the radioactive isotopes formed. This formidable task was beyond the capabilities of a single person, even of a Fermi. Having struck scientific gold, he most generously invited Rasetti, Amaldi, and me to take part in its exploitation.

We were talking about the need for professional chemical help when Oscar D'Agostino showed up at the Physics Institute. He was a graduate of the department of chemistry at the University of Rome and was spending a postdoctoral fellowship in Marie Curie's laboratory in Paris learning radiochemistry. He had returned to Rome for the Easter vacation. We told him our problems, and Fermi invited him to help us. He postponed his return to Paris, and the delay extended indefinitely.

In the neutron work, each of us assumed special duties, although we collaborated on all the phases of the investigation. Fermi was the natural chief, not in the sense that he told us what to do on a detailed basis, but rather in that he set the general guidelines. If there was any problem, we talked it over together, and it is not difficult to guess whose words carried most weight. Once the program was established, each of us took responsibility for some part of it.

From the very beginning of the experiments, we saw that we needed a minimum amount of money beyond the Physics Institute's regular endowment. Fermi had good relations with the Consiglio nazionale delle ricerche. He had been its secretary for physics, and I his assistant secretary. At the present juncture, Fermi asked for help from the CNR and immediately obtained 20,000 lire (then about U.S. $1,000). I doubt whether any scientific grant has ever been more fruitful.

I was charged with procuring what we needed for our work. Luckily, there was no bureaucracy. I could carry our money in my pocket; it was not much, but I could pay cash on the barrel. With this freedom, money multiplied its purchasing power in an astounding way. For chemicals, I turned to a Signor Troccoli, an old and experienced merchant who took pride in stocking a most extensive supply of chemicals. In his youth he had studied in a seminary, and he liked to speak Latin, once in a while offering me some chemical that had been on his shelves for years "gratis et amore Dei." After I explained to him what we were doing, the worthy gentleman helped me in any way he could. However, when, in my ignorance, I asked him for a sample of masurium, he answered, "Nunquam vidi" (I have never seen it). A few years later I realized why. Masurium did not exist.

For some absorption measurements, we needed a gold ingot. I went to the firm of Staccioli, who were dealers in precious metals, and without any difficulty, on the basis of a simple receipt, they loaned me the ingot; I returned to the institute loaded with gold. I purchased necessities that could not be found at Rome through my friend Bakker in Holland.

Work proceeded rapidly. Our group reminded me of a well-rehearsed orchestra, and its conductor, Fermi, got superb music from it. We all outdid ourselves, each achieving more than any of us could have done on his own. The whole was definitely greater than the sum of its parts, Fermi included.

Our communications were published in *La ricerca scientifica,* the bulletin of the CNR, where Amaldi's wife worked, and it is easy to follow the daily progress of our work in it.[22] We sent reprints to well-chosen, strategically located correspondents who could read Italian, and our reports soon attracted the universal attention of nuclear physicists. Corbino kept in close touch with us through frequent visits to our laboratory.

We systematically proceeded to irradiate all the elements we could find, trying to use our sources as efficiently as possible. We prepared them once a week, because the radon they contained had a half-life of 3.82 days. The operation was delicate, but we proceeded cautiously, and nobody got hurt.

Soon we identified the two reactions, neutron capture followed by proton or alpha-particle emission, or, in the usual notation, (n,p) and (n,α). We also found that frequently neutron bombardment produced a radioactive isotope of the target, but we did not know whether this was owing to neutron capture followed by gamma-ray emission or by emission of two neutrons: (n,γ) or (n,2n). At the time we believed that the more energetic the bombarding neutron, the more efficient it would be in producing nuclear reactions. Only months later did we find out how erroneous this assumption was.

Continuing our bombardments by increasing the atomic number of the targets, we arrived at thorium and uranium.[23] The activities we could produce were weak compared with the natural activity of the targets; hence it was necessary before bombardment to remove from them the different radioactive substances they contained in radioactive equilibrium with the primary substance. This was a long and delicate operation, and the substances removed grew again after some time. We thought that in capturing a neutron, uranium and thorium would form a beta emitter that decayed into transuranic elements, for which we anticipated chemical properties similar to those of rhenium, osmium, and iridium. The nuclear processes occur, but the supposed chemical resemblance is false. We erred in the way we extrapolated the periodic system of the elements. Hahn and Meitner fell into the same trap, as did the Joliot-Curies. Only after several years was it realized that transuranic elements form a family similar to the rare earths.[24]

In 1934, at Rome, we proved that some of the activities formed in uranium bombardment were not isotopic with elements between lead and uranium, but we drew the wrong conclusion that they were transuranic. Like other investigators of this period, we noticed that the total activity produced was much larger than that of the products we were isolating, and we should have further investigated its nature. We did not seriously entertain the possibility of nuclear fission, although it had been mentioned by Ida Noddack, who sent us a reprint of her work.[25] The reason for our blindness, shared by Hahn and Meitner, the Joliot-Curies, and everybody else working on the subject, is not clear to me even today.

Transuranic elements presented a difficult experimental problem, full of pitfalls unless one had the right ideas or impeccable techniques. In fact, in December 1938, Otto Hahn and Fritz Strassmann found the solution to the puzzle through an ironclad experiment proving the formation of radioactive barium in neutron bombardment of uranium.[26]

The outpouring of work mentioned above occupied the spring of 1934. During the summer, the Physics Institute was closed, and Fermi went to South America on a lecture tour. Amaldi, his wife, and I went to the Cavendish Laboratory in Cambridge. On the way, we stopped in London to meet Fritz Paneth, from whom I wanted to learn some chemical techniques, as well as Leo Szilard, with whom we had had some correspondence and who seemed to be, in more than one way, an interesting fellow. We made an appointment for a certain time in the afternoon, but nobody showed up. Around 10 P.M., we found Paneth with tears in his eyes and Szilard obviously shaken. We did not talk science. Hitler had attempted a coup in Austria; Chancellor Engelbert Dollfuss had been murdered, and no one knew what would happen next. Paneth was Austrian and Szilard Hungarian; the blow struck close to home. Italian mobilization, ordered by Mussolini, foiled Hitler's plan. The attempted putsch should have forced French, British, and other European politicians to open their eyes. Perhaps what they did not want to see was too ugly to contemplate. They hesitated and missed another opportunity of putting an end to Hitler.

Amaldi and I brought with us the manuscript of a paper summarizing the neutron work done in Rome, which we delivered to Lord Rutherford at the Cavendish Laboratory, begging him to communicate it to the Royal Society. He showed keen interest in our paper, took it home, and returned it the next day with some corrections to the English, saying that he was forwarding it immediately to the *Proceedings of the Royal Society*.[27] I imprudently recommended prompt publication, whereupon he answered, whether in jest or annoyance I could not tell: "What do you think I am president of the Royal Society for?" Rutherford spoke with his pipe in his mouth, and I had difficulty in understanding him. Once in a while he made jokes and laughed aloud at them, but frequently I missed the point. He invited us to tea at his house, where we met Lady Rutherford. We saw him every day in the laboratory, where he went around discussing the current work, recalculating results with a pencil stump he carried in a waistcoat pocket. He was obviously highly respected and listened to, possibly with some awe. I had the impression that when he addressed somebody, the person unconsciously stood at attention. Although Rutherford was most approachable, we talked more with the Dane T. Bjerge and H. C. Westcott, our contemporaries, who were working on neutron-induced radioactivity. Maurice Goldhaber showed us the experiments he and James Chadwick had done on the photodisintegration of the deuteron and invited us for dinner at Caius College, where he lived. Wynn Williams showed Amaldi, who was especially interested in them, several electronic devices, including his linear amplifier. We also met Peter Kapitza, Marcus Oliphant, James Chadwick, John Cockcroft, and several other physicists whom I reencountered later. From the point of view of our immediate work, however, the most interesting exchange was with Bjerge and Westcott; it allowed us to establish a clearcut case of the (n,γ) reaction in sodium, which was to prove important a couple of months later in the discovery of slow neutrons.

At Cambridge we lodged as paying guests in a private home. The food was so bad that I still remember it. Ginestra Amaldi was expecting her first child, Ugo, and her dimensions were growing accordingly. In order to get to the lab, we had to pass between two bollards that formed a barrier in the street. Edoardo decided that when she could no longer pass freely between them, it would be time to go home. He followed this prescription, and Ugo was born a few days after their return to Rome.

A week or two later, Amaldi and I, alone in Rome, resumed the neutron work. One of the first things we did was to repeat the bombardment of Al^{27} in order to verify a (n,γ) reaction similar to the one observed by Bjerge and Westcott in sodium. We found that the period of the isotope formed was indeed that of Al^{28} formed in other reactions, thus demonstrating another case of (n,γ) reaction. We immediately communicated this interesting result to Fermi, who, on his way back from South America, had stopped in London for an international physics conference, and he mentioned our findings at it.[28] In the meantime I caught a cold and could not go to the laboratory for a few days. Amaldi tried to repeat the experiments we had performed a few days earlier and obtained a different decay period. When Fermi got back to Rome, the results were confused, and he scolded us for having irresponsibly given him erroneous data. I resented the accusation because I was sure of our previous experiments, but I could not fathom what was going on. In addition, Edoardo was now finding inexplicable new phenomena.

Within a few weeks, mysteries multiplied. We frequently found irreproducible effects, such as different activations of samples that had been bombarded under what we thought were the same conditions. Later we found out that the difference was that on some occasions we had placed our source on a wooden table and on others on a marble one!

On the morning of October 22, 1934, after several experiments in which a silver foil was irradiated under different conditions, performed mainly by Amaldi and Bruno Pontecorvo, source and silver target were surrounded with paraffin. I was busy with examinations, but around noon I was called to see the strange phenomenon occurring. Paraffin greatly enhanced neutron activation. At first I thought that some counter had gone awry, as occasionally happened, but it did not take much to persuade all present—that is, Fermi, Amaldi, Rasetti, Pontecorvo, and myself, as well as Bruno Rossi and Enrico Persico, who happened to be visiting—of the reality of the effect. We replaced the paraffin with a couple of other substances, but it was apparent that they were not effective.

At this point, rather confused by the morning's observations, we went home for lunch, and after the usual siesta, we returned to the laboratory around 3 P.M. In the meantime, Fermi had come up with an explanation for the paraffin's action. He hypothesized that the hydrogen atoms of the surrounding paraffin slowed down the neutrons from the source by elastic collisions, and that slow neutrons could be more easily captured than fast ones. It is easy to see that in an elastic collision with hydrogen, a neutron loses on average half its energy; the idea that a slower neutron would be more effective than a faster one in producing nuclear reactions was surprising to us. We were used to thinking just the opposite. The effect, however, should be limited to (n,γ) reactions, and not involve (n,p) or (n,α) reactions. If Fermi's surmise was correct, we had to rethink many of our observations, taking it into account.

We rushed to repeat some simple experiments and we recognized at once that puzzles such as the behavior of aluminum could easily be explained. In forming Al^{28}, we had an (n,γ) reaction, enhanced by slowing down the neutrons. In our experiments, we had unwittingly sometimes used slow neutrons and sometimes fast ones, depending on the objects near to our source, obtaining a prevalence of (n,γ) or $(n,2n)$ reactions. With this explanation I gave a sigh of relief, because, after all, Fermi's statement at the London conference, suggested by Amaldi and myself, was vindicated, and Fermi recognized that we had not been careless.

Fermi went to the limit of supposing that neutrons would be slowed down to the energy owing to thermal agitation of the molecules of the medium, about 0.03 eV at room temperature. This process is now called moderation. At the time we tested this hypothesis, which is correct, by trying to slow down the neutrons with paraffin at different temperatures, but that day we could not see any difference.

That same evening, at Amaldi's home, we prepared a short letter to *La ricerca scientifica.* Fermi dictated while I wrote. Rasetti, Amaldi, and Pontecorvo paced the room excitedly, all making comments in loud voices. The din was such that when we left, Amaldi's maid discreetly asked whether the evening guests had all been tipsy. Ginestra Amaldi handed the paper to her boss at *La ricerca scientifica* the following morning.[29]

The discovery of slow neutrons opened up a host of problems and made us reorient our entire research program, including our neglect of the uranium investigation. We measured what we called the coefficient of aquaticity—that is, how much the immersion in water, under standard conditions, would increase the activity—for many substances. This measurement confirmed that the (n,γ) reaction was the only one sensitive to hydrogen. By early November we were sure that the slowing-down hypothesis was correct, and our attention turned more to the slowing-down process than to the substances produced. We tried repeatedly, but unsuccessfully, to see whether cold or hot moderators behaved differently. Ultimately, P. B. Moon and J. R. Tillman in England demonstrated a thermal effect, and we hastened to repeat their experiments.

We soon found that some substances—for instance, cadmium—greatly absorbed slow neutrons and we crudely measured an absorption cross-section for them. We detected the gamma rays emitted on neutron capture, and we started crude measurements of the density of slow neutrons as a function of the distance from a source immersed in water. Finally, we tried to slow down neutrons with substances other than hydrogen and found some effect of inelastic collisions. All this work had produced significant results by December 1934, six weeks after the first discovery.

Of course, we had immediately informed Corbino of our findings, and he at once urged us to apply for a patent on slow neutrons, which he considered potentially of great practical importance. I shall return to this later.

On my return from Germany, I had discussed my romantic difficulties with Riccardo Rimini, my most intimate advisor. He turned out to have similar problems, and talking to each other frankly was a great relief and benefit to both of us. We concluded that as things stood, I should look for another German girl, possibly one living in Italy, and certainly not a Nazi.

I met Elfriede Spiro early in 1934. At the time, she did not know Italian very well, but this was not important; on the contrary, it gave me some advantage, since I spoke German fluently. I wanted to be sure about Nazism; she was prudent and avoided talking politics, but I soon twigged that I did not have anything to fear on that score. She had left Germany because she had been fired from her job, and it did not take much imagination to guess why. I must add that the name Spiro sounded Greek to me, just as Segrè sounded French to her; a bit more Jewish culture would instantly have revealed our roots to both of us.

"La Spiro," as I called her, was handsome, of a somewhat Germanic type, full of vigor, and intelligent. She needed to support herself financially, and this was her first preoccupation. She had no prejudice against any honest work; at first she changed jobs several times, but thanks to her good office training, her neat habits, and her eagerness for work, she soon found a satisfactory job.

Elfriede was born on October 2, 1907, in Ostrowo, then a small German town in East Prussia. Her family, of solid Jewish bourgeois stock, had German feelings and spoke German. Elfriede's father, Max, was a first cousin of the famous chemist Fritz Haber (1868–1934),[30] whose well-known German patriotism was typical of the Spiro family before the advent of Nazism. After World War I, Ostrowo was annexed by a reborn, anti-German, and anti-Semitic Poland, which virtually expelled the Spiros. They then moved to Breslau. In the process, Elfriede's father lost a good part of his fortune.

Elfriede's mother's maiden name was Gertrude Aschert, and Gertrude's brothers were judges and lawyers. Elfriede's older sister married an education professor, Peter Schirbel, an excellent person, anti-Nazi although of "pure German race," and one of the few Germans who contributed, by his behavior, to saving German honor in the horrors to come. Elfriede's younger sister, Lilli, became a nurse, first in England and later in Berkeley.

In Germany, Elfriede had worked as secretary to a socialist politician, Hermann Luedemann.[31] In August of 1933, forced to emigrate, she came to Italy, where her father had some connections, which, however, proved useless. Elfriede was not discouraged; she had spirit, was determined to make a career, and was eager to work. After several positions, she became secretary-interpreter for the "XI congresso internazionale dell'acetilene, della saldatura autogena e delle industrie relative" that took place in Rome in June 1934, a position she held until after the publication of the conference's proceedings, when the organization dissolved.

In the spring of 1934, we were both extremely busy, I with neutrons and Elfriede with the conference, but this did not prevent us from visiting some of Rome's immediate surroundings, including Palestrina, Fogliano, Terracina, and Viterbo. On my return from Cambridge, we found time for a short tour of the Dolomites, where we backpacked from one mountain hut to another. One day we arrived hot and sweaty at a little mountain lake, whose deep transparent water was decorated with small ice floes. Elfriede did not think twice; she donned her swimsuit and jumped in. Not wanting to be thought inferior, I followed her, but almost dropped dead from the cold, while she swam merrily about. At Paneveggio I taught Elfriede to play *bocce,* a game she immediately liked. She remembered this at Recanati when we played the game there on the last day of her life.

On our return to Rome, Elfriede lived on Via XX Settembre, not far from the War Ministry, and I often went to pick her up. One day I saw her at a distance on the sidewalk across the street. An old general was following her with a bouncing gait and seemed to be paying her compliments, which I trust she did not understand. I looked more closely and recognized no less than His Excellency Emilio De Bono, a quadrumvir of the March on Rome and one of the very highest officials of Fascism. Elfriede had no idea of who he was. I teased her about the big fish she had landed.

In October, at the height of the neutron work, we again had a few great weeks. I was alone in Rome, and the city was in all its autumn splendor. We often ate in small old inns on the Via Frattina, now vanished or degraded to tourist traps, where we found excellent fruit, especially marvelous tree-ripened yellow peaches. With luck, after lunch we would even find time for a swim at the Piscina dello stadio. Elfriede's office was on Via S. Claudio, in the same building as the office of my father's company, the Società cartiere tiburtine (SCT). One day, she met my cousin Bindo Rimini in the elevator there, and they recognized each other and introduced themselves on the basis of my descriptions. During the last months of 1934, I explained the importance of the neutron work that kept me so busy to Elfriede and introduced her to my physicist friends.

Unfortunately, Elfriede's job came to an end, and she had to find another one. She and another German girl, a friend and office colleague, found work in a *Landschulheim,* a school for Jewish children, mostly refugees from Germany, who were preparing to go to Palestine. Among the teachers at the school were several who later became distinguished, among them the philosopher P. O. Kristeller, a protégé of the Fascist senator Giovanni Gentile and subsequently a professor at Columbia University in New York. Elfriede took other temporary work too, for instance, as an accountant in a hotel at Sestri Levante, a seaside resort near Genoa that had many German visitors.

The better I knew Elfriede, the more I found in her all the qualities I desired in a wife. I then introduced her to my parents. They had certainly got wind that something was afoot, and they had learned, from previous experience with Angelo, to be prudent and avoid landing in unpleasant situations. Elfriede and I also went to see Ada Rimini, the mother of Riccardo and Bindo, who was most cordial, immediately took a liking to Elfriede, and thereafter frequently invited us to lunch or dinner. Soon my parents, too, started to appreciate Elfriede, and ultimately they came to love her and treated her as a daughter. Elfriede fully reciprocated. Elfriede's parents also came to Italy for a visit; I met them and I hope I made a good impression on them.

Thus we started seriously to think about marrying. I was thirty by then, and she was twenty-seven; we both knew what we wanted and found a solid basis for marriage in our common Weltanschauung. Before marrying, however, I wanted to have a settled position, such as a chair at an Italian university. On the other hand, the past few years had convinced me of the instability of the European situation, and that anything could happen. Although I loved my country, we had to look further afield, whether we liked it or not.

In May 1935 I went to Holland for Zeeman's retirement ceremonies. While there, I held a series of conferences on our neutron work, visiting the Philips Company at Eindhoven, where Bakker had taken a job, among other places. In my speech I told the Philips scientific leaders what I thought the practical applications of nuclear physics were likely to be. My speech must have been convincing, because Bakker wrote to me a few weeks later that Philips had decided to form a new nuclear research group and had recruited him to it; he added that this was owing, in good part, to my visit.

In Holland I also came across the famous German theoretical physicist Arnold Sommerfeld, who had been well paid for lecturing there. He took me aside and told me how glad he was to have the money, saying, "I shall send it to Rutherford for 'displaced scholars' [i.e., Hitler's victims] right away. I can't contribute from Munich, but from here I can." I had some time earlier sent a check, large for me, from Italy to the same organization in London, taking only the precaution of requesting that my name not be mentioned. I would not have dared to do so from Germany.

During the same visit I met my former Hamburg colleague B. Josephy on a streetcar in Amsterdam. In 1933, to escape the Nazis, he had accepted a job in the Soviet Union. We looked at each other with surprise, because I knew he was in Russia and he knew I was in Italy. After a while we decided we were not prey to hallucinations and talked to each other. Josephy told me that in Russia he had had a contract for the preparation of certain hormones. After a few months the authorities, who had not given him the equipment he needed for his work, accused him of sabotage for not having produced the hormones and threatened him with dire penalties for this crime. He had saved himself by referring to his written contract, where, luckily for him, his equipment needs were specified in detail. Since he had not received the promised centrifuges, extractors, and so on, he could not be held liable. Nonetheless, he was fortunate not to have been sent to Siberia, or worse. The Soviet authorities expelled him from the country and sent him back to Germany. "Imagine!" he said. "When I saw the swastika, I rejoiced!" He crossed Germany and was able to exit on the other side; now he was working in a hormone factory in Holland. These episodes helped me to get an increasingly clear picture of the ill wind that was blowing.

It seemed likely that there would be a competition for an experimental physics chair in 1936. The University of Palermo had a vacancy, caused by the death of Professor Michele La Rosa. I had met La Rosa years before: one day Corbino had called me to his study and showed me a photograph of a spectrum, asking what I thought it was. I examined the plate and answered that it was a mercury spectrum, that the plate must have been developed at too high a temperature, softening the gelatine and making the lines unusually broad.

"Could you prepare a similar plate for me by this afternoon?" Corbino asked.

"Certainly," I said.

A gentleman in an elegant blue suit whom I did not know was sitting next to Corbino, who now introduced him to me as Professor Michele La Rosa. I sensed at once that I was in hot water. Corbino added that La Rosa himself had prepared the plate he had shown me and that he thought he had made a discovery. La Rosa was a big shot, and the figure he was cutting was none too brilliant. It was thus important for me to substantiate what I had said. I accordingly made a plate of the mercury spectrum and delivered it to Corbino. A few days later, La Rosa wrote to me a friendly letter, with some bittersweet undertones. He thanked me for having caught his mistake and "hoped that he could at the proper time reciprocate with some similar service." This sounded a bit ominous.

Would I win the forthcoming Palermo competition? Yes, if justice prevailed (in my immodest opinion), but one could never be sure. Hoping for the best, I prepared the necessary documents and filed my application.

During the summer of 1935, I again had an opportunity of visiting the United States with Fermi; my plans were to return to Italy if I won the competition for the Palermo chair; otherwise, I would try to find a job in America and remain there. In either case, Elfriede and I would marry. We were both deeply worried by the political situation, and the difficulties involved in renewing Elfriede's passport and obtaining a U.S. immigration visa were steadily increasing.

When I left for the United States, I gave Elfriede all my liquid assets, to be used in an emergency. At some point the Duce did something, I do not remember what, that convinced me that the best thing I could do was to spend all my cash immediately, and I told Elfriede to buy a typewriter, a Leica, and a gold Longines chronograph for me, and to take the rest of the money and go to the jeweler Settepassi in Florence and buy herself a diamond pin, which she did.

In the United States, as usual, Fermi and I first went to Ann Arbor for a while. Fermi returned to Italy at the end of August, and I accepted an invitation from Dean G. B. Pegram, with whom I had previously corresponded, to transfer to Columbia University, where there was a group in the physics department actively engaged in neutron work, consisting of J. R. Dunning, D. P. Mitchell, G. A. Fink, and Pegram himself.[32] Their neutron sources were similar to ours, but as detectors they used ionization chambers connected to linear amplifiers, which were superior for some purposes to our usual equipment. Furthermore, they had much better machine shops than we had in Rome.

I. I. Rabi, who was also at Columbia, was an interested spectator of the neutron work, although he had strained relations with Dunning. I knew Rabi well because of his molecular beam work, and through him I met his pupils J. Zacharias, S. Millman, N. F. Ramsey, and others. One day a visitor who was well known to me arrived at Rabi's office. He was Moise Haissinsky, who had studied chemistry at Rome and had later gone to Marie Curie's laboratory in Paris. Rabi and Haissinsky could not find a common language, and they asked me to interpret for them, because Haissinsky spoke Italian. I looked at them for a moment and then suggested: "Why don't you try Yiddish?" They were off like a shot and did not need my help.

I took up residence at Columbia University's Casa Italiana and spent most of my time in the laboratory. Soon Rasetti joined me there, and together our group performed several experiments with a neutron velocity selector based on fast rotating wheels. We also built a cadmium wheel trying to verify that the absorption cross section varied as 1/v. We made an error in predicting the outcome of this experiment and got a result different from what we had expected. On second thoughts, I discovered a mistake in our reasoning and explained it to Rasetti, enjoying both his surprise and the opportunity to teach my "Venerated Master," as I used to call him, something.[33]

The political situation was deteriorating badly, and if I had found work in the United States, I might have transferred at once, but jobs were as rare as white flies. The effects of the Great Depression were still being felt and there was also a certain amount of more or less open xenophobia and anti-Semitism in America. For instance, I was dismayed when Rasetti showed me a letter from a renowned chemist at the University of Chicago offering him a job provided he was not Jewish.

I decided to await the outcome of the University of Palermo competition in New York. In 1935, a repetition of the 1933 competition was hardly possible. By now Fermi was a power to be reckoned with, and one could also expect to find at least one other physicist of the younger generation among the judges. The decision was unexpectedly delayed, and only at the end of October did I receive Fermi's cable telling me that I had placed first, followed by Antonio Rostagni and G. Todesco. (In hindsight I think Amaldi and Gilberto Bernardini should have been chosen, but the setback for them was, as it had been for me, a blessing in disguise, especially for Amaldi.) In addition to Fermi, the judges were A. Campetti (chairman), Laureto Tieri, U. Crudeli, and Antonio Carrelli. Among them, only Fermi and Carrelli, a minority, knew modern physics.

On November 16, 1935, I embarked for Italy. The Ethiopian Campaign was raging, and the luxury Italian liner was empty except for a few dozen Italians who had volunteered for the African war. The ship ran into a major Atlantic storm, and many trucks secured on the deck were crushed by the waves like matchboxes. On our arrival at Naples, the poor volunteers disembarked, kissed their native soil, listened to a band playing in their honor, exchanged Roman salutes with Fascist bigwigs, and endured suitable rhetorical speeches. I watched, more and more convinced that they were crazy.

On my arrival in Rome, I contacted Elfriede in Florence, but had to proceed immediately to Palermo to take up my duties.

Chapter Five

On My Own: Professor at Palermo (1936–1938)

Scent of Orange Blossoms

Non sien le genti, ancor, troppo sicure
a giudicar, sì come quei che stima
le biade in campo pria che sien mature;
ch'i' ho veduto tutto 'l verno prima
lo prun mostrarsi rigido e feroce;
poscia portar la rosa in su la cima;
e legno vidi già dritto e veloce,
correr lo mar per tutto suo cammino,
perire al fine a l'intrar de la foce.

(Let not the people be too self-assured
In judging early, as who should count the rows
Of green blades in the field ere they matured.
For I have seen how first the wild-brier shows
Her sprays, all winter through, thorny and stark,
And then upon the topmost bears the rose;
And I have seen ere now a speeding barque
Run all her sea-course with unswerving stem
And close on harbour go down to the dark.)
Dante, *Paradiso* 13.130–38 (trans. Laurence Binyon)

Marriage and transfer to Palermo signaled significant changes in my life. From being a young man living in his parents' home, I now became the head of a new family; from being a subordinate in the Physics Institute in Rome, I became chief of an institute of my own in Sicily. At the University of Palermo I was a young, but important, tenured professor, and my career seemed established, inasmuch according to the Italian law then prevailing, further advancement occurred mostly by seniority. I wanted to give the best of myself. I hoped to set an example of renewal and modernization in teaching and also to initiate some meaningful research in a new Italian center. I felt liberated from the need to write papers for my advancement; only science counted. Similarly, our new family would be ours alone; I loved my parents and tradition, but the family Elfriede and I established would differ from theirs in many ways.

At the beginning of my stay in Palermo, I lodged in the Pensione Lincoln, on Via Archirafi, near the Physics Institute. The pension was comfortable in its simplicity. The institute was located in a new building, with very large rooms and much wasted space. The existing apparatus dated from the nineteenth century. To offset this, there was a bronze head of Professor La Rosa, my predecessor. The personnel consisted of a middle-aged assistant, who seemed to me unretrievable for useful work, an old mechanic, competent within his limitations and full of good will, and some more than adequate janitors.

On the floor above the Physics Institute, occupying territory that in theory belonged to physics, was the Mineralogy Institute. Since I did not need more space, there was no conflict. On the contrary, the professor of mineralogy, Carlo Perrier (1886–1948) was a nice fellow, a true Piedmontese gentleman, and an anti-Fascist.[1] He was a bachelor, about twenty years older than I, and well versed in classical mineralogy and analytical chemistry. Soon we became close friends, and this friendship later brought its fruits. He also efficiently guided me through the shoals of Palermo's university politics.

My first priority was to organize the important service courses for engineers; my second, to provide instruction on more advanced physics (*fisica superiore*), which had also been entrusted to me; my third, to start some research.

I amused myself by inspecting old teaching apparatus, as I had done in Rome once with Amaldi when we were still students. At that time, we had discovered several pieces of equipment dating from the second half of the nineteenth century, among them a gadget for demonstrating conical refraction, which required some thought before we could figure out what it was. At Palermo I found pieces going back to the times of Augusto Righi, Damiano Macaluso, O. M. Corbino, and other of my predecessors. The library was devoid of modern books and journals. On the other hand, I had a beautiful office with elegant furniture, and a letterhead that possibly went back to King Umberto I (assassinated in 1900), which I enjoyed using. For the rest, the Physics Institute was a desert.

At the first faculty meeting, with about a dozen professors sitting around a table, I could see that there were no big fights afoot. The mathematicians Michele de Franchis and Michele Cipolla were authorities in their fields. The botanist Montemartini was confined to Palermo because he was notoriously anti-Fascist, and the zoologist Giardina, although now very old, had once been brilliant. The chemists did not seem exactly at the level of their great predecessor, Stanislao Cannizzaro, and neither did the astronomer appear to be the equal of his great predecessor, Giuseppe Piazzi. All were good professors, however, with whom it was easy to agree provided one maintained polite behavior and due respect for turf.

I clearly stated that I had no intention of being a bird of passage. I would do my best to improve physics and I would not spend day and night planning how to contrive a transfer, as many professors from the mainland used to do. When my Sicilian colleagues perceived that I truly meant what I said, they helped me in whatever ways they could and adopted me as one of them. Thus my university relations were excellent.

Palermo was not, in fact, one of the minor posts usually conferred at the start of a university career, such as Camerino, or Sassari, but neither was it one of the major seats in which one landed at the end of a meritorious career, such as Rome, Bologna, Pisa, or Turin. At Palermo there were a good many Sicilians, for whom it was the seat of choice; some notorious anti-Fascists, such as Perrier and Montemartini, who were not in the good graces of the minister and would not be transferred even if they wanted to be; and some young professors at the beginning of their careers.

As soon as possible after my return from America, I had joined Elfriede in Florence, and we started making detailed plans for our imminent wedding. Elfriede ordered linens for our home from the house of Pini in Florence. She bought an elegant dress at Zecca in Rome and stocked up on top-quality household items and clothing, destined to last a long time. This fitted our philosophy, as well as that of my parents. However, when the bills arrived, some were pretty stiff. Imprudently, my father or I (I do not remember who) made some comments on this. Elfriede immediately started crying; her bitter and unusual tears startled me even more because they showed a surprising misunderstanding. No criticism of her had been implied; on the contrary, everybody was satisfied that she had done very well.

Elfriede and I decided to marry on Sunday, February 2, 1936. To our great regret, Elfriede's parents could not come to the wedding, but they visited us later when we were settled in Palermo. I went to the Rome synagogue to make arrangements for the wedding ceremony and told the rabbi that I wanted the simplest and cheapest wedding available, the more so as the parents of the bride could not attend. The rabbi winced, and I added that I found it inappropriate to spend money on ceremonies when there were so many tragic situations that needed help. To dispel any doubts in his mind I added: "How much does a luxury wedding cost?" He told me, and I gave him the sum, saying that he should arrange the simplest possible ceremony for us, as I had requested, and spend the difference for German refugees. This was the agreement. On the day of the wedding, however, the Temple was full of flowers and tapestries with great pomp. The rabbi gave us a short homily. "See! Adonai. . . . Before yours, there was a luxury wedding ceremony and there was no time to change the decorations. Thus you too will have a luxury wedding." A reception at the old Hotel de Russie followed. It was attended by friends and relatives, including Corbino, Levi-Civita, and my physicist friends.

From Rome, in terrible weather, we went to the Hotel Vesuvio in Naples, and as the rain persisted, we went to visit my friend Carrelli, professor of physics at Naples, who showed us a splendid calcite crystal, a present of Fresnel to Melloni, from his museum. Bad luck had it that it slipped from Elfriede's hand and was chipped in one corner. Our embarrassment is hard to describe.

At Palermo we lodged at the Hotel Excelsior in Piazza della Libertà. Papà had commanded me not to return to Pensione Lincoln but to find the best possible accommodation. The Excelsior was then an excellent hotel, with a first-class chef and an able manager, who was stuck in Palermo because he was suspected of anti-Fascism. He took a liking to us and treated us as his protégés, giving us the best rooms of the hotel and keeping them always at our disposal.

Before our departure from Rome, my father, unbeknown to me, had taken Elfriede aside and given her a small sum, telling her that she should use it for postage stamps to write to both families. The money would have sufficed for writing by special delivery all her life and more. Elfriede deeply appreciated the gesture.

This and similar episodes must be seen in relation to my wish to live within my professorial salary of about two thousand lire a month. My father, who was more practical, decided to add a substantial monthly supplement to my salary. When I refused to accept this, he instructed Bindo Rimini: "Go to your cousin and tell him he is not only a fool, but also rude." My father was quite right. The sum was trifling for him; it pleased him to give it to us, and it helped to make our life more pleasant. Furthermore, neither Elfriede nor I was lazy or spendthrift. After a while, I realized that instead of being haughty with my father, I should be grateful and thank him.

Immediately after our arrival in Palermo, we started exploring its surroundings. It was an exceptionally cold spell; there was even snow on some of the mountains, a most unusual condition. Later, however, we became fully acquainted with the extraordinary beauty of that part of Sicily.

On our first vacation, we decided to go on a true honeymoon trip, skiing in the Dolomites. At that time there were no ski lifts, and one climbed using sealskins; but we really enjoyed our avocation of cross-country skiing. We went around the Sella group carrying our rucksacks and sleeping in small hotels or huts. In one of them, our room remained quite cold in spite of an electric heater. I examined it and changed the connections of its resistors from series to parallel, quadrupling the heat output. Elfriede admired the power of physics, but the next day the innkeeper made a scene because we had used too much power.

Finally, we returned to Palermo to stay for a longer period. We started by making an official round of visits to the dean and the rector, who was most cordial. The dean was not at home, but his wife was, and she received us in a friendly way. We noticed her conspicuous beautiful and brilliant red hair and began a polite social conversation. After suitable platitudes, she offered us some *karkade,* an infusion of an Ethiopian plant that in those times of sanctions by the League of Nations was supposed to replace tea. It was deep red and had a flavor new to us. Caught by surprise, we found it hardly drinkable. A look between us showed us that we had had the same thought: perhaps it was used to dye our hostess's hair.

No less than mine, Elfriede's life had changed radically with her marriage and coming to Palermo. She was no more "La Spiro," but Signora Segrè. However, we had not changed our fundamental habits of first working hard and then finding our recreation in the mountains or in touring. In the beginning we did not own a car, but we soon acquired one and drove it down from Rome to Palermo. I had thus repeated a good part of the itinerary I had covered with Rasetti in 1929, but we could not enter the Palazzo Cimbrone at Ravello because Greta Garbo and Leopold Stokowski had rented it and locked themselves up in it.

The Palermo of 1936 was a beautiful city; despite its location, it was not provincial. At the beginning of the century, it had enjoyed a great cultural and architectural flowering. It had shops comparable to those of the greatest Italian cities, an excellent opera house, and magnificent villas flanking the Viale della Libertà, not to mention the antiquities, Arab, Norman, and baroque, that testified to its millennial history. All told, one could recognize a capital, perhaps slightly Bourbon, of the Kingdom of the Two Sicilies. Surrounded by Monte Grifone, Monreale, Monte Pellegrino, and Mondello like precious stones set in a ring around a central diamond, the city offered splendid outings. In the spring we could smell the scent of orange blossoms, which became more pungent at sunset. War and uncontrolled population influx have ruined Palermo, as they have most Italian cities.

While very busy organizing the Physics Institute, I started teaching the elementary experimental physics course, performing many demonstrations with apparatus that had been out of use for perhaps fifty years. I also prepared a few instruments in the hope of being able to begin some research. As a start, I built one of our standard ionization chambers and ordered a Perucca-type electrometer and other equipment needed for radioactive work. I hoped somehow to secure long-lived radioactive isotopes for study. I asked my physicist friends to suggest names for a couple of vacant assistant professorships automatically placing former pupils of the Scuola normale at Pisa high on my list. I thus met B. N. Cacciapuoti and Manlio Mandò, whom I was subsequently able to hire. Years later, the one became a professor at Pisa, and the other, after a long period as a prisoner of war in India, at Florence; Mariano Santangelo, an able young student at Palermo, became professor at Modena.

Among the students was a young lady, Ginetta Barresi, related to the Crocco Family, famous in Italian aviation. She was an unusual person; most intelligent, with deep Sicilian roots, sincerely religious and learned in Catholic doctrine. In those days a woman physics student was a rarity, and in Palermo she was the only one. Ginetta had no qualms about the matter; she studied her chosen subject proficiently and if people wondered, she let them wonder. Her unusual culture extended to literary subjects and was always very solid and well digested, never superficial. She became our dear friend and helped us admirably in the difficult times that were to follow.

In keeping with tradition, I started writing lecture notes for my course in experimental physics. I completed and published the first volume,[2] and I started the second but could not finish it before my dismissal. For *fisica superiore,* I taught electricity. I introduced written examinations, a novelty in Italian universities. The attempt produced a certain ferment, but ultimately the students became resigned to this innovation, although it was dubious whether written examinations were legal. A typical question for such exams was: Calculate the weight of a mercury sphere of 3 cm radius. Unfortunately, the answers were not edifying.

Since the majority of physics students became high school teachers, I thought it would be useful to write a book on "elementary physics, from a higher point of view," modeled on the similar ones for mathematics, edited by Felix Klein in German and by Federigo Enriques in Italian. I worked seriously on the project, writing a detailed program for the work. It was to consist of a series of articles, and I looked for collaborators, and for a publisher. This last was to be Sansoni of Florence, who belonged to the Gentile family. The racial laws put an end to my endeavors, but Giovanni Gentile, Jr., continued the project until he died prematurely in 1942. After the war, Gilberto Bernardini resumed the initiative, and the first volume appeared in 1947. Bernardini's preface summarizes the history of the book. I believe that the idea has still some merit.[3]

In 1936 I could not yet assemble the minimum equipment necessary to start research at Palermo, but I took advantage of vacations to do some experiments in Rome. I found Amaldi and Fermi deeply engulfed in their fundamental investigations on the slowing down of neutrons in hydrogenous substances. I had the impression that they did not want to waste time even with an old friend like myself. I spoke to Wick, who was in Rome, and did something by myself with the instruments and sources available.

All told, the school year 1936 passed quickly and pleasantly. For the summer I thought of going to have a look at the United States with Elfriede. At the beginning of the summer, it turned out that she was pregnant, but since she had no complaints whatever, we decided to go anyway. Later, with a small child, it would be much more difficult to travel, hence this would be our last chance, at least for several years, to visit America. Moreover, we were disturbed by the steady downhill trend of events. Although we personally had prospered, we were convinced of the precariousness of the situation, and this was one more reason for keeping in contact with American physicists and for showing up in the United States.

For me, the natural place to visit was Columbia University; I had been there before, had rapidly done good work there, and had struck up a friendship with the Columbia neutron physicists. I knew their instruments, and we had common scientific interests. I thus wrote to Dean Pegram proposing to go there, and on July 2, 1936, we landed in New York. Amaldi, too, came to New York in the same period.

We had not, however, reckoned with New York's hot, humid weather and with the suffering it would bring to a pregnant woman. Elfriede could not sleep well for the heat; she got up at night to take showers to cool off; it was clear that the heat was not only unpleasant but unhealthy. Thus, as soon as possible, we departed for better climates.

Otto Stern had extolled the future of Ernest Lawrence's cyclotron to us in previous years. In 1935, when we were in Ann Arbor, Fermi and I had corresponded with Lawrence. At that time, I do not remember for what reason, he offered Fermi a millicurie of radiosodium. Doubting Berkeley's radioactivity measurements, Fermi replied suggesting that Lawrence had perhaps made a mistake and actually meant a microcurie, a thousand times less. In answer, he received a letter containing a millicurie of radiosodium. We were dumbfounded. By then I was sure I wanted to go to see the cyclotron. Later, when I knew the Radiation Laboratory from the inside, I could imagine the effect that Fermi's letter must have produced and Lawrence's reaction.

At Rome we had discussed the possibility of building such a machine, and we had even tried to locate a magnet similar to the one used for the 37-inch cyclotron in Berkeley (Marconi had used it in the radio station at Coltano many years earlier; in fact, I believe it was the one I had seen there as a child before World War I). Ultimately, however, the plan came to naught, and in 1936, cyclotron and climate attracted me to Berkeley.

We left New York by train and stopped for a few days in Ann Arbor, Michigan, where Elfriede started feeling better; then we crossed the continent on a famous train called "The Challenger," reaching Berkeley in three days. We had two objectives: learning about the cyclotron and visiting California and the West. We rented a Ford car for one month, hoping that it would also serve Elfriede to learn to drive. I then plunged into the Radiation Laboratory for several days.

Lawrence was most cordial.[4] It was my first meeting with him and I was not used to his personality, so different from that of any other physicist I knew. He invited us to his home for dinner, where I ate avocado for the first time. (I did not like it then, but I do now.)

Besides Lawrence, I spoke with Edwin McMillan,[5] Don Cooksey, Franz Kurie, Philip Abelson,[6] and others I do not remember. J. Robert Oppenheimer also invited us for dinner.[7] I gave a couple of lectures on neutrons and made a detailed tour of the Rad Lab, speaking extensively with Abelson, then a graduate student working on uranium. I told him that there was undoubtedly a mystery in uranium bombarded by neutrons. With a source as powerful as the cyclotron, the Berkeley researchers had the inestimable advantage of being able to generate an activity that was large compared with the natural activity of uranium. This should enable them to see things hidden from everybody else. As a first step, I proposed they see whether there was a difference between the activity produced by fast and slow neutrons. Of course, I had no idea of the nature of the mystery presented by uranium, but I knew it was there. Abelson worked a little on the subject and gave me the decay curves he obtained, but at the time he went no further.

I also renewed my acquaintance with Count Lorenzo Emo Capodilista. I had known him since his student days in Florence, where he had worked with the local cosmic-ray group. He had lost his mother when still very young and cherished his wealthy grandmother, a Mrs. Parrish of Philadelphia, who was a true lady, remarkable for her vigor, intelligence, and imposing appearance. Lorenzo had a heart of gold and was a wise man and a true gentleman, but not a rabid physicist like most of his colleagues. Possessing independent means and extended interests beyond physics, he tended to enjoy life. Elfriede liked him at once, too, and he later became one of our closest friends.

In visiting the Rad Lab, I noticed that there was a lot of radioactive metal scrap lying around. Nobody knew what it contained. I asked for some samples to take with me to Palermo. The radioactivities undoubtedly had long periods, and I would be able to study them at leisure upon my return. With luck, there might be something interesting. I took several pieces of metal that had belonged to cyclotron parts. Lawrence was very kind and generous in giving me this material; he said he was happy I could use it and glad to be able to help so penurious a place as Palermo.

When I had concluded my scientific visits, Elfriede and I drove off to see the marvels of the West. We were convinced that this was a unique opportunity to do so, and some fifty years in fact elapsed before I returned to some of the places we visited. Other places, on the contrary, became frequent destinations when we settled in Berkeley.

Rasetti, as usual, had lectured us on the places we must see, and his guidance from afar was very useful to us. We went to Yosemite and Death Valley, continuing to some truly wild deserts in Navajo country, but missed the Canyon de Chelly, which Rasetti had rated a must. We visited Boulder Dam still half empty, and the Utah national parks, bought cactus seeds for Montemartini and Palermo's botanical garden, and went to Mount Wilson Observatory and to the movie studios in Hollywood. Happy with our trip, we returned to Berkeley, where I collected my radioactive material. On October 10th, 1936, we landed in Naples.

On my return to Palermo I immediately started work on the material I had obtained in Berkeley. The instruments I had prepared the previous year were perfectly suited to my program; furthermore, I had built a chemical hood and had found glassware and chemical instruments in the lab, which perhaps went back to the time of Cannizzaro. I was thus able to start chemical separations using the usual radiochemistry techniques.

I soon discovered that I had taken with me a true mine of radioactive substances. The cyclotron had been used for bombarding a little of everything, although mainly phosphorus; no special precautions had been taken, so I found many different substances that had vaporized from the target. In addition to phosphorus, a preliminary survey revealed radioactive cobalt, zinc, perhaps silver, and other activities I could not ascribe to any known element.

I first recognized a large quantity of P^{32}, with a half life of about two weeks. I thought immediately that this might be useful for biological experiments, but naturally, not being a biologist, I did not know specifically what to do with it. I explained the tracer technique, then relatively new, in detail to my colleague Camillo Artom, professor of physiology, and offered him the radioactive phosphorus and the necessary technical help with radioactive measurements. Artom at once grasped the technique and the possibilities it offered, and immediately thought of some interesting applications to phospholipid metabolism. Thus began a fruitful collaboration, which produced good results. Having learned a minimum of physiology and biochemistry, I tried to make a rudimentary mathematical model of a mouse, describing its metabolism by suitable coefficients. Some of the ideas went back to Volterra's old studies, which I had read as a student.[8] I believe that this type of investigation, greatly refined and expanded, has developed into a fashionable endeavor.

In February 1937 I received a letter from Lawrence containing more radioactive stuff. In particular, it contained a molybdenum foil that had been part of the cyclotron's deflector. I suspected at once that it might contain element 43.[9] The simple reason was that deuteron bombardment of molybdenum (atomic number 42) should give isotopes of element 43 through well-established nuclear reactions. My sample, the molybdenum deflector lip, had certainly been intensely bombarded with deuterons, and I noted that one of its faces was much more radioactive than the other. I then dissolved only the material of the active face, in this way achieving a first important concentration of the activity.

By now I was more sophisticated than I had been in Rome in 1934, and I knew that the "masurium" announced by I. W. and W. K. Noddack in 1925 was probably a mistake. Among other reasons, nuclear systematics raised strong suspicions about its stability. I thus had to prove that I really had in hand a new element, created artificially and devoid of stable isotopes. The methods for such an investigation had been pioneered long ago by D. I. Mendeleyev and Marie Curie. One predicts the chemical properties to be expected for the new substance by criteria similar to those used by Mendeleyev, and then one tries to verify the predictions by radiochemical methods, taking into account that the behavior of trace amounts of a substance can be different from that of matter in bulk.

For this investigation I enlisted the cooperation of Carlo Perrier, who had more experience in chemistry than I. First we separated the activity we were studying from all known elements to make sure that it was not isotopic with any of them. Next we established several of the chemical properties of element 43. Separation from rhenium was the most difficult problem, but in the end we succeeded in two different ways: by precipitation as a sulfide in a very acid solution and by distillation in a current of gaseous hydrochloric acid. All this work was most amusing and of obvious importance.[10]

By following the radioactive decay of our samples and by measuring the absorption in aluminum of the electrons emitted, B. N. Cacciapuoti and I found three decay periods: 90, 80, and 50 days. Looking back on the data fifty years later, I see that in effect we had only two radioactive isotopes: technetium 95, with a period of 61 days, and technetium 97, with a period of 90 days. They are both nuclear isomeric states with complex electronic radiations, obtained by deuteron bombardment of several molybdenum isotopes.

In this work we had discovered the first chemical element created by man.[11] Perrier and I decided not to name the new element at the time, although we received suggestions for names celebrating Fascism or Sicily, such as Trinacrium (from Trinacria, an ancient Greek name for the island), which we did not like. Moreover, for us to avoid controversy with Walter Noddack and Ida Tacke-Noddack, they first had to retract their claims, or these had to fall of their own weight, as later happened. We also knew that many more elements had been named or announced than truly existed. Haste in naming did not seem like good style to us.[12]

Georg von Hevesy, who knew the Noddacks' work at first hand, wrote to me explaining its weaknesses. Hevesy, a Hungarian educated with all refinements of the old Austrian Empire, was one of the greatest living chemists and a close friend of Niels Bohr. He and Fritz Paneth had invented the radioactive tracer method, and Hevesy and Dirk Coster had discovered the element hafnium, using X-rays as an analytical tool.[13]

The Noddacks were chemists, highly respected for their discovery of rhenium, which they detected in several ores in 1925. In the same paper they had announced the discovery of two elements: element 75, which they named rhenium (from Rhenus, the ancient name for the Rhine), and element 43, which they named masurium (from Masuria, the easternmost part of East Prussia, where German armies had repeatedly defeated the Russians in World War I). Rhenium was soon confirmed, and the Noddacks prepared it in macroscopic amounts, but they did not make any further mention of masurium. In 1933, when I bought all the elements available in Rome for our neutron work, I found a sample of rhenium, but not one of masurium.

In 1937, after receiving the letter from Hevesy mentioned above, I had some doubts about the Noddacks' results and decided to visit them and to obtain firsthand information on their work. About September 20, on my return from Copenhagen (see p. 122), I stopped in Freiburg, where the Noddacks had their lab. Professor Walter Noddack kept me waiting for a while, but ultimately he received me. I did not see his wife.

I showed Noddack the proofs of our Lincei paper giving the properties of element 43 and asked him whether his results agreed with ours. "Yes," was the answer. I asked him whether he had found something on the chemistry of 43 beyond what we had, and he said, "No." I asked him how much masurium they had, and he answered about 1 mg, which to me seemed unlikely. He told me he had sent it to Francis Aston at the Cavendish Laboratory for isotopic analysis, which surprised me. I asked to see some of his X-ray plates, with the characteristic spectrum of 43. He answered that unfortunately the plates had accidentally been broken and hence were not available. When I asked why he had not made more plates, I could not obtain a clear answer. By then I was thinking that either they were deluding themselves or they had doubts about their results and hoped that further work might resolve them; in the meantime they did not want to prejudice the issue. In any case it was unlikely that they had clear-cut results. Having formed this opinion, I took my leave.

I was surprised when a couple of weeks later Noddack, his wife (if I remember correctly), and a cohort of assistants showed up at my lab in Palermo. I showed them what we had. These are the only personal contacts I remember having had with the Noddacks.

After the war, when nuclear reactors produced macroscopic amounts of element 43, I had the satisfaction of seeing, not only that we had made no mistakes, but also that we had found the main properties of the new substance. Only then did Perrier and I give it the name *technetium* to commemorate the fact that it was the first artificial element.[14]

One day Fermi came to visit me at Palermo and told me he thought our work on element 43 was the best piece of work in physics in the preceding year. Since Fermi did not make such statements merely to please, or without due consideration, I was elated.

The prime necessity for further work was the supply of radioactive substances. I asked Lorenzo Emo to send me more material from Berkeley, which he did, with Lawrence's permission. When I received a letter from Berkeley, I measured its radioactivity before opening it. I also sent to Berkeley a collection of test tubes containing several substances to be put near the cyclotron target where they would be neutron-irradiated. Among them I included some purified uranium and thorium, because I was aware of the uranium mysteries, some ammonium nitrate, in the hope of finding C^{14}, and sundry other materials. The cyclotron produced so many neutrons that if my samples were simply kept in a box near the target, I could obtain precious material that would keep me busy for quite a while; or at least I so hoped.

The beginning of the year 1937 was darkened by an unexpected tragedy. Corbino caught pneumonia and died in a few days, on January 23. His death was a severe blow. He was only sixty-one, and we had all counted on his wise counsel and guidance in the difficult times we anticipated.

I immediately saw what the consequences of Corbino's death would be, and I was soon proved right. When I went to Rome for the funeral,

I found that his post, which should logically have gone to Fermi, had become the target of obscure cabals. The end result was Lo Surdo's appointment as director of the Physics Institute. I could not have imagined a worse choice. Among other things it ensured hostility, in place of benevolence, toward Fermi's group, which, in my opinion, was exerting a most salutary influence on Italian physics.

Another surprise followed shortly after: Amaldi was appointed professor at Rome. He had competed successfully for a chair at Cagliari, in Sardinia, but had renounced the appointment in order not to have to leave Rome. Immediately afterward, he was called to the University of Rome. The whole deal had very negative implications for me. I came from the same stable as Amaldi, had been first assistant to Corbino, and had seniority over Amaldi; nor could it be said that his scientific work overshadowed mine. Obviously my chances of returning to Rome and rejoining the group were vanishing. Amaldi's appointment at Rome also meant that Fermi and Rasetti either could not or did not want to put up a fight for me. Fermi, as a matter of principle, avoided losing battles, and the whole development signaled to me that my chances of being appointed to a better chair than Palermo were slim indeed. Although the idea of remaining at Palermo for a long time was not disagreeable per se, I was concerned for the future of my research. It was not easy for me to imagine how I would be able to continue to do interesting work in Sicily. Ultimately, however, Amaldi's appointment did me little harm and turned out to be a stroke of good luck for Italian physics.[15]

Having assessed the situation at Rome, I put out some feelers for other chairs, with discouraging, even humiliating, results. For instance, I still regret having asked His Excellency Professor Nicola Parravano, *accademico d'Italia,* to communicate our note announcing the discovery of element 43 to the Accademia dei Lincei as a gesture of appeasement. On this occasion, I saw manifest signs of anti-Semitism, and they were not the first. Anti-Semitism had always been endemic in Italy, but it had not prevented talented people from making their way. Now one felt, however, that the disease was getting worse.

My father bought us a brand-new modern apartment on the Piazza Francesco Crispi in Palermo, which had windows overlooking the beautiful Giardino Inglese. It was furnished for us by the Florentine firm of Gori and with some pieces designed by an architect friend of mine, which I had brought from Rome. After fifty years of service, I can still admire their quality in my California house.

We were expecting a child in March, and Rasetti's mother and the Amaldis helped us to find exactly the help we needed: Lella, a woman from Abruzzo, who had never been to school, but had uncommon intelligence and personality, and a sweet nursemaid from Poggio, where the Amaldi family had an estate. Both women excelled in their work, were of sterling honesty, and affectionate; they remained Elfriede's lifelong friends.

On March 2, 1937, Gori came from Florence to assemble our furniture; he wanted to do it personally. At about 3 P.M., when he had just finished his work, Elfriede told me that she thought it might be better to go to the hospital, and about two hours later our son Claudio was born. My colleague the professor of obstetrics at the University of Palermo was in attendance, although once in a while he fell asleep. A few days later Claudio developed a sizeable lump on his neck, much to our horror. The pediatrician, a German doctor chosen by Elfriede, who had formerly been her colleague at the *Landschulheim* in Florence, made an alarming diagnosis, but suggested we show the child to the university's pediatrics professor. The latter, a very elegant Sicilian gentleman, whose looks reminded me of Freud, briefly examined the infant and then said: "Do not worry. It is nothing serious. His neck has been pulled at birth. All he needs is to sleep for a few days with his head tilted and he will be all right." This turned out to be the case.

With the 1936–37 school year approaching its end, we prepared for our summer vacation. Since Claudio was only a few months old, we could not travel far, and we rented a house at Alba di Canazei in the Dolomites, where we occupied one floor and Amaldi another. We also arranged lodgings in the immediate vicinity for the families of Bakker from Holland and of Bernardini from Florence. Unfortunately, this was to be the last vacation I was able to enjoy in the old-fashioned style familiar to me from my childhood. We collected large amounts of wild raspberries, from which we made jam, and of edible mushrooms (which I learned to identify from a German booklet), thus commencing two lifelong culinary hobbies.

In the middle of the summer, I was called to the colors and had to attend a military training school in the ancient seaport town of Civitavecchia, north of Rome, for several weeks. While there, I received a telegram from the rector of the University of Palermo urgently recalling me, because Il Duce, Mussolini, was about to visit and all the professors had to be present. I took the telegram to the colonel commanding the school and applied for leave. The colonel looked at me intently and asked: "In this season is Palermo very hot?" I understood at once the meaning of the question and answered: "It is terribly sultry." To this the colonel responded: "Answer that you are serving in the army and that leave has been denied." I must add that the colonel gave me leave every weekend to join my family at Alba di Canazei, where the weather was good.

While at Civitavecchia, in the deep of night, I received a telephone call with the news that my father, who was at Tivoli with my mother, had been taken gravely ill. Shortly thereafter Bindo Rimini arrived by car and took me to Tivoli, where I found my mother, Riccardo Rimini, and Marco. My father was in a coma, and according to Riccardo, an excellent doctor whom we all trusted, there was little hope of his surviving. A few hours passed, and the situation was unchanged. Somehow rumors of my father's state spread, and people from the paper mill and city authorities made discreet, concerned inquiries. Somebody even started thinking about funeral arrangements. No signs of improvement appeared.

In the afternoon, the patient, still in a coma, passed a lot of wind, and then loudly and clearly spoke some famous lines from Torquato Tasso's *Gerusalemme liberata* (my translation):

> The raucous sound of the Tartarean bugle
> Calls the inhabitants of the eternal shadows.

My mother, who was at her husband's bedside, almost fainted. We all rushed in, and to everybody's amazement, my father regained consciousness. In a few hours he was greatly improved. For about a week he slightly dragged one leg in walking, but soon he totally recovered, without visible trace of what had happened in either body or mind. We had been terribly scared. My father's comment was: "Now I know what there is in the beyond: nothing."

Before the summer vacation, Bohr had invited me to one of his annual conferences in Copenhagen, showing that our work at Palermo had not escaped his attention; I was highly pleased and immediately accepted. On the train to Copenhagen, I met Hans von Euler and several other young physicists proceeding to the same conference, which thus began en route. They explained some of the mysteries of the latest cosmic-ray observations, harbingers of what were later called muons, to me.

At Copenhagen, the meetings were extremely strenuous. In such company, one tried to absorb as much as possible, and thus one had to concentrate without interruption for many hours at a time. I was exhausted by the end of each day.

Bohr's residence and lifestyle impressed me; they were truly princely in the best sense of the word. We also made some of the usual excursions, but continued talking physics all the time. I spoke on the new element 43.

On my way back I stopped briefly in Hamburg. From there, on September 15, 1937, I wrote as follows to Riccardo Rimini:

> . . . Yesterday evening the Congress ended, with a humorous, but rather moving, feast. We acted in a sort of variety show summarizing Bohr's recent travels around the world. Through the jokes one could feel the respect and almost veneration that everybody feels for Bohr. I could not approach him very much, but I understood that he is one of the most remarkable personalities produced by mankind, and that he hovers in heights incomparably higher than those reached by common mortals, be they even Fermis. Also morally and from a human point of view he must be superior to others. Immediately after the feast I left with [Werner] Heisenberg [winner of the Nobel Prize in Physics in 1932] and his wife. Heisenberg . . . has been a pupil of Bohr's at Copenhagen for three

years, and he has done his best work there. Bohr said a few words of good-bye to him and his wife that well-nigh made the company shiver, and everybody was clearly shaken.

I had a bad train trip, because we were continuously disturbed by customs agents, police, and similar characters. Next morning I arrived here and lodged where you see from the letterhead [the Hotel Continental]. I slept a little and then phoned "I." We made an appointment, I believe by chance, on the spot where we first met. It was then a winter evening; it is now an autumn morning. While waiting I thought of many things of that time, about progress made and changes since then in myself and in the world. I do not go into detail so as not to annoy you. In any case you know too many things not to be able to more or less guess my thoughts.

"I" is physically not much changed, although she is with child. Morally she seems to me to have become rather stupid. She has entered the mentality of the local lower middle class, which is rather unappetizing. The husband is a lawyer and notary and is 40 years old. She has lost some of the vivacity and flexibility she had in years past. She is *simpatica,* but completely, irremediably, foreign to me, and I have no reason to see her any more in the future. Although we did not say it, the long pauses interspersing our conversation clearly spoke for themselves. In any case I would not even have gone to bed with her with any enthusiasm. Other ties and interests are now at the forefront and perhaps now, for the first time, Claudio had signaled his existence. In any case, as you know, I dearly love Elfriede, and although I am perfectly aware of the possibility of separating persons and affections, a possibility that, although frequently contested, in certain cases exists, I do not want foolishly to hurt her.

The city of Hamburg and Germany in general after such a long absence have a curious effect on me. Although the exterior aspect has somewhat changed, I could not say that the country looks different, in spite of the abundance of soldiers, each stiff as a ramrod. The shops, with the exception of the booksellers, are the same, and so are the public places, but the whole looks to me like a shell without the animal. For me, who knew Germany as the freest country, as a fountainhead of culture for a physicist, as an unprejudiced country for girls, full of new ideas and with a lively intellectual life, it gives the impression of a total void. Void, void, and nothing else. . . .

In any case the result of this whole trip and of this experience is

> rather to turn me to the future, and now Bohr and his discourses are more alive, or, better, more important to me than memories of 1933.

As already described, I also stopped on my way home at Freiburg to see the Noddacks.

Shortly after the Copenhagen conference I attended a congress held at Bologna to celebrate the 200th anniversary of Galvani's birth, but all I remember of it are visits to Ravenna, previously unknown to Elfriede and myself, and the general dismay at the announcement during the conference of Rutherford's death.

All told, 1937 had been a good year for us both personally and scientifically, although clouded by Corbino's death. Elfriede and I got along together better and better. She proved to be an excellent wife, ever equal to the often difficult demands placed upon her. She took care of Claudio with good sense, helped measure radioactive phosphorus in the lab, ran the household, acted as a secretary, read on her own account, and grew intellectually; in short, she was an excellent companion in every respect. She had also become attached to my parents, who abundantly reciprocated her feelings.

Our weekend trips had shown us a good part of Sicily. Agrigento and Selinunte (the ancient Selinus) on the southwest coast of the island in particular appealed to my imagination, although when we drove to Selinunte, the local boys scratched the paint of our car and spat on our rucksacks out of pure spite, while alternately begging or vainly trying to sell to us some fake old coins. Hiking in the Bosco della Ficuzza, we saw wild peonies for the first time. I did not know what they were, but guessed, remembering them from Chinese porcelain. At the foot of Monte Pellegrino, we found fourteen different kinds of wild orchids in an area of about five acres. Once in a while, we went to Mondello to buy live lobsters or swim.

In spite of everything, I was worried. I knew that the discovery of element 43 had been a stroke of luck, not likely to be repeated at Palermo, and doubted I could develop a sustained research program without radioactive sources and better instruments than the simple ones I had built. The university had assigned me two hundred thousand lire, a substantial sum, but a good part of it was needed for a machine shop and for other indispensable plant; the future was not all rosy. By nature I was inclined to what I used to call "physics without apparatus," in which new ideas make up for the simplicity of the techniques. This attitude derived both from my indifferent ability as an instrument builder and from my education in Rome, where theory prevailed over technique. However, there were limits to what could be done this way. I tried in vain to obtain some money from the Rockefeller Foundation and from the Italian Consiglio delle ricerche.

To invigorate physics in Palermo, I wanted to establish a chair of theoretical physics. There was no scarcity of young candidates who could brilliantly fill it; first among them Gian Carlo Wick and Giulio Racah. I did not consider Ettore Majorana because by then he had become a recluse and never left home. I could count on a good choice because Fermi's opinion would be decisive.

I discussed the subject with the rector of the University of Palermo, the jurist Professor G. Scaduto. He was most cooperative and promised to help me however he could, but was worried that the new professor might regard Palermo merely as a springboard and might not stay long enough to exert a truly beneficial influence. Scaduto wanted a commitment on this point.

The subsequent competition had a peculiar history. Initially, I had expected that the three winners would be Wick, Racah, and Giovanni Gentile, Jr. I never dreamed Majorana would enter the competition, because he had lived in seclusion for several years. Completely unexpectedly, however, he did. The consequence was clear: the three winners would be Majorana, Wick, and Racah; Gentile would be left out. In a theoretical physics competition, the opinions of Fermi and Enrico Persico would be decisive, and both would honestly recognize merit.

Then something unprecedented happened. The appointment committee (Fermi, Lazzarino, Persico, Polvani, Carrelli) met on October 25, 1937, and put forward a most unusual suggestion. It proposed to appoint Majorana as a professor for "exceptional merit" independently of the Palermo competition, and to suspend further deliberations until the minister had acted on this proposal.

I believe, on good grounds, that in order to avoid a defeat for his son, Gentile's father, a former minister of education and still a power in Italian politics, had conceived this plan and suggested it to the committee. With the competition held in abeyance, Majorana was appointed professor at Naples based on exceptional merit. A law allowed for this procedure in special cases involving illustrious persons, and had been used, for example, in the case of Marconi. After Majorana's appointment, the competition was reinstated, obviously without Majorana's candidacy. The three chosen were Wick, Racah, and Gentile. To my delight, Wick came to Palermo not long thereafter. Needless to say, at the time I was completely in the dark about the maneuvers mentioned above.

This was not, however, the end of the story. After a few months in Naples, where he had started his course in theoretical physics, Majorana wrote a suicide note to his colleague Carrelli and took a boat for Palermo. From there he wired Carrelli that he had changed his mind; he also mailed him a letter on the writing paper of the Hotel Sole at Palermo, dated March 26, 1938, saying:

> Dear Carrelli,
>
> I hope my telegram and the letter arrived simultaneously. The sea has rejected me and I shall return to the Hotel Bologna [in Naples] tomorrow, perhaps traveling together with these lines. However, I want to give up teaching. Do not think of me as a girl in an Ibsen play, because the case is different. I am at your disposal for further details.
>
> Affectionately, E. Majorana.

It is easy to imagine Carrelli's alarm and dismay on receiving these communications. As Majorana did not show up in Naples, Carrelli contacted Majorana's family in Rome, as well as Fermi. Ettore's brother, Luciano Majorana, who had also been my schoolmate, rushed to Palermo and came to see me; together we tried to trace Ettore's moves through the police. We found only that he had been at the Hotel Sole, as was clear, anyway, from the writing paper he had used. Fermi immediately alerted the government, and Mussolini personally ordered the chief of police at Palermo to use all his resources to find Majorana. To no avail. He had reembarked from Palermo for Naples, but after boarding the ship, he vanished without a trace. In all probability, he jumped overboard and was lost at sea. His body was never found.[16]

On my return to Palermo in the fall of 1937, Perrier and I renewed our investigation of element 43, but the cream had already been skimmed, and results were harder to get. Nonetheless, we succeeded in finding interesting novelties. I had set my hopes for the future on the package, previously described, sent for irradiation at Berkeley. I also started building a linear amplifier to detect the alpha particles I expected from the transuranic elements I hoped would be present in irradiated uranium.

In the meantime I had been asked to join the Rotary Club in Palermo. Italian Rotary Clubs are very different from their American counterparts. At Palermo, the club's membership was restricted to important local civic leaders. Furthermore, the club was definitely not Fascist. My father urged me to join, and knowing me well, strengthened his arguments by offering to pay the substantial monthly fee.

At the Rotary Club I met several interesting and important persons, both visitors and local residents. I remember especially the inspired face of the composer Don Lorenzo Perosi, which could have served as a model for a sculptor representing "Genius." My election to the club was another sign that Sicilians liked and accepted me. One of the members was the excellent rector of the university, scion of an illustrious family of lawyers. We were friends, but not intimates. One day, however, at the Rotary Club, when I went to greet him with a handshake, he surprised me by embracing me with open arms, whispering in my ear: "Watch out. You have behind you the secretary of the Fascio"—the highest local Fascist authority. Mussolini had just forbidden shaking hands as an un-Fascist gesture.

In 1938 Elfriede returned to Germany for a visit. It was the last time she saw her parents. When she got back to Palermo, I met her ship at the pier with a bunch of roses. They did not suffice, however, to counterbalance a scary piece of political news: Hitler's visit to Mussolini, of which the poet Trilussa (Carlo Alberto Salustri) so appropriately wrote:

Roma di travertino
Rifatta di cartone
Saluta l'imbianchino
Suo prossimo Padrone.

(Rome of marble splendor
Patched with cardboard and plaster
Welcomes the housepainter,
Her next lord and master.)

The allusion is to patch work ordered by Mussolini along the route to be followed by him and his guest.

I decided to spend the summer of 1938 in Berkeley in order to study short-lived isotopes of element 43 that could not survive the time it took to get from California to Palermo.

For these summer forays, I used a scientific strategy I had successfully tested years earlier in Amsterdam at the time of my first visit to Zeeman's lab. I prepared a detailed plan of work, rehearsed the techniques I would use, and knew exactly the instruments I needed. With such preparations, once on the spot, it was easy to obtain good results rapidly. In this specific case, I knew how to isolate element 43 from a molybdenum target, and I knew what to measure in the new isotopes, and how.

At the time Claudio was about one year old, and it was not expedient to bring him to the United States for a few months. We thus decided that he and Elfriede would stay in Italy, first in the Alps to escape the summer heat and then at Tivoli. I would return in October for the beginning of the school year.

In 1938 it was very difficult to get U.S. visas. U.S. consulates would not give one a visa unless one's Italian passport was specifically validated for the United States, and the Italian government would not validate a passport for the United States unless it already contained a visa. In theory, this precluded obtaining even a tourist visa. Immigration visas involved additional quota difficulties, practically excluding Italians and Poles. The last fact was important; Elfriede fell under the Polish quota, although she had never been a Polish citizen. Rasetti had, however, told me that the U.S. immigration law then in force contained a Section 4(d) that permitted entry, irrespective of the quota system, to artists, priests, and professors of a recognized university. At the time this did not concern me, because I only wanted a tourist visa, but it became vital later.

Under these circumstances I went, with my passport, to see an important official of the appropriate department. As a last-minute inspiration, I also stuck Elfriede's passport in my pocket. Our conversation proceeded approximately as follows:

"Commendatore, I am professor of physics at the university and I would like to go, for the summer, to study in California. I have a return ticket and I would like to obtain the validation of my passport."

"You know that I cannot validate it without a previous U.S. visa."

"Yes, I know; however, with this system nobody can move any more."

"Ah! You are the new physics professor?"

"Yes."

"The nasty one! I have a nephew who is very scared by your exam he has to pass in October."

"Commendatore, what is your nephew's name? Tell him not to worry."

The commendatore gave me the name, and I added, "I shall remember it; tell your nephew he has passed the exam."

With this, the commendatore took my passport, stamped, and signed it. I concluded: "Many thanks for your kindness; I sincerely appreciate it and shall not forget it. However, I leave here my wife and a child. One never knows. Couldn't you validate their passport too?" And I pulled out of my pocket the other passport, which was immediately validated. I still regret having been unable to repay the good commendatore's kindness; he may well have saved Elfriede's and Claudio's lives. Unfortunately, however, I obtained a U.S. visa only for myself, and not for Elfriede and Claudio.

Before departing for America, I went to Tivoli to take leave of my parents. It was the last time I saw them. Papà took me aside and said to me: "You are right in going. If I were half a century younger, I would do the same." These are the last words I heard him speak. Elfriede and I stopped in Rome and went to see *Aïda* at the Terme of Caracalla, but we were not in a cheery mood.

I embarked for the United States at Naples on June 25, 1938.

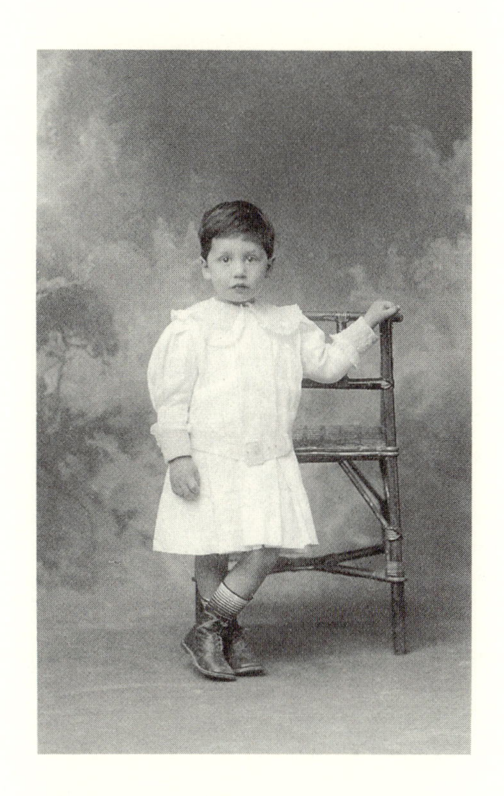

1. Emilio Segrè (Pippi) at age 3, 1908.

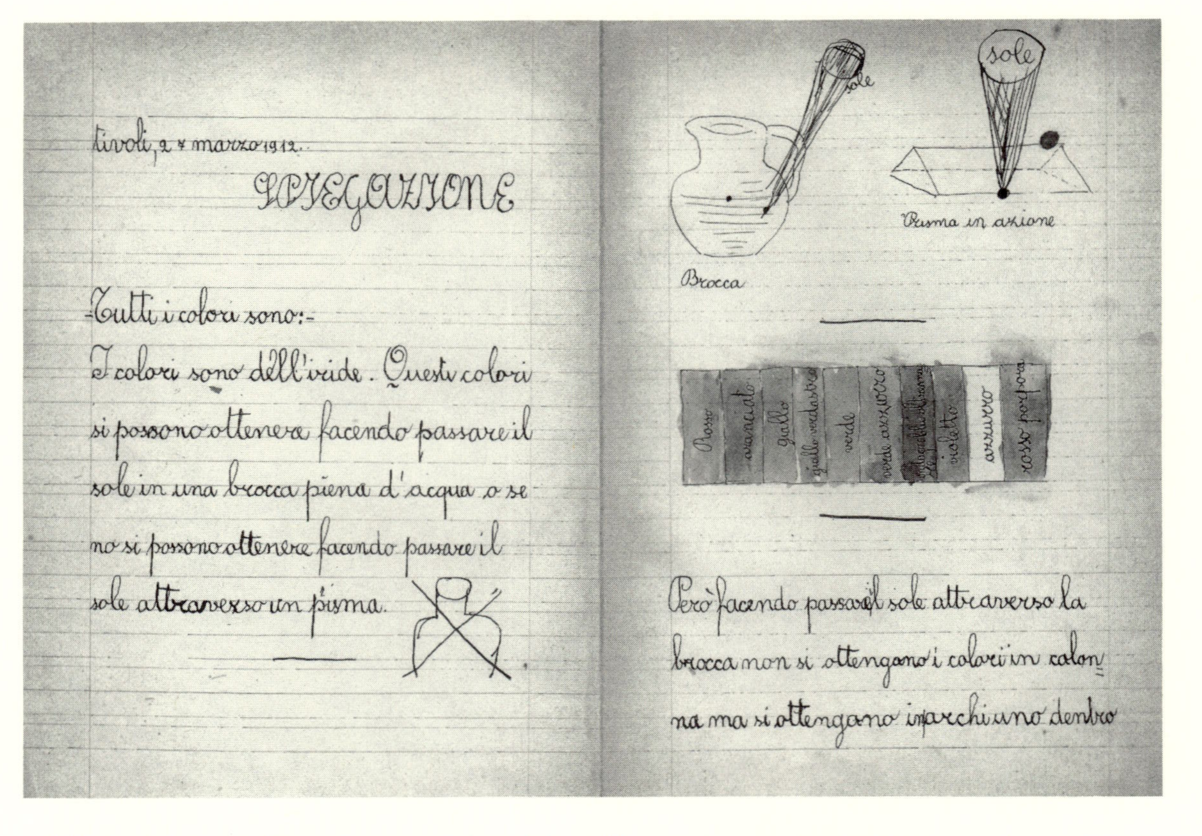
Tivoli, 27 marzo 1912.

SPIEGAZIONE

-Tutti i colori sono:-

I colori sono dell'iride. Questi colori si possono ottenere facendo passare il sole in una brocca piena d'acqua o se no si possono ottenere facendo passare il sole attraverso un prisma.

Brocca

sole

Prisma in azione

Rosso | arancato | giallo | giallo verdastro | verde | verde azzurro | azzurro | violetto | azzurro | rosso porpora

Però facendo passare il sole attraverso la brocca non si ottengano i colori in colonna ma si ottengano in archi uno dentro

2. “All the colors are: The colors are of the rainbow. These colors can be obtained by passing sunlight through a pitcher filled with water or by passing sunlight through a prism. However, by passing the sun through the pitcher one does not obtain the colors in columns but in arcs, one within [the other].” Emilio, age seven.

3. Segrè family on vacation in Castiglioncello, 1917. Back row from left: Egle Segrè (Gino's daughter), Gino and Claudio Segrè (uncles), Angelo Marco and Marco Claudio Segrè (brothers), and Giuseppe Segrè (father). Front row from left: Emilio between Bice and Fausta Segrè (Gino's daughters), Cina Segrè (Gino's wife), and Amelia Segrè (mother).

4. and 5. Giuseppe Segrè and Amelia Treves Segrè, 1937.

Amelia Treves in Segrè

6. Emilio in artillery lieutenant's uniform at the Physics Institute at Via Panisperna, on the day of Sir Chandrasekhara Venkata Raman's visit, circa 1930.

7. International Physics Conference at Como, 1927. Emilio is at the far left. (Lawrence Berkeley Laboratory)

8. Enrico Fermi, about 1928. Photo by G. C. Trabacchi.

9. Franco Rasetti (nicknamed "Cardinal Vicar"), Enrico Fermi ("Pope"), and Emilio Segrè ("Basilisk") in academic dress, 1931.

10. Edoardo Amaldi, Franco Rasetti, and Emilio Segrè on a hike, 1931. (Lawrence Berkeley Laboratory)

11. "The Group of Rome," 1934. From left: Oscar D'Agostino, Emilio Segrè, Edoardo Amaldi, Franco Rasetti, and Enrico Fermi.

12. Elfriede Segrè, circa 1937.

13. Copenhagen Physics Conference, 1937. Front row, from left: N. Bohr, W. Heisenberg, W. Pauli, O. Stern, L. Meitner, R. Ladenburg, J. C. Jacobsen. Second row, seated, from left: V. Weisskopf, C. Moller, H. Euler, R. Peierls, F. Hund, M. Goldhaber, W. Heitler, E. Segrè . . . Third row, seated, from left: G. Placzek, C. von Weiszacker, H. Kopferman . . . Standing: H. D. Jensen, L. Rosenfeld, G. C. Wick.

14. M. Stanley Livingston and Ernest O. Lawrence with the 27-inch cyclotron, 1932. (Lawrence Berkeley Laboratory)

15. Dinner at International House in Berkeley, circa 1939. Seated from left to right: Emilio Segrè, J. Robert Oppenheimer, and Chien-Shiung Wu. (Lawrence Berkeley Laboratory)

16. Segrè conducted his first experiments with spontaneous fission in this abandoned forester's cabin in Pajarito Canyon, Los Alamos, in June 1943. (George Farwell)

17. Group leaders at Los Alamos, late 1943. From left: Herbert Anderson, D. Froman, Enrico Fermi, Henry Herman Barschall, R. R. Wilson, Joe Fowler [?], John Manley, Seth Neddermeyer, L. D. P. King, Egon Bretscher, Emilio Segrè, and Hans Staub.

18. Segrè group at Los Alamos, late 1943. From left: Clyde Wiegand, G. A. Linenberger, M. Kahn, Owen Chamberlain, George Farwell, J. Miskel, Ann Kahn, Bill Nobles, J. Jungerman, Emilio Segrè, and Martin Deutsch.

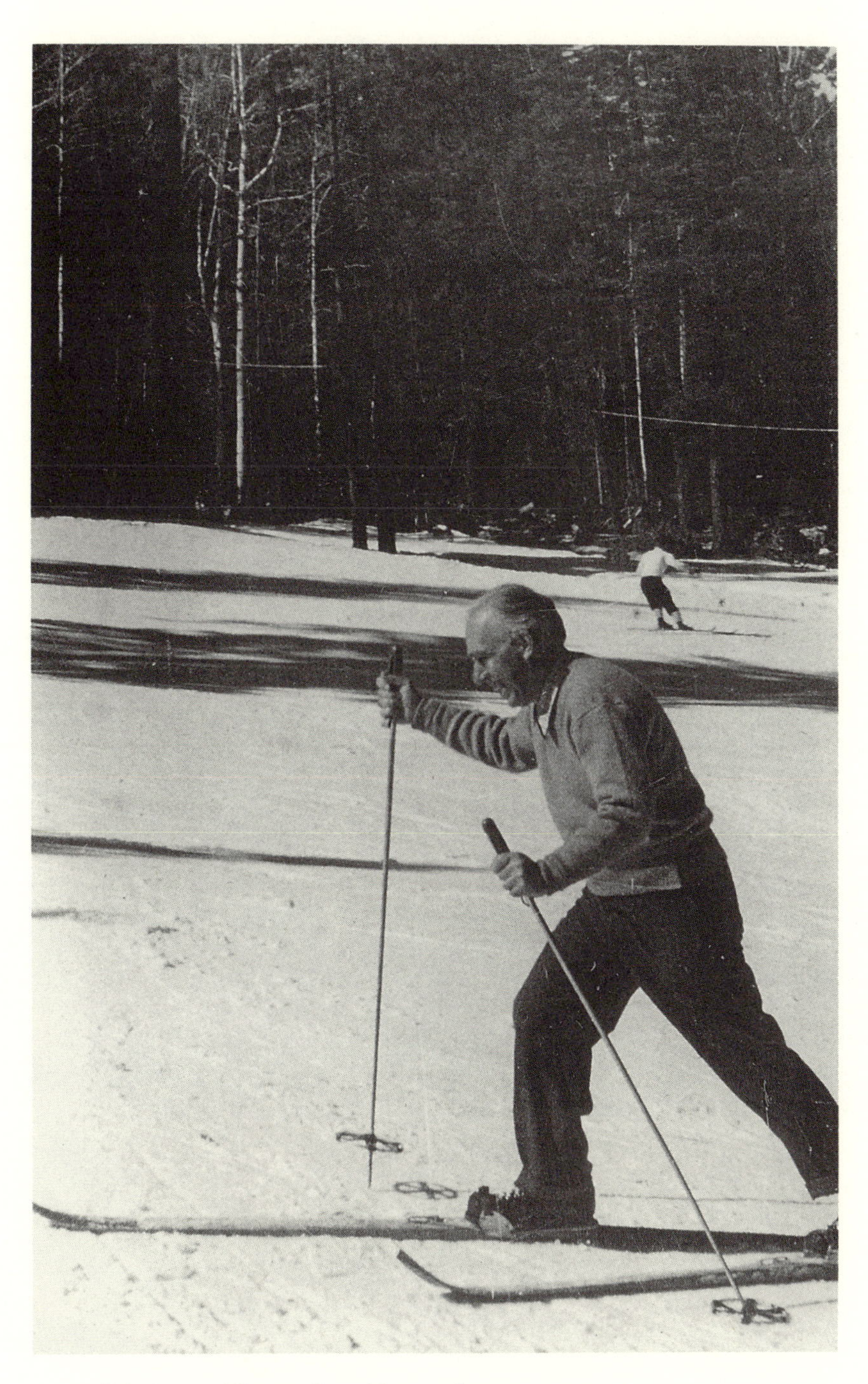

19. Niels Bohr skiing at Los Alamos, January 1945.

20. First atomic bomb explosion at Jornada del Muerto near Alamogordo on July 16, 1945. (Los Alamos Scientific Laboratory)

21. Fermi and Segrè at Los Alamos on V-E Day, May 8, 1945.

22. The Segrè family in Berkeley, 1946. From left: Elfriede, Fausta, Claudio, Amelia, and Emilio.

23. Emilio Segrè, spring of 1952. The equations on the board indicate the reactions that produced elements or isotopes discovered by Segrè; technetium, astatine, and plutonium-239.

24. In front of the antiproton equipment, 1955. From left: Emilio Segrè, Clyde Wiegand, Edward Lofgren, Owen Chamberlain, and Thomas Ypsilantis.

1 H																	2 He
3 Li	4 Be											5 B	6 C	7 N	8 O	9 F	10 Ne
11 Na	12 Mg											13 Al	14 Si	15 P	16 S	17 Cl	18 Ar
19 K	20 Ca	21 Sc	22 Ti	23 V	24 Cr	25 Mn	26 Fe	27 Co	28 Ni	29 Cu	30 Zn	31 Ga	32 Ge	33 As	34 Se	35 Br	36 Kr
37 Rb	38 Sr	39 Y	40 Zr	41 Nb	42 Mo	43 Tc	44 Ru	45 Rh	46 Pd	47 Ag	48 Cd	49 In	50 Sn	51 Sb	52 Te	53 I	54 Xe
55 Cs	56 Ba	57 La	72 Hf	73 Ta	74 W	75 Re	76 Os	77 Ir	78 Pt	79 Au	80 Hg	81 Tl	82 Pb	83 Bi	84 Po	85 At	86 Rn
87 Fr	88 Ra	89 Ac	104 Rf	105 Ha	106	107	108	109	(110)	(111)	(112)	(113)	(114)	(115)	(116)	(117)	(118)
(119)	(120)	(121)	(154)	(155)	(156)	(157)	(158)	(159)	(160)	(161)	(162)	(163)	(164)	(165)	(166)	(167)	(168)

LANTHANIDES	58 Ce	59 Pr	60 Nd	61 Pm	62 Sm	63 Eu	64 Gd	65 Tb	66 Dy	67 Ho	68 Er	69 Tm	70 Yb	71 Lu
ACTINIDES	90 Th	91 Pa	92 U	93 Np	94 Pu	95 Am	96 Cm	97 Bk	98 Cf	99 Es	100 Fm	101 Md	102 No	103 Lr
SUPER-ACTINIDES	(122)	(123)	(124)	(125)	(126)									(153)

25. The periodic table, showing Segrè's discoveries: technetium (no. 43, Tc), astatine (no. 85, At), and the light isotope of plutonium (no. 94, Pu), Pu-239. (Lawrence Berkeley Laboratory)

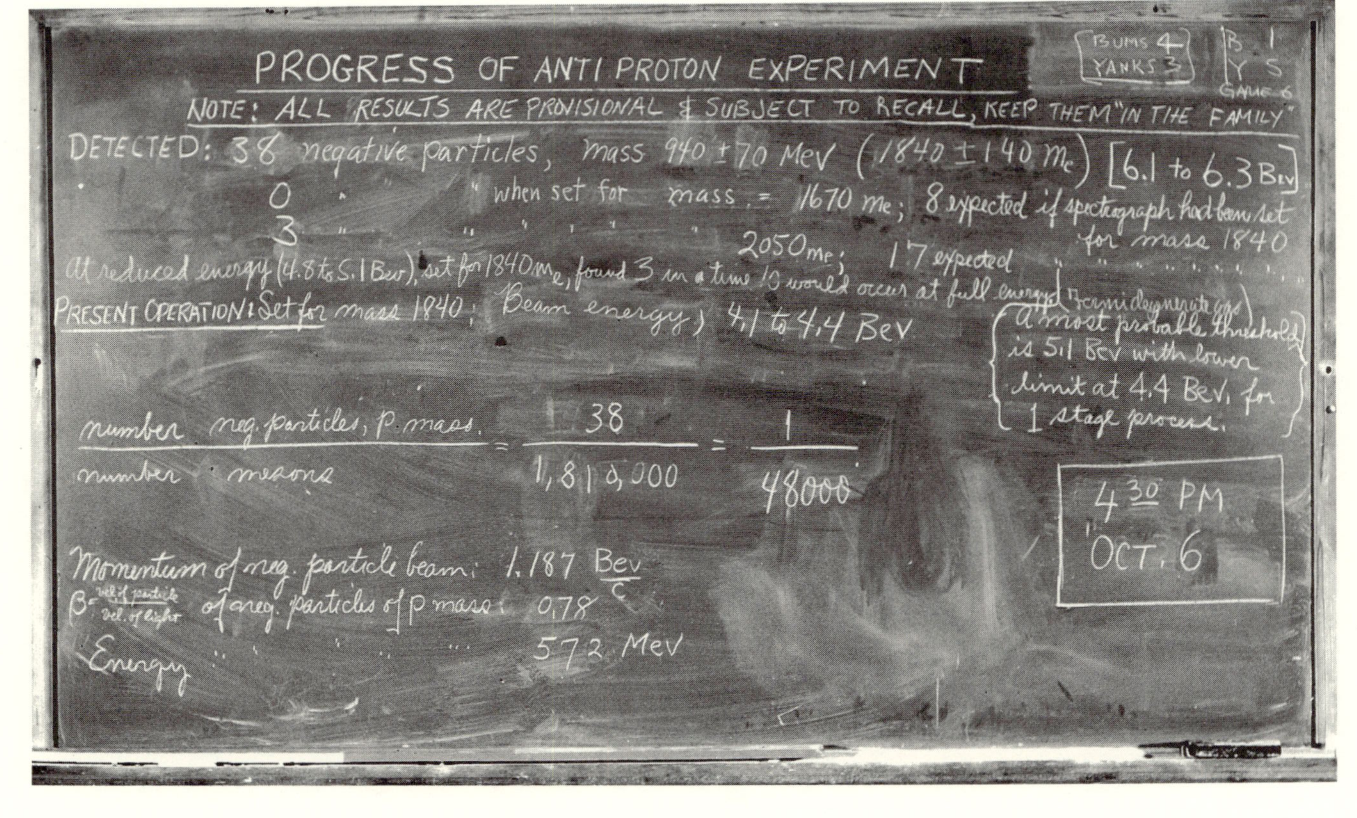

26. The Rad Lab's noticeboard for 6 October 1955 with progress of the antiproton experiment. By then 38 "events" had been "observed," or 1 for every 48,000 unwanted mesons.

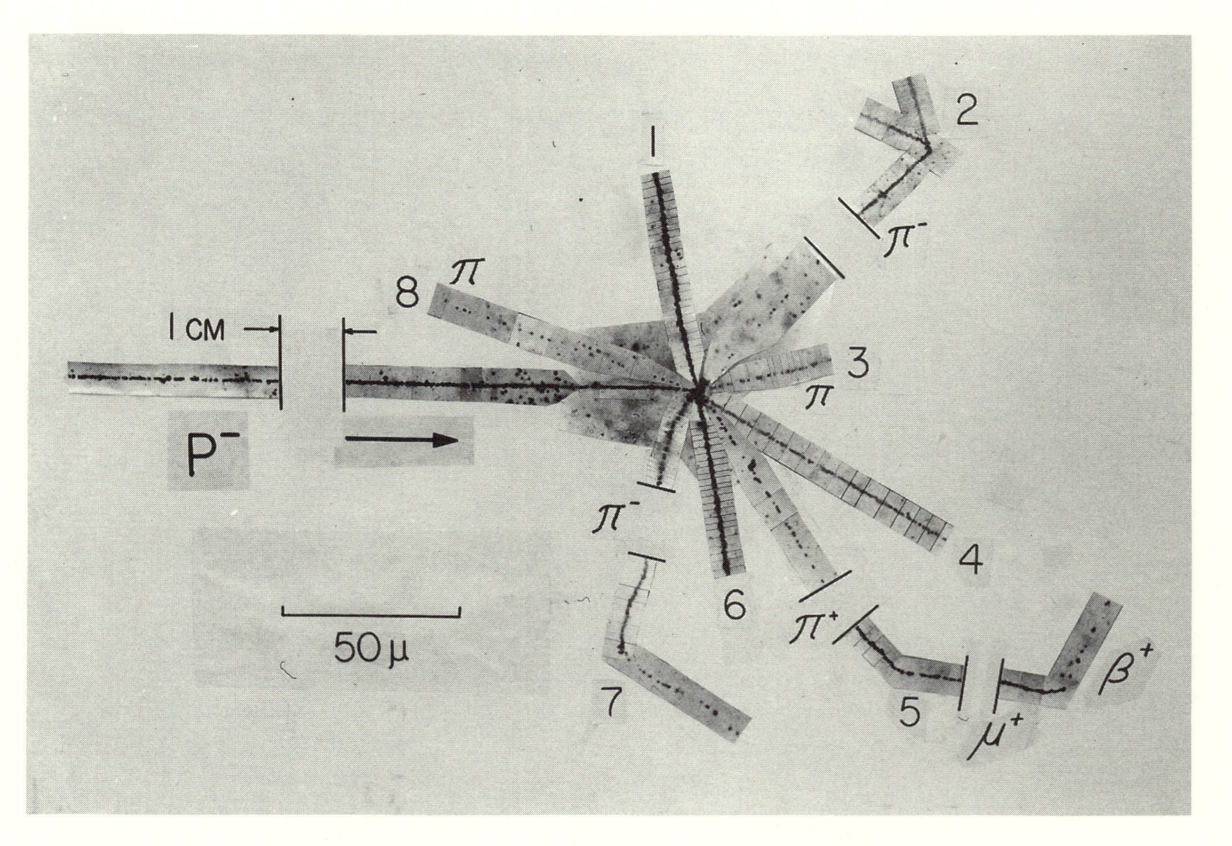

27. Antiproton Emulsion Star, 1955.

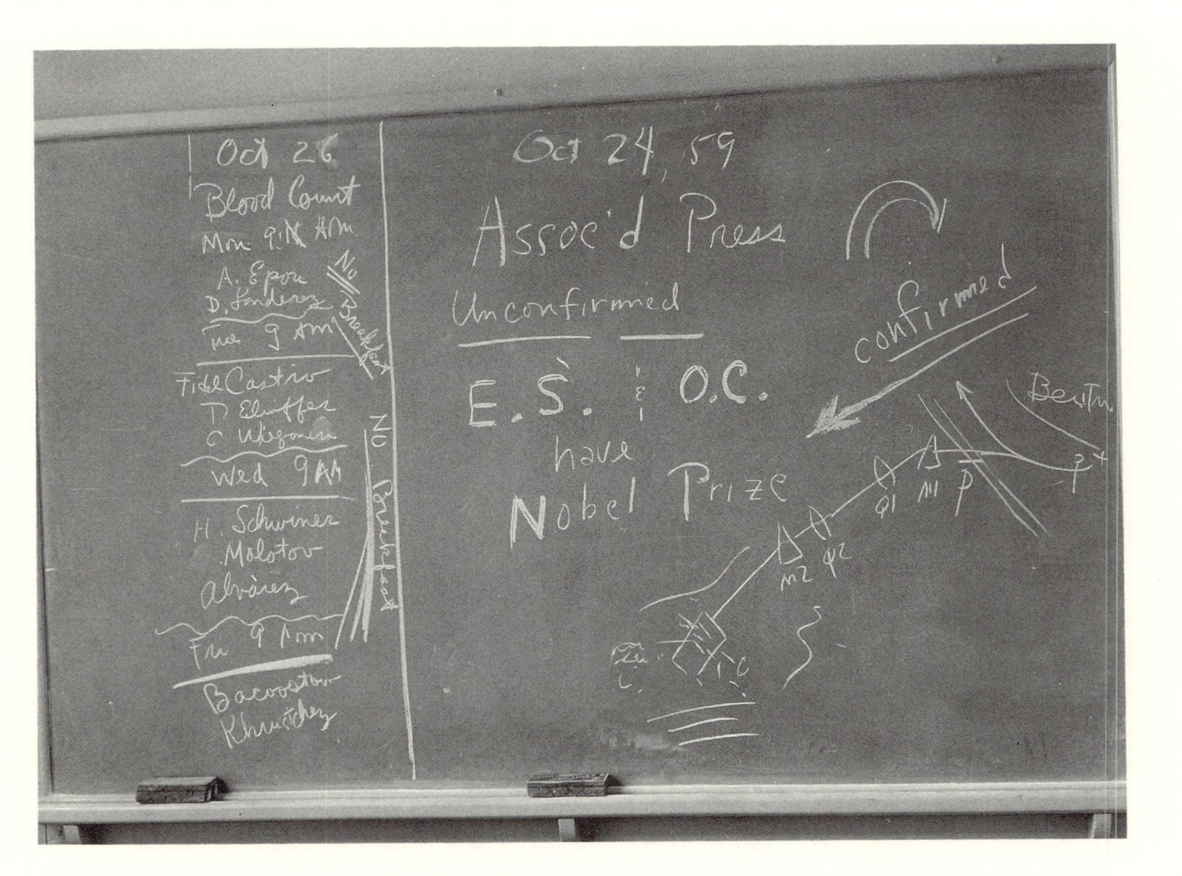

28. Blackboard at the Lawrence Lab, October 24, 1959, announcing the news of the day. At left is the schedule for routine blood tests, with some humorous annotations. (Lawrence Berkeley Laboratory)

29. Emilio Segrè receiving the Nobel Prize in physics from King Gustav of Sweden, 1959.

30. Emilio Segrè's Nobel medal.

31. Cover of *Time* magazine, January 2, 1961. © 1960 Time Warner, Inc. Reprinted by permission.

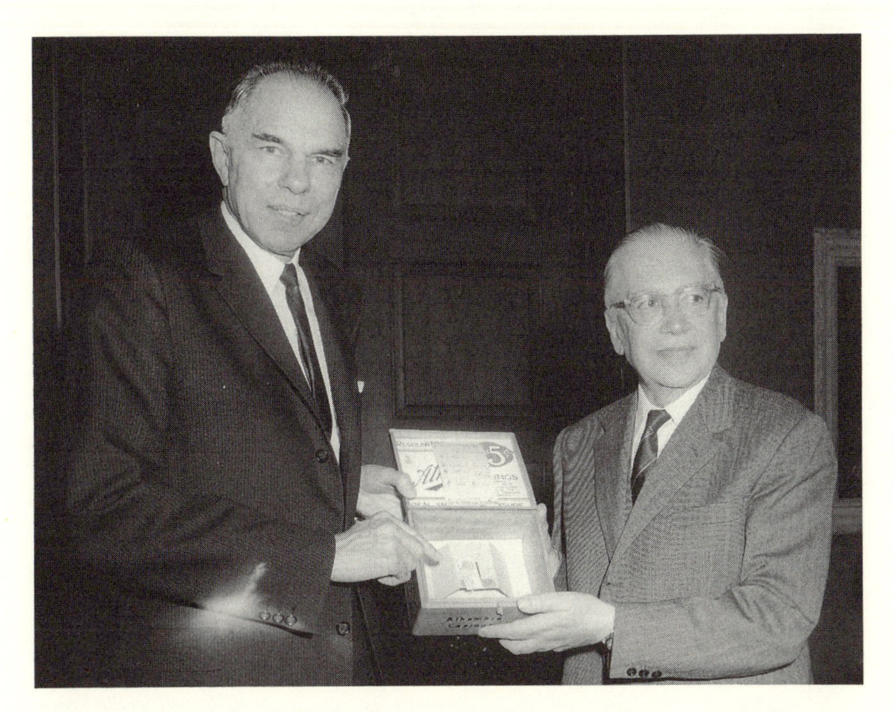

32. Glenn Seaborg and Emilio Segrè present the first sample of plutonium-239 (their collaborative discovery), in the cigar box in which it had been stored for a quarter of a century, to the Smithsonian Institution in Washington, D.C., March 28, 1966.

33. Segrè family, 1982. Back row from left: Amelia's husband Joseph Terkel, Rosa and Emilio Segrè, Amelia, Claudio, and Zaza Segrè holding son Joel. Front row from left: Amir Terkel, Gino and Francesca Segrè, and Vivian Terkel.

34. Rosa and Emilio near Locarno, Switzerland, 1987. Photo by Robert Schira.

35. Emilio Segrè, 1980. Photo by M. Balibrera. (Los Alamos Scientific Laboratory)

36. Emilio at home, doing his daily chinning exercise, 1981. (Courtesy of the *San Francisco Examiner*)

Chapter Six

In the New World: Refugee at Berkeley (1938–1943)

Smell of Cyclotron Oil

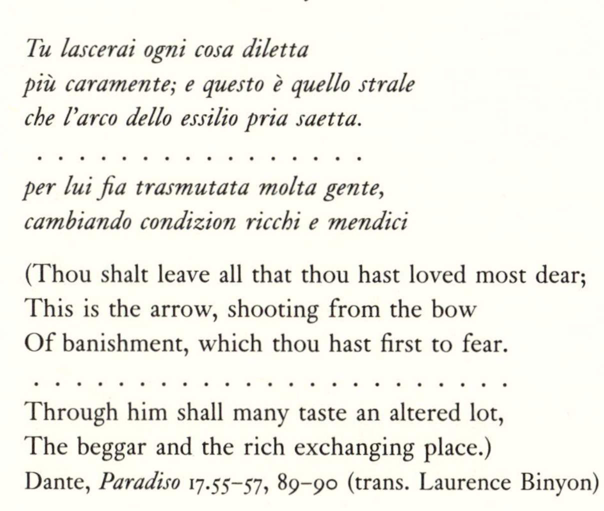

Tu lascerai ogni cosa diletta
più caramente; e questo è quello strale
che l'arco dello essilio pria saetta.

. . . .

per lui fia trasmutata molta gente,
cambiando condizion ricchi e mendici

(Thou shalt leave all that thou hast loved most dear;
This is the arrow, shooting from the bow
Of banishment, which thou hast first to fear.

. . . .

Through him shall many taste an altered lot,
The beggar and the rich exchanging place.)
Dante, *Paradiso* 17.55–57, 89–90 (trans. Laurence Binyon)

I landed in New York on July 13, 1938, expecting to return to Italy in the autumn for the beginning of the school year. Instead, nine years were to elapse before I revisited Italy. By that time, having lost my Italian job, I had built a second career, participated in great historic events, won a superior university position, and become a U.S. citizen and a Californian. Fate had been kind to me, although I had had my share of "blood, sweat and tears." I am grateful that I was spared both morbid homesickness and the terrible disasters that befell our parents and many of my friends. In my early American years I was lucky to have in science an absorbing occupation that both gave me pleasure and helped me rebuild my life.

On my arrival at New York, I visited my Columbia University friends and then, after a few days, left for Berkeley. I wanted to execute my work program as fast as possible. Leo Szilard, the Hungarian physicist I had met a few years earlier in England, came to see me off at Grand Central Station on my departure for San Francisco. He inquired about my plans, which I detailed to him. When I told him I expected to return to Palermo in October, he said that would be impossible because of what Mussolini might be expected to do; Italy might adopt Hitler's racist politics, and in any case, Hitler might start a world war soon. With these cheerful thoughts, I started on the first leg of my trip, from New York to Chicago.

At Chicago I bought a newspaper and read a short but chilling news item on the new charter of anti-Semitism in Italy, the *Manifesto della razza* (race manifesto), which obviously Mussolini had encouraged, even if it bore only the signatures of his minions.[1] I ruminated for the rest of the trip on its contents and implications. I decided not to mention the subject to strangers in Berkeley, however, until I knew more.

I was thus catapulted from my apparently secure job in Palermo into a precarious new situation. It is true that I was not unknown as a scientist, but my prestige was not sufficient to place me in the category of those for whom it was easy to find a job. Furthermore, I had arrived a few years later than contemporaries like Bethe, Bloch, and Teller, and the job market was more saturated. Strangely, the first shock did not affect me very much emotionally. The blow was not unexpected, and all in all I was too busy trying to rebuild my life to brood aimlessly. The result was that I did not feel the sudden rending from my previous life as an irreparable wound, as happened to many émigrés.

On my arrival in Berkeley, I immediately tried to look up my friend Lorenzo Emo, but he was away for a few days. The owner of the house where he lived had another room for rent and I took it. In that house

I found a little book of Admiral Nelson's letters to Emma, Lady Hamilton, his lover. In them Nelson often complained about seasickness and signed with his Neapolitan ducal title, Bronté. Thus, in our friendly intercourse, I nicknamed Lorenzo "Bronté" or "Brontuzzo."

Lorenzo was of the greatest help to me in my early days at Berkeley. First, he explained to me the university minuet that one danced at Berkeley, as one does in all universities; then he guided me a little among the various personalities I met. Above all, he was a trusted friend with whom I could share my worries about what was happening in Europe. Between us we could talk with open hearts, sure of mutual understanding. In those weeks our acquaintance grew into a solid, intimate friendship that lasted for life.

Lorenzo, a handsome man, a man of the world, a count, and of independent means, was most popular among physicists and their wives, both at the university and at the Radiation Laboratory. Although he looked to be a *farfallone amoroso* (amorous butterfly), he was an astute observer of people and shrewdly appraised them. The first people I met through him were Don Cooksey (Lawrence's alter ego), Francis Jenkins and his wife Henriette, and Robert Brode and his wife Bernice.[2] In addition, Lorenzo introduced me to Dr. Giacomo Ancona, a physician from Florence living in San Francisco, who we discovered had examined me in 1926 when he was a young assistant to the professor of medicine at the University of Florence, Frugoni. I also struck up a close, lifelong friendship with Ancona.

I looked around the lab and saw at once that, while the cyclotron was a wonder, the methods used to measure radioactivity left much to be desired. I then tried to secure an ionization chamber of the type I had used in Rome and Palermo, and finding that Franco Rasetti, on a short visit in 1935, had left parts of such an instrument in Berkeley, I completed it. I was used to coupling the ionization chamber to an electrostatic electrometer, but when I asked Lawrence to buy such an instrument, he denied my request. He said that it was old-fashioned stuff, and that I should build a DC amplifier with an FP 54 tube. I had no choice and started working. Fortunately, Lee DuBridge, the future president of Caltech, then a professor at Rochester University, was vacationing in Berkeley. He was an expert on DC amplifiers and kindly offered his help, rapidly assembling the rest of the instrument. The resultant simple apparatus was used later for many investigations. It is now a museum piece at the Smithsonian Institution in Washington, D.C.

The more familiar I became with the Rad Lab, the more surprised I was; it operated very differently from any other laboratory I had been in. There were many students, but they seemed to me to be left to themselves, without scientific guidance. I was simple-minded or imprudent enough to mention this to Lawrence, and I even offered to help in guiding students. He answered me rather coolly that the Rad Lab and America were not a German university with a big boss dominating the students; that here students were free and learned by themselves, and similar tales.

The truth was that Lawrence's interest centered on the cyclotron and on building the Rad Lab's diverse activities; his knowledge of and interest in nuclear physics were limited. Students, in practice, served as cheap labor for the building and tending of the cyclotron and any move that might divert them from this task was frowned upon. It was difficult for me to understand the scientific policy of the Rad Lab. The cyclotron was a unique device, with seemingly infinite potential, but the main concern of those who controlled it was apparently to make the machine bigger and put it to work in areas outside of physics; there was little thought given to making proper use of what was on hand for nuclear studies.[3]

In hindsight I now believe Lawrence's attitude was more far-sighted than I then deemed. The concentration of effort was necessary in order to develop the machine, and Lawrence tried to do what he knew best. His early efforts in the nuclear field had not been a success and may have left him with unpleasant memories. The cyclotron in 1938 was still a relatively simple machine and could be developed primarily by empirical procedures, without a deep understanding and an elaborate theory of its functioning. To some extent, the methods used initially at the Rad Lab resembled the approach to technical development of an Edison or a Marconi; Lawrence fitted the same slot: brawn prevailed over brain. Financial support was what limited the development of the machine, and Lawrence spent most of his time raising money from foundations, from the university, and from anywhere else he could get it.

In the Rad Lab there were men of greatly differing ability, but all were young. L. W. Alvarez and Ed McMillan were obviously first-class scientists.[4] Emo had described the first to me as a little fascist leader, fawning to the Duce, but mean to his equals or inferiors. McMillan, according to Emo, was very clever, but lazy. They were the only ones who also had University of California appointments. Of the others, R. L. Thornton, M. D. Kamen, and R. R. Wilson were more than able. Thornton, in addition, was a true gentleman, brought up English-style, who combined technical prowess with superior human qualities. Kamen worked as a physicist and a chemist to earn a living but his heart was in music, and if he could have afforded to do so, he would have become a professional musician.[5] Wilson had his roots in the Wild West, but made a serious effort to civilize himself. At Berkeley, he and McMillan were the first to make a serious experimental study of the fine points of the workings of the cyclotron and to try to understand them theoretically. Outside of Berkeley, Hans Bethe, Morris Rose, and L. H. Thomas had applied sophisticated theory to accelerators, but had less contact with the realities of the problems.

At a somewhat lower level, there were Paul Aebersold, David Kalbfell, Robert Cornog, Philip Abelson, and many other graduate students or young postdoctoral fellows. They had sufficient ability, but, I believe, would have profited from more guidance. In addition, there was a whole group of medical doctors using radiation as a clinical and research tool. They were privileged because money came to a large extent from foundations nourishing optimistic hopes of medical applications, especially cancer cures.

Don Cooksey had a unique position. He came from a rich, cultivated family and had met the young Lawrence at Yale, where Cooksey and his brother Carleton were both members of the physics staff. Cooksey had limited scientific ambitions, but an immense love for and confidence in Lawrence. He devoted his considerable technical ability, his refined manners, his generosity and human understanding, to helping Lawrence in every way. Lawrence reciprocated by making Cooksey his most intimate confidant and advisor.

Soon after my arrival in Berkeley, I met Glenn Seaborg at the Faculty Club. He had obtained his doctorate the previous year and was a research assistant to G. N. Lewis, a famous chemist and one of the most important members of the Berkeley Faculty. Seaborg was keenly interested in anything happening around him and kept his ears and eyes open. Almost immediately he offered me his cooperation, which I gladly accepted. I found that he knew analytical chemistry according to the system propounded by A. A. Noyes and W. C. Bray,[6] rather than the more classical schemes I was familiar with. It seemed to me he knew everybody in the Chemistry Department and could find anything we needed. At the time Seaborg had a heavy load helping Lewis in his organic chemistry research, as well as in anything else that Lewis's fertile mind came up with. On the side he worked on nuclear problems for several hours every day and compiled data for an important review article he was preparing, as well as for a table of isotopes.[7]

Seaborg could do all this because he had iron discipline, a lively and highly systematic mind, immense persistence, exceptional endurance for work, and a sincere, open-minded interest in science. He was a superb organizer, but was not too strong in physics or instrumentation. At the beginning of our collaboration, I probably remembered my much earlier experience with Bakker at Zeeman's laboratory in Holland, when I, a newcomer, and he, the local boy, greatly helped each other and became steadfast friends. Little did I recognize Seaborg's unbridled ambition and his unshakable determination to succeed and to be preeminent.

As a general guideline for myself, I decided to fulfill my obligations as a citizen of the Cyclotron Republic but, at the same time, to try to preserve my individuality and to avoid being absorbed completely by the Rad Lab. I also wanted to use the immense opportunities offered by the cyclotron as much as possible, which I considered a contribution to the Lab. I soon found students willing to work with me. The first was Chien-Shiung Wu, who joined me when she realized that, in so doing, she would be able to learn something. She was very handsome, and very elegant in her Chinese dresses. When she walked on campus, she was often followed by a swarm of admirers, like a queen. She was a fiend for work, almost obsessed by physics, highly talented, and very shrewd, as well as witty. Many years later she became world-famous for the experimental demonstration of the nonconservation of parity. I admired her and liked her, and we remained friends for life.

Alex Langsdorf, Jr., a postdoctoral fellow, also started working with me. Disappointed with Lawrence, he had begun building a continuously operating cloud chamber, but this project ran into difficulties, and Alex joined me. He was the son of a well-known professor of electrical engineering at Washington University in St. Louis; while we worked, he diligently instructed me in things American, and we became steady friends.

A few days after my arrival, I started my first investigation, with Seaborg, a search for short-lived technetium isotopes. This was the natural continuation of my work at Palermo, the reason for which I had come to Berkeley in the first place. One of the new radioactivities, obtained by deuteron bombardment of molybdenum, immediately presented an unexpected and interesting nuclear phenomenon; it was a case of nuclear isomerism—that is, of a nucleus possessing long-lived excited states. We did not dream then that this isomer would in time become a mainstay of nuclear medicine as a powerful diagnostic tool. Today it is used by thousands of practitioners, and its applications are the basis of a multimillion dollar industry.

However, what interested us at the time was the phenomenon of nuclear isomerism. In 1936 C. F. von Weizsäcker had proposed a theoretical explanation of isomeric states, attributing to them a high angular momentum that forbids transitions to lower, small angular momentum states. It followed from this, according to theory, that there should be plenty of internal conversion electrons. Seaborg and I looked for them and found them, reporting our results in a letter to the *Physical Review* on September 14, 1938. A few days later, however, on the advice of J. Robert Oppenheimer, who for reasons unknown had told him (but not us) that we were mistaken, Lawrence demanded that we wire the editor of the journal to stop publication of our letter. I protested within the limits permitted by my position, and the letter was resubmitted to the *Physical Review* on October 14, 1938, after Bruno Pontecorvo had reported getting results on rhodium similar to ours.[8]

This was one of my first contacts with Oppenheimer, then professor of theoretical physics at Berkeley, later famous for his part in building the atomic bomb, for his political activity, and for his unjust victimization.[9] At the time, he was considered a demigod by himself and others at Berkeley, and as such he spake in learned and obscure fashions. Besides, he knew quantum mechanics well, and in this he was unique at Berkeley. He taught it in none too easy a fashion, which showed off his prowess and attracted a number of gifted students. His course later formed the basis of Leonard Schiff's well-known treatise on quantum mechanics.[10] Oppenheimer's loyal disciples hung on his words and put on corresponding airs. Just as we in Rome had acquired Fermi's intonation, in Berkeley Oppenheimer's students walked as if they had flat feet, an infirmity of their master's.

Oppenheimer and most of his acolytes followed the political line of the Communist Party of the United States, which was highly uncritical and simple-minded. Although most of these young people were not members of the Party, some were members of auxiliary organizations and later fell victim to cruel persecutions. Oppenheimer and his court did esoteric things; they read very highbrow books, cooked and ate unusual food, and during the summer went to a mountain ranch in New Mexico leased by Oppenheimer, beautifully located near Santa Fe. Their physics was valid, but often they attacked problems prematurely, or problems beyond their capabilities, resulting in indifferent success. The best of Oppenheimer is perhaps his astrophysical work on neutron stars, which many years after his death proved truly prophetic. His research on cosmic rays and atomic and nuclear problems embodied many good and even prescient ideas, but was often inconclusive. Oppenheimer's beneficial influence on the development of theoretical physics in the United States was considerable, and several of his pupils have achieved major results, as well as deserved fame.

Oppenheimer and his group did not inspire in me the awe that they perhaps expected. I had the impression that their celebrated general culture was not superior to that expected in a boy who had attended a good European high school. I was already acquainted with most of their cultural discoveries, and I found Oppenheimer's ostentation slightly ridiculous. In physics I was used to Fermi, who had a quite different solidity, coupled with a simplicity that contrasted with Oppenheimer's erudite complexities. Probably I did not sufficiently conceal my lack of supine admiration for Oppenheimer, and I found him unfriendly, even if covertly, for a good part of my career, except when he wanted me to join his team at Los Alamos.

At Berkeley, in 1938, I had the impression that Oppenheimer regarded me as a great Fascist. I was a Fascist Party member, as every Italian state employee was required to be by law, but it did not take much acumen to figure out that I could not be a Fascist at heart. I could not, however, conceal my skepticism when I heard him repeat, with the faith of the true believer, the nonsense originating from Stalin's Cominform.

Talking politics with American colleagues, I found an incomprehension of things European that was appalling to me. My partners in conversation had many different opinions, but most seemed convinced that what happened in Europe did not concern the United States, and that if the Americans minded their own business, they could avoid entanglements in European quarrels. It was, fundamentally, the isolationist thesis; they did not grasp Hitler's nature and his plans of world domination. These plans were the products of a deranged mind, but the disease had spread to a whole nation as powerful as Germany, and it was not something to trifle with. There was also a good measure of optimistic skepticism about Hitler's true intentions, which in effect proved much worse than the most pessimistic forecasts.

Others, although not many, followed the Communist Party line, like Oppenheimer, and deemed that the European quarrels were caused by capitalist imperialists, and that Holy Communism would avoid them. Which side of the bad capitalists won was irrelevant. If they weakened themselves sufficiently in their internecine strife, Mother Russia would later establish the millennium. This senseless view suddenly changed when Hitler and Stalin signed their nonaggression pact. Hitler suddenly became palatable!

My guts rebelled when I heard such talk. I strove to persuade isolationists that things were not as they hoped and believed. For the Communist true believers, there was nothing to be done, because their "ideas" were more religious feelings than political reasoning. Fortunately, President Roosevelt saw farther and more clearly than the majority of Americans. Anyone who was in the United States in the years immediately preceding World War II knows the difficulties Roosevelt faced in opening the eyes of the Congress and of the American people on the subject of Hitler.

I followed the news from Italy with increasing alarm, but I could relieve my feelings and unburden myself only with Emo. The Fascist *Manifesto della razza* had been followed by legal measures that left no doubt about the final purpose of the campaign. I recognized more and more the foresightedness and wisdom of having preserved a certain amount of money abroad. By the end of July, I had decided to forget Palermo and to summon my wife and son to California.

In the meantime, to escape the summer heat, Elfriede had taken Claudio to Frassenè, a resort in the Alps, where she was shortly joined by her parents from Germany. The publication of the *Manifesto della razza* followed a few days later. Elfriede was less alarmed than I, because Italian newspapers published a version of the facts that played down the probable consequences. When I wrote asking her to pack up and come to America, however, she agreed immediately.

In the following months, Elfriede had to face a heavy burden, first alone in Italy, and then in America. To start with, in Italy, she was confronted with sticky bureaucratic problems that required infinite patience, while the sultry summer weather and the effort of taking care of a one-year-old child further sapped her strength. She had good help, but the proceedings were morally debilitating and physically exhausting. All this emerges from her letters of the time.

In the meantime, the Czechoslovakian crisis was progressing and seemed likely to precipitate a world war. I thus urged Elfriede to speed up her departure, cabling her: "Vieni immediatamente con tuo comodo" (Come immediately, at your leisure), counting on her realizing that the first two words were the key ones.

Finally, Elfriede had to return to Sicily, lock up our apartment, and board a ship. She arrived in Palermo dead tired and sick with digestive troubles. There she was met at the dock by my student Ginetta Barresi, whom she had told she was coming. Ginetta was accompanied by her father, a colonel in the army, and by two other gentlemen. They asked without further ado what Elfriede needed. She explained the state of her passport—that it had the necessary Italian validation, but no U.S. visa—and said she wanted to close up our apartment and depart as soon as possible. The answer was immediate and to the point. "The next ship leaves on . . ." and they gave the date. "You get some rest now, and we'll take care of EVERYTHING." I do not know whether Elfriede went to our apartment or to the Barresis' villa at Mondello; in any case, they brought her food and, three or four days later, in time for embarkation, her visa, the ticket, and everything she needed. As far as closing the apartment and forwarding its contents were concerned, Ginetta asked for instructions and in due course took care of everything to perfection. Moreover, Ginetta visited my parents whenever she was in Rome and sent us news of them. "I am pleased my compatriots are showing themselves to be decent people. Having gotten to know Ginetta and her like better, however, I shall become a Sicilian by choice," I wrote to my parents on August 3, 1939, responding to a letter of theirs in which they took comfort in the signs of friendship and esteem they were receiving in Tivoli. From Ginetta we had proofs of friendship in "heroic degree," as the Church says of virtue.

In 1938 and 1939 my parents continued to live more or less as they usually did, at least on the material side. They traveled a good deal, vacationing at Forte dei Marmi or some other familiar resort when Rome was too hot, and spending long periods at Tivoli. In almost all their letters, they mention old friends, above all Signora Rasetti and Amaldi, Bernardini, and other physicists who went to visit them and tried to cheer them up. In these letters, my father's great moral strength, clarity of mind, and farsightedness are also apparent. He never loses sight of the essential point. In rereading my parents' letters half a century later, I am impressed.

It was an anxious time. I was afraid that hostilities might break out while Elfriede and Claudio were on an Italian ship, and that they might be detained at Gibraltar for the duration of the war. Furthermore, in view of the circumstances, I had cabled certain instructions concerning my funds, and I received an answer in a commercial code that was incomprehensible without the key. I confided this to Emo, who, half jokingly, half because he did not know what to say, indulged in the blackest hypotheses on the mysterious meaning of the words of the cable I had received. "Homgo Homil," it began, and Emo said that this obviously meant "All is lost!"; for many years these words remained our private joke. Finally, I found a European apprentice at a bank in Berkeley who recognized the code and translated it for me. It was nothing important, but at the time, Emo and I were so upset by the events that we lived on milk shakes only; we could not eat because any solid food disagreed with us.

At the worst of the Czechoslovakian crisis, Elfriede, in blissful ignorance of everything, was at Gibraltar on an Italian liner, but the situation cleared up, at least temporarily, and she was able to proceed to New York. There she caught a train, and since it was her birthday, October 2, I sent her a telegram, using one of those long and pompous standard texts delivered on a special birthday form, all for twenty-five cents. "The darkest clouds have a silver lining," it declared among other things. On receiving the telegram en route in a style so unnatural for me, and unaware of Western Union's bargain services, Elfriede thought I had lost my mind. Finally, in early October, we were reunited, to our great mutual joy.

During the summer I had also started to become acquainted with some of the attractive features of life in the United States under the guidance of Brode, Jenkins, and their wives. As one of the first lessons, the husbands and their friend S. K. Allison, a physicist from the University of Chicago,[11] took me for a long hike in Yosemite National Park. We visited the high mountains near Vogelsang Pass, from where, in a couple of days, we descended into the Valley, walking along the Merced River. Jenkins had brought along an extra rod and he showed me how to fish for trout. Within an hour I was hooked on the sport, which delighted me for many years. My three physicist companions sometimes marveled at my ignorance of things American, and I still remember Jenkins's laugh when I told him that to inform myself about America, I had read Defoe's *Moll Flanders*. From it, however, I had learned that Quakers like himself were good, kind people. He and his wife liked to tease us when we could not understand *New Yorker* cartoons they showed us. On the other hand, he was very surprised when I pointed out to him that if there were counties, one expected to find also counts. He had never thought of the connection between the two.

Jenkins, Brode, and Allison, all three of them children of professors, were much more cultivated than Lawrence and some of our other colleagues, and they were also better-natured and more easygoing. The Jenkinses remained our best and closest American friends until their sad, early deaths. We often visited them informally in the evening, and their tales, example, and help with the problems of daily life contributed substantially both to our morale and to our adaptation to American customs.

Don Cooksey was another dear friend who valiantly helped us to adapt to America. He owned a ranch in Trinity County in northern California, not far from Forest Glen. One reached it after a walk of about forty-five minutes in a pine wood along the South Fork of the Trinity River. The ranch's log cabins were in a clearing near a small placer-mining operation. The cabins had been built around 1906 by a couple who had cared for Don, who was orphaned at an early age. These people had willed the cabin to their charge. The gold-mining claim had never been important in itself but helped in securing a deed from the Forest Service. A man who lived permanently at the camp with a couple of horses helped Don with the upkeep of the place. At the camp, one lived very simply, spending a good part of the time on or near the river. There was complete freedom: one could do nothing, read one of the interesting books stored in the cabin, swim, fish, float on an inner tube, walk around, or even, if one really wanted to, pan for gold. The trout fishing was good, and in the water one could see huge salmon that came to spawn in the river. The company was congenial: it consisted of the owners, and, in turn, Thornton, Kamen, McMillan, and several others connected with the Rad Lab. Whenever Cooksey's observant eye saw somebody in need of a restorative vacation, he invited him to his camp. There we were treated with exquisite courtesy, which we tried to reciprocate by helping with the necessary chores. Photographs show the kind of life we enjoyed. Such vacations made deep impressions and left all those who had the good fortune to share them with pleasant memories.

Before Elfriede's arrival, Henriette Jenkins and I went looking for a furnished house to rent. We found one at 2532a Piedmont Ave. in Berkeley for forty-five dollars a month, and as soon as she saw it, Henriette urged me to take it. I signed the rental agreement at once. The house was a wooden cottage in a yard behind some larger homes. It had a kitchen, living room, a couple of bedrooms, and a small porch. The whole was in indifferent shape and old, but in a convenient location and sufficient for us, at least temporarily. There was enough furniture so that we could manage until our own arrived from Palermo. When it arrived, with some effort we fitted it in. The landlady was a widow who tried to be helpful and accommodating. With a little good will, we could even take in a guest; and in fact Felix Bloch came to stay with us when he was working at the cyclotron with Alvarez. He slept on the porch in his sleeping bag, and Claudio used to wake him up in the morning, calling in Italian: "Bloch! Lavora!" (Bloch! Work!). Amaldi too spent some time with us and happened to be with us when World War II started.

With an eighteen-month-old baby to take care of, Elfriede also had to learn what food to buy, and where; where to find clothes; how to get medical help if necessary; and how to cope with a thousand details of daily life that were different in the United States. Above all, in Italy we had had a maid and a nanny; here everything fell on Elfriede's shoulders. She soon learned how to cook excellently, something she had never done before. No wonder she acquired the nickname "Tuechtigona" (an italianized augmentative of the German *tüchtig*, able and hard-working), which she fully deserved on all counts. How strenuous and demanding all this was emerges from our correspondence with my parents, where one often finds hints of tiredness.

In the beginning we did not have a car, but university, Rad Lab, and shopping centers were all within a fifteen-minute walk from home, so we did not need one. I did the shopping and I knew that my carrying strength sufficed up to five dollars worth of groceries. In May 1939, we bought our first American car, and a few days later Elfriede got her driving license, increasing our mobility.

From the moment of our arrival, we had had an immediate and important problem: obtaining an immigration visa. My tourist visa was for six months, so it lapsed in January 1939. In order to change one's visa, it was necessary to go to a U.S. consul outside the United States, and the most convenient place seemed to be Mexicali in Mexico. Immigration restrictions were steadily increasing as Hitler exacerbated his persecutions and more people tried to flee Europe; there was no time to waste. I remembered Rasetti's instruction in the subtle and marvelous points of Section 4(d) of the immigration law then in force, but when I tried to obtain some help from a service in San Francisco that was supposed to help immigrants, I concluded that it was better to do everything myself (one of Fermi's fundamental rules).

I asked the good Dean Pegram of Columbia University for a letter of recommendation to the consul, and he wrote a masterpiece. I asked for help in Berkeley too, but with little success. The university officials I contacted were polite but seemed exceedingly scared by the idea of incurring (nonexistent) responsibilities and limited themselves to the minimum they could not deny. However, armed with Pegram's recommendation and with my knowledge of the law that fitted my case perfectly, I took a bus and in two days arrived at Mexicali. Elfriede and Claudio were already in Berkeley with a tourist visa, but they could not possibly come with me at the time, and I thought it better to make two trips to Mexicali rather than wait until we could all go together.

On November 16, 1938, I obtained the visa that solved my immigration problems, and in February 1939, we returned to Mexicali to exchange Elfriede and Claudio's tourist visa for an immigration one too. We hoped we would not have to produce Claudio, but the consul insisted on his presence. As soon as Claudio entered his office, however, he started crying and yelling loudly, and the consul, at a loss for what to do, said, "Take him away immediately," which we did. Thus, with two trips to Mexicali, we settled our immigrant status. Later I read in a newspaper that the consul at Mexicali had been convicted of selling visas!

On November 29, 1938, we celebrated our first American Thanksgiving, with Emo and Bloch as our guests. I thought that we had every reason for celebrating, and that we pretty well resembled the early colonists who had originated the custom. Fermi used to say, whenever somebody bragged about his Pilgrim ancestry, or when he heard about the Daughters of the American Revolution, that he and other newly arrived refugees were the true, new pilgrims, who understood and appreciated some American ideals better than the American-born. About the same time, I received the news of the departure of my Rimini cousins from Italy for Montevideo, in Uruguay.

On February 2, 1939, we celebrated my father's eightieth birthday. We had written to him in time.

On May 10, 1939, my brother Angelo arrived in New York with his wife and two children. At first he tried to work at Columbia University in his field, economic history, and initially he had some success at this. Subsequently, however, he quarreled with everybody and gave up teaching, devoting himself entirely to painting. He exhibited in New York and had a certain critical success, but no financial reward, and did not persist in trying to sell his work. Although I invited him repeatedly to visit us in California, he never budged from the East Coast.

Angelo lived on family money, administered by me, according to my father's policies. I was cautious in the use of this money. I thought it should first be at my parents' disposal if they wanted to emigrate, and this took precedence over everything else. Second, one had to consider emergencies that might occur at any moment. I thought my family and I should live on my earnings alone as long as we could do so without great sacrifices and without endangering our future.

As soon as Angelo arrived at New York, he started pestering me with demands that ran counter to my instructions and policy. I gave him a monthly allowance of $150; it was not much, but it should have sufficed under the circumstances; in emergencies, he could ask for more. Angelo greatly upset me by threatening to complain to our father. Fortunately, a trusted friend of the family was out of Italy for a business trip, and through him I was able to communicate with Papà, who let me know that he approved of my conduct. Later, when I started earning a little more at Los Alamos, I sent some money of my own to Angelo, whose son needed an operation. In thanks I received a letter from Angelo saying that he considered my gift restitution of funds I had wrongly appropriated.

In 1938 after I had worked at the Rad Lab for a few weeks, Lawrence asked me if I could extend my stay beyond the month of October, the date I had given him for my return to Palermo. The offer was a godsend, but I sensed that precipitous acceptance was not to my advantage. I thanked Lawrence profusely and asked for some time to answer. Lawrence then went to Alaska on vacation, and I began to fear that he might change his mind, and that by trying to be too smart, I had destroyed my chance. Fortunately, on his return Lawrence renewed the offer of a salary of $300 a month for six months. My title was to be "research associate," a nondescript qualification that could apply to persons at very different levels. It had, however, one definite connotation: it implied a temporary job and did not commit the university or the Rad Lab beyond the term of the contract. At Christmas we were invited to Lawrence's home for dinner; oddly enough, he took the opportunity of telling me on this occasion that the lab was short of money, and that if necessary he would ask me to return part of my salary.

In 1939, $300 a month was a good salary, and it got me out of the woods for some time, but after six months, in July 1939, Lawrence, who by then must have realized my situation, asked me if I could return to Palermo. I answered by telling him the truth, and he immediately interjected: "But then why should I pay you $300 per month? From now on I will give you $116." I was stunned, and even now, so many years afterward, I marvel at Lawrence's impulsiveness; he did not think for a second of the impression he conveyed. With a minimum of reflection and diplomacy, he could have saved his $184 a month without cutting a horrible figure in my eyes. However, although I have not forgotten his conduct, I now see it in a different light than I did then.

I did not know what the salaries of other members of the Rad Lab were at the time. It was I who, with $300 a month, was the exception, and salaries around $116 were not rare. If I had known this and how Lawrence behaved with Americans who were unemployed because of the depression, I might have viewed the episode somewhat differently. I was somewhat older and better established as a scientist than some excellent Americans were, but they too got meager salaries, although they had academic positions and guaranteed careers.

In any case, with a wife and a child, a salary of $116 was scant, but it was not totally impossible to live on it. With $200 a month I could make ends meet, sparingly, but without deprivation, and I could bring my salary to that level by using my private funds, the existence of which was known only to Elfriede and to my friend Emo, whose discretion was absolute.

Elfriede and I constantly thought of our faraway parents. We wrote at length and regularly, at least once a week. Neither transatlantic airmail nor usable telephone connections with Italy existed, but the mails were better organized then, and a letter did not take much longer to reach Italy than it does today. My mother and Elfriede wrote at greater length than my father and I did, but he added at least a few lines, sometimes insignificant, sometimes important, to every letter. Each side tried to reassure the other, so the letters sounded a little more optimistic than they should have, and definitely more optimistic than those I exchanged with my cousins in Uruguay. When I complained, as I sometimes did, about the precariousness of my position, my father insisted that there was only one way of improving it: do good physics. If I succeeded in this, recognition would not fail to come, and career problems would take care of themselves. The times demanded prudence in writing; my parents interspersed praises of the Duce and other phrases intended for the eyes of a possible Fascist censor in their letters, although I doubt they would have fooled him.

The letters mirror Italian Jewish life in those tragic years, as one by one the young people emigrated and the older generation were left increasingly alone. At the same time, to their credit, many good old friends appeared, seeking to comfort my parents with frequent visits: Rasetti and his mother; Amaldi; Ginetta Barresi; the Salvati family and others in Tivoli, among them both important citizens and humble people; Ada Rimini and other relatives; my cousin Renzo Ravenna, the mayor of Ferrara, who came to visit his aunt and uncle and his close friend Italo Balbo, in Rome for a meeting of the Fascist Gran Consiglio. The letters report a great variety of other news: marriages, births, deaths, the recipe for some cakes I liked, urgings to Elfriede not to overdo things and to get the rest she needed, encouragement to use family money, indirect news of Elfriede's parents, as well as of Bindo and Riccardo, and so on. There is even a report on an exchange of letters with E. O. Lawrence in which they congratulated him on his Nobel Prize and received a friendly answer. On July 6, 1940, my father writes: "I am eager to know your arrangements with Lawrence, but in any case I am happy if scientific concepts absolutely prevail, an opinion shared by M" (this sentence means that I should, if necessary, use family funds to support myself). On April 9, 1941, he comments on the friendliness of a Carabinieri officer who had come to withdraw my Fascist Party card. Other news: the U.S. consul, who had rented our Palermo apartment, is leaving. On May 7, 1941, my mother "is memorizing our letters" and father "is working to prevent idleness from wearing out his spirit."

Correspondence with Germany and with Elfriede's parents was much more difficult and dangerous, and thus is much less informative.

All regular communications stopped in December 1941, with the entry of the United States into the war.

At Berkeley I had realized that there was only one salvation: to do good physics. With that weapon I might perhaps save myself; without it, I would be thrown out without mercy. This simple estimate of the situation was supported by my father's advice and inspiration and by prudent, but well informed and extremely welcome, encouragement from Fermi. Fortunately, I did not lack ideas, and the Rad Lab, for all its defects, offered unique opportunities for experimentation.

Half a century later, I see that my personality did not allow me any other way of survival; someone else might have made an easier life for himself than I did by being less touchy, less proud, more able to dissimulate, better at public relations, and simultaneously less timid and and less critical. I have by now attained some slight knowledge of myself, and I know how unpleasant I can be. However, if I am a curmudgeon, I have paid for it.

The 60-inch cyclotron was under construction; the magnet was there, but not much else yet. The sight of such a big, powerful magnet suggested to me the possibility of improving on my previous studies of the quadratic Zeeman effect. Jenkins was a spectroscopist, and we collaborated in this work.[12] Such research continues today because with bigger magnetic fields and much better spectroscopic resources, physicists keep finding new interesting details. The subject is now part of the study of what today are called Rydberg atoms.

In January 1939, having clearly understood the relation between nuclear isomerism and conversion electrons, it occurred to me that it might be possible to separate nuclear isomers with a method similar to that devised by Szilard and T. A. Chalmers for isolating products of the (n,γ) reaction. I communicated my idea to Seaborg and told him that I needed a chemist who could synthesize a suitable organic molecule containing Br^{80}. Seaborg found me Ralph Halford, who knew how to prepare a suitable organic bromine compound. We imprudently spoke of our project at the Faculty Club in the presence of Willard Libby, later famous for his carbon dating (I have been told that the suggestion came from Fermi in a Chicago seminar). At the time Libby was Seaborg's great rival and an assistant professor, while Seaborg was slightly his junior. Without telling anybody, Libby went to his lab and applied my method to tellurium isomers. He then wrote a letter describing the result to the *Physical Review* and was about to send it, when Seaborg got wind of the fact. A row developed, and I asked Libby to wait twenty-four hours before sending his letter, so as to allow us to finish our work, which was half done. Under strong pressure, Libby acquiesced, and the two letters to the *Physical Review* thus appeared side by side.[13]

It was an unusual experience for me. I do not think that Libby would have gotten away with such behavior in Europe at the time, certainly not in Rome. Criteria of intellectual property were more elastic in America; on the other hand, at least at the Rad Lab, there was great generosity in the exchange of instruments. It is difficult to pass judgment. Subsequently several Berkeley radiochemists, including Joseph Kennedy and Seaborg, eagerly pursued this method for separating radioactive isomers.

At the end of 1938, there were great hopes in Berkeley that Lawrence would win the Nobel Prize. I thought that if it did not go to Otto Stern or somebody older, it would go to Fermi. (The year before in Palermo, to my great surprise, I had received a nomination form, and I had nominated him.) I did not, of course, know that Bohr had confidentially told Fermi that he would be chosen that year if he wanted to be, or that Fermi was preparing to emigrate. When the official news that the physics prize for 1938 had been awarded to Fermi arrived, I was delighted and wrote to congratulate him, adding: "The only sadness is the thought of the various people of the old guard who would have rejoiced to be near you now that the reward of so much work, so many hours of labor, as well as *c.i.f.* (*con intuito formidabile*, 'with formidable intuition,' a joking acronym we used for statements by Fermi that were true, but that he could not prove), is here, and they are prevented by an inscrutable destiny." Lawrence acted with good grace, but he was clearly disappointed. He asked me whether I thought he would get the prize next year. I said I did, and this indeed came to pass. By chance, on November 14, 1939, the day of the announcement of Lawrence's prize, he was having dinner at our house, and we opened a bottle of champagne, as reported in a letter home.

At the beginning of 1939, the news of the discovery of fission by Otto Hahn and Fritz Strassmann reached Berkeley. The experiments were immediately repeated, but not by me. I did not like to rush into a competition, and although I had been among the first to work on uranium, I continued my investigations on other subjects. On my arrival in 1938 I had renewed my conversations of 1936 with Philip Abelson, who clearly indicated that he considered uranium his property, not troubling to hide his feelings about foreign interlopers. I therefore left uranium to him until the discovery of fission. Abelson then recognized that he had seen X-ray lines belonging to the tellurium K-series, but had misinterpreted them, a rather crude error, by which he missed making a great discovery. However, uranium also fooled Fermi, the Joliot-Curies, Hahn and Meitner, and other eminent scientists, all of whom made gross errors, and my own mistakes on the subject stand out. If Abelson had reason to be angry with himself, some time later I did too.

After the discovery of fission, Joliot-Curie and McMillan independently devised a clever physical method for separating the fission products from the product of the (n,β) reaction occurring in uranium. As a result of bombardment of a thin uranium layer, the fission fragments emerge from the layer and can be collected on a suitable adjacent foil, while the products of the (n,β) reaction stay in the uranium foil. It was thus possible to confirm Hahn and Meitner's finding of a 23-minute activity due to U^{239}. In addition the uranium layer showed an activity, with about a 2-day period, and I started studying it chemically. I suspected it was a beta-decay product of the 23-minute activity and hence an isotope of element 93 (neptunium) of mass 239, but I did not expect that element 93 might be chemically similar to a rare earth. Everybody up to then believed that element 93 would be similar to rhenium, and this was one of the errors that had produced great confusion in the interpretation of all the results of uranium bombardment. I established chemically that the activity I was studying behaved similarly to a rare earth and then convinced myself that a fission fragment constituted by a heavy rare earth might stay in the uranium layer. I discussed this with Felix Bloch, who concurred, but the responsibility for the error is totally mine. On second thoughts, I should have realized that my interpretation was, to say the least, suspect.

I tried to make a stronger sample and to see whether the 2-day activity could be interpreted as the radioactive daughter of the 23-minute activity. In this I had bad luck. The ionization chamber I used had a window too thick for the beta rays of Np^{239}, which are unusually weak. Furthermore, I went to the movies at a critical time when I should have been in the lab measuring the activity, although this fact was not of paramount importance.

In conclusion, I erred and did not recognize the genetic relation between the two activities. The resulting paper is fundamentally wrong, but it contains an important truth: the similarity between element 93 and the rare earths.[14] This similarity had even been considered in the literature, but the relevant papers had escaped me, and not only me. Shortly afterward, the problem of the 2-day activity was cracked by McMillan and Abelson, who discovered neptunium. After their discovery, in a letter home, dated June 4, 1940, I observed:

> I do not know whether I have ever written to you of my misadventure with element 93. After so much work, discoveries and undiscoveries, last spring I had it in my hands for several weeks and did not recognize it. On the contrary I have even published a short note affirming that the substance they have now proved to be element 93 was a rare earth. Altogether it is an ugly blunder, combined with my having lost, in a most stupid way, the opportunity for a rather interesting discovery. Now there is nothing else I can do about it. Let us hope that the thing will not be proclaimed to all comers by my friends.

In other studies on fission products with Alex Langsdorf and Chien-Shiung Wu, we inter alia found Xe^{135}, which is a tremendous neutron absorber. This last fact was discovered only later, when xenon poisoned the first nuclear reactors.[15]

In spite of my scientific activity, Lawrence must have come to the conclusion that I was too expensive. I contributed to the exploitation of his machine, but not sufficiently to its development or operation. Hence I could not aspire to a permanent position at Berkeley. Lawrence told me all this very forthrightly in December 1939 and demanded that I use at least half of my time in the service of the Rad Lab, probably intending the cyclotron. I may have believed that I was already using all my time for the benefit of the lab. He also urged me to find a job elsewhere and suggested I try an oil-prospecting company in Tulsa, Oklahoma. "In industry they are no great shakes and you will pass for a good physicist," he added. So saying, he wrote me an excellent recommendation. Lawrence's intentions may have been good, but his diplomacy was not. Remembering also the salary cut I had suffered, I reluctantly started thinking of leaving Berkeley.

I remembered, however, how my dear friend George Placzek consoled himself for having passed from the state of a wealthy gentleman to his present penury: "See! I am at Cornell University. I have an excellent salary: $1,000 a month, but expenses kill me!" He then listed imaginary expenses: $150 a month for having escaped the Nazis; $150 for living in a good climate; $100 for the use of the library and for having access to seminars and to worthy colleagues, and so forth. "I am left with only $120 per month, but it could be worse," Placzek concluded.[16]

By attending meetings, writing letters, passing the word around, and so on, I strove to find a more stable place than the one I held at Berkeley. Some of the answers were friendly, but some were chilling, like that of S. K. Allison, who pointed out to me that not even his student Skaggs (a completely unknown character) had found a job. The idea that I came after Skaggs in the mind of so knowledgeable a friend and gentleman as Allison scared me. Among the encouraging and friendly letters I received, I remember one from James Franck, the great experimental physicist of Göttingen, who had been dismissed by the Nazis and was then working at Chicago. Since Franck was a major figure in physics and was also known for his good heart (besides his courage), I expect he received many appeals for help. I wrote him and he answered with a solicitous and encouraging letter dated September 1, 1939. Finding insufficient what he had dictated to his secretary in English, he added by hand in German, "In your case I am truly optimistic, because people do not let slip a man of your ability." The encouragement of such a person helped me. Later Franck visited us in Berkeley and I met him many times in subsequent years, the last time at a meeting of Nobel laureates in Lindau, Germany, in 1962.

I also tried to mobilize Fermi's help, but I had the impression he was unwilling to bestir himself on my behalf. I now believe his unconcern was more apparent than real, because he preferred to appear indifferent rather than to say that he could not do much. More important, I believe that seeing me doing good work at Berkeley, he reasoned, like my father, that my position in Berkeley was undoubtedly scientifically most advantageous, and that it would be an error to exchange it for a permanent job in a minor university.

Few machines were to be as productive of important discoveries in nuclear physics as the 60-inch cyclotron, which started working at the beginning of 1940. I immediately suggested that since it could accelerate alpha particles enough for them to penetrate potential barriers of heavy elements, one could form isotopes of the missing element 85 by bombarding bismuth with alpha particles. Robert Cornog has described what followed:

> One Monday night at a meeting of the Radiation Laboratory group, Emilio Segrè described his plans to make element 85 by bombarding bismuth with alpha-particles accelerated in the 60-inch cyclotron. After the meeting, as Dale Corson and I walked together across campus, we talked of Segrè's proposed experiment. Unaware that Corson was already preparing to do some alpha-particle bombardments with the 60-inch cyclotron, I said: "You know, Dale, I have a lump of bismuth." "And I have a linear amplifier," Dale countered.
>
> The next morning, bright and early, Corson and I bombarded bismuth with alpha-particles from the 60-inch cyclotron. We saw gobs of giant pulses when we placed the bismuth in front of our linear amplifier. I was elated but felt guilty as sin to have poached what I felt was Segrè's experiment, so I went directly to see Segrè. "Emilio, would you mind much if I had a try at that bismuth experiment you described last night?" I asked. After a short pause Segrè replied: "No. There are other experiments that I can do." Now came the sticky part. "It's worse than that, Emilio. Dale Corson and I have already bombarded bismuth. We got giant pulses on his linear amplifier." Segrè paused somewhat longer. "I have only one request. Let me do the chemistry."
>
> It was a day or two after these events that Luie [Alvarez] was especially articulate and direct. He suggested that I work either on discovering element 85 or on discovering the stability of hydrogen-3 and helium-3, but not on both! So Corson, MacKenzie and Segrè discovered element 85 and Alvarez and Cornog discovered hydrogen and helium of

> mass 3. All these events notwithstanding, Segrè continued to let me use his electrometer, an instrument which now resides in the Smithsonian collection.[17]

Today, as we proposed, element 85, the last of the halogens, is called astatine. As in the case of the other halogens, the name refers to one of its outstanding properties: its instability. Astatine chemistry is complicated, and as all its isotopes are short-lived, it can be studied only on the tracer scale. Chien-Shiung Wu and I also tried to form the last cisuranic missing element, of atomic number 61, by suitable bombardment of rare earths, but at the time the chemical separation of rare earths presented difficult problems, subsequently overcome by elution on resins. We certainly produced isotopes of element 61, but we could not prove it to our satisfaction, and the element was later discovered by Glendenin and Marinsky.

From January 30 to February 20, 1940, Fermi came to Berkeley as Hitchcock Lecturer. I was delighted to see him (it was the first time we had met since leaving Italy), and we resumed our habit of taking long walks together, talking mostly physics. Fermi explained to me his latest studies on the stopping power for ionizing particles and on its relativistic rise. We tried to verify some of his predictions experimentally using the electrons of P^{32}, but the results were inconclusive. On the other hand, with the new 60-inch cyclotron we fissioned uranium with alpha particles, reporting our findings in the last paper we coauthored (although certainly not the last we discussed in the planning stage or to interpret results).[18] At that time Fermi did not mention his studies of the chain reaction to me. As I have emphasized, he was always reserved, and in this case he had weighty reasons for being more than merely cautious.

As was to be expected, Fermi often spoke to Lawrence, but when he started talking physics, Lawrence usually changed the subject, possibly because it was uncomfortable for him. Fermi doubted Lawrence knew or understood much physics, and thought he was rather full of himself. "It's a real problem when people must play the great man but are not up to it," he said. Fermi also attended Oppenheimer's seminars; coming out of one of them once, he said: "Emilio, I must be getting senile. I went to a learned theoretical seminar and could not understand anything except the last words, which were 'And this is Fermi's theory of beta decay.' "

Some of the conversations I had with Fermi concerned beta decay and its ramifications, and it occurred to me that it might be possible to alter the decay constant of a K-electron capturer by chemical means. Since the decay constant is proportional to the electronic density at the nucleus, by subtracting electrons from an atom, it should be possible to alter that density. I estimated the effect and found out that it would be small but probably observable. The best substance to try was Be^7, because of its atomic and nuclear characteristics. On second thoughts, I also concluded that the same mechanism should alter the internal conversion and, as a consequence, the decay periods of my old friends the nuclear isomers. Generally the technique opened up a small chapter of nuclear physics, on altering nuclear processes by chemical means. This had been tried by simple-minded methods by the founders of nuclear physics, but since they lacked the necessary insight, their results had always been negative.

In pursuit of this idea, Chien-Shiung Wu and I tried preparing radioactive BeO and asked the Brush Beryllium Co. to convert some of it to metal. However, after a while more urgent work forced us to drop the project for the duration of the war. I did not publish anything on the subject then, because I disliked publishing ideas of experiments without having performed them. It was most fortunate that we suspended the work, because we did not know of the toxicity of beryllium and would unwittingly have incurred deadly dangers. After the war I completed the investigation with C. E. Wiegand, who was then my student.[19]

In 1940 Lawrence was planning a 184-inch cyclotron to reach the then-enormous energy of 100 MeV. Lawrence expected to overcome relativistic difficulties by putting a million volts on the dees and reaching the final energy in only fifty revolutions. It was a typical brute force method, which might have pleased Admiral Farragut of "Damn the torpedoes—full speed ahead!" fame, but I do not know if it would have succeeded. Lawrence was convinced that with enthusiasm, hard work, and persistence one could overcome every obstacle, or that somebody would find a way out by a new invention, as indeed occasionally happened, particularly in this case. However, ingenuity may circumvent nature's laws, but not violate them.

I realized that if I were to survive at the Rad Lab, I had better contribute to the great 184-inch cyclotron project, so when asked, I eagerly started helping William Brobeck, the exceptionally able engineer who headed the project. Brobeck was a rich young man, the son of a prominent San Francisco lawyer, and in love with his profession. He had a profoundly salutary influence on the Rad Lab because he introduced sound engineering practice in place of the physicists' rough-and-ready way of doing things. Machines were planned with reasonable safety margins, and good engineering techniques replaced improvisation. Due attention was paid to gaskets, welding, stress distribution, and choice of materials. Brobeck introduced all sorts of important preventive maintenance routines. In short, he injected the art of engineering into accelerator development. As a consequence the functioning of the machines greatly improved, the wasting of time owing to breakdowns markedly diminished, and overall efficiency increased substantially. Brobeck was highly respected by the ablest physicists, who understood that his demands were well-founded and abundantly repaid the work and expense of complying with them. From 1937 until 1956, when he left the Rad Lab, Brobeck was involved in all its projects.

In 1940 Brobeck was planning the 184-inch cyclotron magnet, and I helped him in the building and testing of a model of it. In the beginning, it seems, Brobeck had a certain diffidence about working with a foreign physicist, unfamiliar with American engineering units and methods. However, we very rapidly found ourselves to be congenial and greatly enjoying the collaboration. Each learned from the other. I was pleasantly surprised in finding that Brobeck had advertised my work to Lawrence, who was impressed, since the praise came from an unexpected source. A letter home of May 22, 1940, says I was working full-time on the magnet.

At this time I ran into another serious poisoning risk, besides that of beryllium. I used to fill my ionization chamber with methyl bromide to enhance its sensitivity to gamma rays. The commercial gas came in a small canister, which I used every so often to refill the ionization chamber. Fortunately, for no conscious reason, I always performed the operation in the open air, on a balcony. When the cylinder was empty, I called the salesman to get a new one. "How did it go?" he asked. "Have you killed all the rats?" I was surprised by the question, and he told me that he knew of only one use for the gas: as a fumigant for rats. I shuddered. There was no indication of toxicity on the cylinders, and I had not known that the gas was poisonous. Possibly my having handled it in open air saved my life.

As mentioned, soon after Fermi's visit, Lawrence recommended I apply for a job with an oil-exploration firm in Tulsa. I went, to explore the possibilities. By then it was spring, and I decided to go, not only to Tulsa, but also to Washington, D.C., for the annual meeting of the American Physical Society, and to some other place to advertise my existence. In Washington, I saw Rasetti, Rabi, Lee DuBridge, and others; in New York I visited my brother Angelo.

In May 1940, I spent a week in Tulsa, where I met two young men about my age: Serge Scherbatskoy, a Russian aristocrat whose father had been a czarist general, and his partner, a Jew from Russian Poland named Neufeld (who I fancied might have escaped some pogrom commanded by Scherbatskoy's father). They were trying to prospect for oil by radioactive methods; first they had used gamma-ray diffusion on rocks surrounding a drilling hole. Now they wanted to extend their method to include the use of neutrons, which they hoped might help identify hydrogenous material.

They offered me a good salary to join them as their physicist. I looked around carefully and concluded that the only reliable method for interpreting the logs they would get was to build an artificial well in the lab and to observe the behavior of neutrons under conditions similar to what they might find in nature. A calculation without experimental verification was possible, but unreliable. We discussed these technical problems and other details, but after a few days, I decided it was not a job for me.

My stay in Tulsa coincided with the end of the phony war period, and while I was there I heard that Germany had invaded Belgium and Holland, starting an active offensive war in the west. I remember I was having breakfast when the radio gave the news. I almost choked.

From 1930 to 1946, for over sixteen long years, my youth was dominated by the specter either of war or of political disaster. So much worrying must perforce have influenced my character and made me something of a pessimist. I have never been able to live simply from day to day, and the anticipation of events, which unfortunately often came to pass, and from which I did not know how to, or could not, defend myself, has caused me much suffering. Conflicts born out of such frustrations have embittered me greatly and for a long time.

When I turned down the Tulsa job, Scherbatskoy asked me about Bruno Pontecorvo, who then was in Paris with Joliot-Curie, and I warmly recommended him. The oil-exploration firm, Wells Surveys, then decided to offer him a job and cabled him. Thus Pontecorvo, escaping on a bicycle from Paris, about to fall to the Nazis, and in imminent peril of his life, suddenly found himself with an assured job in America. A true miracle!

My trip on the whole had not been successful. I returned to Berkeley because there I could do good physics and I had not found anything better elsewhere. My situation, however, became increasingly precarious until it took an unexpected turn for the better.

In the summer of 1940, we were visited at Berkeley by the head of the physics department at Purdue University, Karl Lark-Horovitz.[20] He was of Austrian origin, of a touchy and sullen character that made him many enemies, but at heart a very decent man and a good physicist. At Purdue he had created an excellent physics department and ruled it autocratically, but with good results. He recognized quality in science and helped whoever he thought merited it. When I became acquainted with him, I felt I could openly tell him my personal problems. He invited me to go to Purdue for a limited period. He could not secure a permanent job for me, but thought that even a temporary appointment would greatly improve my position at Berkeley. Thus in the fall of 1940, Elfriede, Claudio, and I took the train for West Lafayette, Indiana. At Purdue, we settled in at Union Hall, which was strongly heated and extremely dry, so much that in walking on its insulating carpets, one got highly electrified, and I could amuse Claudio by pulling sparks from his nose. I had also prepared some experiments I could easily and rapidly carry out, and I did my best to elicit appreciation for my lectures. Lark-Horovitz's plan worked to a fault. Things went well, and I was pleased.

At Berkeley, when they perceived that I was really leaving, Lawrence and the head of the physics department, Raymond T. Birge,[21] found the money to pay me and wired me with the offer of a lecturership. It was not a tenured position, but it was great progress. I had, more or less, returned to the position with which I had started in 1938, before they knew I could not return to Italy. I asked for a contract with the University of California, not with the Rad Lab, of which I had had enough. I knew that Birge kept his word, whereas Lawrence's intentions and capabilities could change at any time. In any case I would have access to the cyclotron, and I hoped that relations with Lawrence would improve, because having good work done by me in the Rad Lab at no expense to himself, he would be happy and sweet, as indeed happened. Thanks to Lark-Horovitz, I had $300 a month and a much improved position.

My new boss, R. T. Birge, was a very fussy man, a true hair-splitter, and this had served him well, allowing him to ferret out many scientific errors of long standing and greatly to improve our knowledge of universal constants. To better understand his peculiarities, one has to read his history of the physics department at Berkeley.[22] It reveals his many good qualities, which, all in all, more than made up for his weaknesses. These flaws, on the other hand, are easily recognized reading between the lines. In some respects, he was a narrow-minded man, with prejudices against foreigners, especially Chinese, women, and anyone who spoke with an accent. His perfectionism sometimes caused him to lose sight of what was important and run after minute details. Once I was invited for dinner at his home, with Fermi, Harold Urey, and other important people, all in black tie and the wives in long dresses. A magnificent roast turkey arrived at the table, and Birge, at the head of the table, carved it with perfect art, but it took him almost an hour, and we all felt we had to wait until the end of the carving operation before we started to eat, so that the turkey on our plates was by then almost cold.

Birge had a lively sense of humor, which I appreciated, and did not mince words, saying forthrightly what he thought, or, rather, often thinking aloud. His comment to a teacher at a summer school was typical. This professor arrived and found that he had been assigned a small classroom that was possibly insufficient. He complained to Birge, who answered, "Let me see; last year Fermi taught this same course, and we had to change classrooms twice, because they were always too small, but for you this one will be sufficient."

As a department head, Birge had some good, clear ideas; for instance, he early realized the great future in store for the Rad Lab, fostered it, and always kept in mind that it was to the reciprocal advantage of the department and of the Rad Lab to cooperate and to avoid fights. Furthermore, in the early years of his career as head of the department, he made excellent appointments, which greatly contributed to Berkeley's later success. He remained head of the department for over twenty years (1933–55), although at the end he lost contact with active research. Nevertheless, for a period after the war, he still managed to exert a beneficial influence with the advice of younger faculty members.

In any case, in 1940, I found myself much better off dealing with Birge than with Lawrence. For this reason I strove to be employed by the university rather than by the Rad Lab. Furthermore, I liked teaching and I much preferred the cultural aspects of physics to the engineering ones, although I operated in both and recognized their interdependence. All told, I was a born professor and felt more at home at the university than in the Rad Lab, and I established good relations with Birge, who, I believe, had a good opinion of me, despite my imperfect English pronunciation.

So much for Birge; I return to my story. From Purdue University, before reverting to Berkeley, I went to visit Fermi at his home in Leonia, New Jersey. At Berkeley in January 1940, he had hardly mentioned nuclear energy release. At Leonia things were different, and we went into it in depth. In very cold weather, we hiked along the

Hudson River, which was dotted with small icebergs. On the bare trees, to my surprise, I saw hanging many used prophylactics; the winter absence of leaves made this strange vegetation conspicuous. Our thoughts, however, were elsewhere. It was expected that U^{238} by neutron capture and beta decay would form 94^{239}, of which no one knew anything. One could, however, speculate by analogy to U^{235}, that this even-odd nucleus might undergo slow neutron fission. If this was the case, 94^{239} could perhaps replace U^{235} as a nuclear fuel or explosive. It all depended on unpredictable cross sections, decay periods, and the number of neutrons emitted per fission. The prospect was obviously of great importance, but one needed information on this hypothetical isotope.

In a favorable case one might expect an entirely new source of a material fissionable by slow neutrons, independent of the separation of uranium isotopes. Further vistas were immense, including reactors producing the new isotope and its use as a nuclear explosive. On the other hand if the nuclear properties of the new isotope should turn out to be unfavorable, that whole approach would come to nothing. The only way of answering these momentous questions was by direct experiment. One had to make enough of the new substance to measure the desired unknowns. This was possible with the help of the cyclotron. We made several calculations and estimates and found that, with some luck, the plan was feasible. It was imperative to try. Similar ideas also occurred to others, among them Alvarez, L. A. Turner, and Egon Bretscher, but I was unaware of this, and I believe Fermi was too.[23]

To accumulate sufficient material, one needed a substantial cyclotron bombardment, and since this would commit the machine for some time, we required Lawrence's approval. By chance he was then in New York, and on December 14th, 1940, Fermi, Lawrence, Pegram, and I discussed the matter in Pegram's office at Columbia University. Lawrence immediately gave his assent, and a few days later, I returned to Berkeley and started the work. From the very beginning of 1941, I realized that I could not carry out the work as fast as needed alone, and I asked Seaborg to help.

On January 10, 1941, I wrote to Fermi: "Here I found Lawrence who had received from Washington (over Abelson's signature!) a letter concerning our experiment and the suggestion to entrust it to Seaborg. We all agreed that it is not a one-man job, and thus Seaborg and I will carry it on, except that (with Lawrence's concurrence) we will bring in somebody else to help if necessary." Those brought in were J. W. Kennedy, a recent Ph.D., with whom we had previously worked on isomerism, and A. C. Wahl, a graduate student.

Kennedy was a Texan of Irish ancestry, lively, intelligent, very shrewd, and utterly honest. He was a year or two junior to Seaborg, who had introduced him to radiochemistry and collaborated with him on several investigations of radioactive isotopes. Slowly I found out that it was easier to deal with him than with Seaborg. Kennedy's openness inspired confidence, whereas I came to feel that Seaborg had secret personal plans. Kennedy and I noted the unparalleled documentation he was accumulating, way beyond what was needed for current work, with a mixture of admiration and puzzlement.[24]

Seaborg also made every effort to acquire and maintain strict control over our group's communications with the outside world. In this he was helped by the difficulties I had as an alien, by my uncertain position, and above all by my meager ability in public relations, as well as by Kennedy's junior status. As time went by, Kennedy and I, alarmed by what we saw, ended by associating very closely in self-defense.

At the time, Arthur Wahl was a rather innocent student, out of rural Iowa, and ignorant of the world. At least at the beginning, he was strongly influenced by Seaborg. He was a thoroughly good and honorable fellow, but too inexpert and ingenuous to understand the game he was being drawn into. Later he opened his eyes, associated strictly with Kennedy, and came to Los Alamos with him. After the war, Wahl, together with other Berkeley chemists who had originally been students or collaborators of Seaborg's but were averse to further ties with him, went to Washington University in St. Louis, Missouri, which had appointed Kennedy as head of a revitalized chemistry department.

The history given by several of the published documents pertinent to that period is not always complete, and there are occasional disagreements.[25] The authors wrote under the demands of secrecy regulations, patent requirements, and the wish to protect their own personal scientific claims to the utmost. Thus there are occasional disagreements. The possibility of Nobel Prizes being awarded did not simplify matters.

I give here some chronological points: in June 1940 McMillan bombarded uranium with deuterons and by chemical methods he and Abelson had previously devised, separated from it a beta activity due to neptunium and let it decay. After the beta activity had decayed, he found that the residue emitted alpha particles, possibly due to element 94.

In December of 1940, after McMillan had left Berkeley to go to radar work, Seaborg, Kennedy, and Wahl, with McMillan's agreement, started work on deuteron-bombarded uranium. In their bombardment, they obtained a mixture of several isotopes of 93, and this multiplicity of isotopes complicated further work. The beta activities of element 93 decayed, leaving alpha active residues, most likely due to element 94. Some preliminary chemical trials showed that the alpha activity could be chemically separated from most elements. I believe that this was all that was known at the beginning of our common work.

The primary purpose of the investigation by Kennedy, Seaborg, and myself was to measure the fission cross section of Pu^{239}. In order to obtain an isotopically pure sample of this substance, it was best to bombard uranium with slow neutrons. Those gave results that were easier to interpret and also gave larger amounts of Pu^{239} than those obtainable by deuteron bombardments.[26]

At the beginning of our work, we simply chemically separated neptunium from bombarded uranium, using a rare earth carrier, and let it decay. The samples thus obtained, however, were too thick for accurate investigations of alpha and fission processes.

In the meantime, Kennedy, Seaborg, and Wahl continued developing the chemistry of plutonium and its separation from neptunium, but they did not tell me their reagents, having received orders to be prudent with an alien. It took until the end of February 1941 before Wahl found a clean way of oxidizing 94 to various valence states. By then Kennedy, Seaborg, and I already had samples of Pu^{239}, co-precipitated with cerium fluoride, in relatively thick layers.

Around March 1, 1941, we performed a big neutron bombardment of uranium to prepare a substantial sample of Pu^{239}. It was a dangerous operation considering the large amount and the explosiveness of the ether we used, as well as the radioactivity of the bombarded uranium and its fission products. Everything went well, however, and we survived to tell the tale. A first thick sample was ready by April for first rough measurements. In May the sample was thinned by Wahl, using the separation methods he had developed, to a total weight of about 200 micrograms (almost totally rare earth carrier), and we got accurate results.

We all worked on all phases of the investigations, but Wahl performed the bulk of the microchemical operations (I performed some chemical separations without knowing the reagents I was given!). Kennedy, Seaborg, and I performed the bulk of the physical measurements. In the following months, Wahl further developed the chemistry of neptunium and plutonium and laid the foundation for much of the later work.

We inferred the mass of 94 we had in our sample from the measured beta activity and the half-life of its mother substance, 93. From the mass of 94 and its alpha activity, we calculated its decay period. For the slow neutron fission cross section, direct comparison with a known uranium sample in the same neutron flux gave the best results. From all this substantial work, we gathered, by May 1941, that the slow neutron fission cross section of 94^{239} was about equal to that of U^{235}, and its decay period about twenty-five thousand years. Later measurements confirmed all these data. About May 25, 1941, Seaborg, Kennedy, Lawrence, and I signed a letter to Dr. Lyman J. Briggs, chairman of the government's Advisory Committee on Uranium, reporting these results.

We had thus gathered information of capital practical importance, demonstrating the feasibility of using 94^{239} as a nuclear fuel or explosive, and we had opened up a new way of tapping nuclear energy, avoiding the need for isotope separation. We fully realized the importance of our results, and Kennedy, Seaborg, Wahl, and I wrote a letter to the *Physical Review* reporting them. In compliance with the policy of voluntary secrecy then prevailing for results of possible military importance, however, we requested that publication be withheld until better times, and the report eventually appeared only in 1946.[27]

In spite of Lawrence's concurrence, at the beginning of our experiments it had been difficult even to obtain the necessary uranium. I had to write to Fermi at Columbia University asking for it. In a letter of January 11, 1941, I wrote him: "For the uranium work, Cooksey, who holds the purse strings, is not very lavish. It would be most useful if you could send us 5 kg of pure uranyl nitrate. Please let me know if there is any way of getting some money to buy the ether and several vessels for the first extraction after the bombardment. Such difficulties may slow us down." The government had no part in this phase of our work. It woke up after the facts, in June 1941, and asked us to file a secret patent application. We complied, to our advantage, as we shall see in due course.

I do not know to what extent Lawrence initially realized the importance of our findings, although he had helped, indirectly, in obtaining them. I tried to discuss them with him, but I doubt he appreciated the weight of their implications. He told me to talk to his friend Alfred Loomis, a multimillionaire banker and amateur physicist of great intelligence who was visiting the Rad Lab.[28] I hesitated because of security, but Lawrence reassured me by saying that Loomis was cleared for every technical secret concerning defense, and that furthermore he was a cousin and close friend of Secretary of War Henry Stimson's. Loomis understood everything I told him promptly and completely. I believe he helped to open Lawrence's eyes, although it is possible that Lawrence had fully grasped what I told him, and simply wanted Loomis to hear the news directly from the horse's mouth.

It was in this period that Kennedy and I, surprised and confused by some of the occurrences during our work, opened up to each other. We talked at length while walking around the physics building innumerable times.

Lawrence wanted a clean and correct situation and did not care much about personalities. His position is shown in a letter dated May 31, 1946, to Harold Urey, in which he wrote: "Sometime I should like to talk to you about this whole business [of transuranics], as I think that the work of Abelson, McMillan, Seaborg, Segrè, Wahl and Kennedy should be recognized some day in Nobel awards; perhaps the Nobel Prize in Physics to McMillan and Abelson and another one in Chemistry to Seaborg and associates. At any rate sometime I should like to get your ideas and opinions along this line." There were also other forces at play, including G. N. Lewis, the famous Berkeley chemico-physicist, a commanding authority in the chemistry department, and above all Professor W. M. Latimer, who was well known for his xenophobia and anti-Semitism. They communicated with Seaborg only, and naturally they were inclined to accept whatever they heard from him.

Seaborg used his uncommon organizing talents to foster the plutonium study. He enlisted in his operation all the chemistry graduate students he could contact, hiring young chemists left and right and planning large-scale investigations. In short, he was busy building an empire. If I compare this operation to that of other emperors I have known: Lawrence and Fermi (absit iniuria verbo), I immediately perceive a great difference. Lawrence was the chief of a great enterprise he had created from scratch, and he identified himself with the Rad Lab and its successes. He was devoid of jealousy and generous in attributing credit. Although no more than a mediocre scientist himself, he perhaps mildly looked down on his fellow scientists and their squabbles, which he disliked. He justly felt that their successes, if obtained in the Rad Lab, would always extend to him too.

Fermi, a very great scientist, was primarily interested in physics and did not especially like administrative work. He was perfectly honest, open, and scientifically generous, as befits one who had so much to give. He was emperor because subjects and equals in rank recognized him as such, and as such trusted and liked him. Seaborg, a great organizer, and a man of exceptional stamina and capacity for work, but not an exceptional scientist, had unbridled personal ambitions and was determined to get ahead by any means. The simplest was to hire a great number of young collaborators who could not overshadow him, and to take a small part of the credit for their work. Many of these students later had distinguished careers of their own (a very incomplete list of them might include Jack Gofman, Fred Leitz, Ray Stoughton, Morris and Isidor Perlman, Gerhart Friedländer, René J. Prestwood, S. G. English, and Stanley G. Thompson). Taking 10 percent of the work and reputation from each of fifty young men, he could become deservedly famous. The method had been previously used on a large scale by illustrious organic chemists, and similar methods now prevail in large physics collaborations.

What was unusual in Seaborg was the long-term planning he diligently applied to everything. In 1941 he would say: in 1946 I shall be a dean; in 1948, chancellor of the University of California; in 1955, senator for California, and so on, and he never lost sight of his aims. In 1938 he always dressed in a blue suit, with a tie, differently from his colleagues, because he thought that these clothes would help him become a full professor, a small first step in the grand design. Ultimately, he devoted much effort to public service as chairman of the Atomic Energy Commission and many other organizations, receiving more than fifty honorary degrees and collecting pictures of himself with a number of presidents of the United States and other such figures.

Seaborg's designs were fostered by his marriage to Helen Griggs, Lawrence's secretary, an attractive young woman, superior in her profession. She may have helped him by giving him access to documents she controlled. Some disappeared from Lawrence's files when Seaborg and his wife transferred to Chicago in 1942. Some were particularly important to Kennedy and me because they contained data on the plutonium work. Kennedy, alarmed, guessed where they might be and asked Lawrence to enquire from his former secretary whether she had taken them to Chicago in error. The documents were promptly returned, with apologies for the mistake. In the process, however, the original documents had suffered some alterations. In the letter on the discovery of 94^{239}, some of the names under the document had been cut out. When I complained to Gregory Breit, the secretary of the Advisory Committee on Uranium, he answered in a letter dated October 26, 1942:

> I have your letter of October 17th. It is indeed true that your letter to Dr. Briggs written jointly with Seaborg, Kennedy and Lawrence is in the

> official files of the Reference Committee on Nuclear Physics and Isotopes and that it has been circulated as Report A-33. The report is registered for publication in the *Physical Review.*
>
> In circulating the contents as a report, it has been thought advisable to omit your and Kennedy's names. This was purely a formality concerning the report and not the authorship of the paper in the *Physical Review.* At the time the matter of clearance was a difficulty and the letter was transmitted to the NDRC in fulfillment of financial obligations as a report from the official investigators at Berkeley. I should like to assure you, however, that so far as the *Physical Review* is concerned the authorship of the communication has not been changed. It is in fact largely for this reason that the original letter has been transferred to the files of the Reference Committee.
>
> The authorship of this communication is kept secret for the duration of the emergency so as not to disseminate information regarding who is concerned with important work in the section.

This curious letter hardly explains the matter.

At the beginning of 1941, I found it advantageous to buy a house rather than to continue renting. I was able to use family money for the purpose, and for about $8,000 I bought a house at 1617 Spruce Street in Berkeley, about fifteen minutes' walk from the campus. It was a great improvement on our previous house and had a garden with apricot and apple trees, in which we were able to grow excellent vegetables and flowers (although my heart was not in gardening). Our immediate neighbors were Dean Mulford of the School of Forestry and his wife, who were a good deal older than we were. Mrs. Mulford once told us that the FBI had gone to them asking about us; she resented the intrusion, but had given them a glowing report.

On Sunday, December 7, 1941, I was working in the garden when I heard over the radio of the bombing of Pearl Harbor. We were surprised by the fact in itself and by the unpreparedness of the United States. We realized at once that the losses were serious, but a few days later we heard President Roosevelt detailing the construction programs by which he intended to repair the blow. I remember I thought they were like Mussolini's humbug, and that it would never be possible to build so many ships, airplanes, tanks, and so on, in the time allotted by the president. I was wrong, and actually the goals were surpassed. Even after living three years in the United States, I underestimated its industrial potential and what it could accomplish once mobilized.

The reaction to the attack was strong and even hysterical, as in the deportation of Axis citizens. The Italian declaration of war on the United States, which followed at once, interrupted direct communications with my parents. Our letter #129 was returned to sender with a stamp indicating the suspension of service. It was a great sorrow to us, but little compared with the tragedies that were engulfing the world.

The Pearl Harbor attack immediately unified a country that seemed still divided, full of isolationists and of friends of peace at any cost. The effect on the Rad Lab was equally radical. Lawrence woke up to the possibility of nuclear weapons and devoted himself and the Lab to the separation of uranium isotopes, which he decided to do by means of colossal mass spectrographs. Typically, he chose the most direct and "brute force" method and pursued it with indomitable energy. He used the magnet of the 37-inch cyclotron for a model mass spectrograph and started studying it. The 60-inch cyclotron was left intact because it was needed for other programs, especially studies on transuranics, but the magnet of the 184-inch cyclotron, which was under construction, was converted to mass spectrography too.

In all this frenzied activity, one obtained currents depositing uranium ions into suitable receptacles. The hope was that isotopes would be separated according to plan, but initially there was no isotopic analysis of the product. I believed it was imperative to set up an adequate analytical method to know what was happening. I found a solution to this problem by weighing a sample, measuring its alpha activity, and measuring its fission cross section for slow neutrons. Crudely speaking, the mass gives the amount of U^{238}, the alpha activity that of U^{234}, and the fission cross section that of U^{235}. I thus organized an analytical lab, with Lawrence's enthusiastic encouragement and support. Kennedy and several graduate students, who were immediately hired by the Rad Lab, joined in this enterprise.[29] Unemployment among physicists was by now on the wane.

One day I was trying to build a power supply for an electronic apparatus on my own. A student started looking at me, and after a while, with a half-disgusted expression, asked whether he could help me. I was happy to accept, and within half an hour he provided me with a much better power supply than I could ever have made. The student was Clyde Wiegand, and this started a collaboration and friendship that lasted for the rest of our careers.

With the entry of the United States into the war, the migration of physicists to war work, which had started long before Pearl Harbor, greatly increased. Alvarez, McMillan, Jenkins, Brode, and others joined war projects under way elsewhere and left Berkeley. At the same time the university was swamped by students undergoing military training who needed accelerated physics instruction. In this emergency, Birge asked me to teach some "peace" physics courses to the few regular students who were left. I eagerly accepted the assignment because I liked teaching and because I thought it might help me to win a permanent position. The salary too was slightly higher than what I had been receiving up to then. Thus I found myself teaching various branches of physics at an upper division or graduate level: physical optics, quantum mechanics, spectroscopy, thermodynamics, atomic physics. I knew the subjects adequately from experience, and I did reasonably well, acquiring much credit.

In one of the optics courses there was a student who amused himself in finding flaws in the lectures. His objections, always polite, were often well taken and showed a critical and alert mind. I appreciated the young man, who obviously was interested in the course and used his head, and I made friends with him. He was Owen Chamberlain, the son of an eminent radiologist. I must likewise have impressed him favorably, because he became my graduate student and remained my associate for many years.

The outbreak of war between the United States and Italy made me an "enemy alien," and soon the government decreed that enemy aliens must leave the Pacific Coast—first the Japanese, then the Germans, and lastly the Italians. If they did not go voluntarily, they were to be rounded up in relocation camps. In the meantime, they must register, obey a curfew, and surrender their radios and arms. The government started executing these orders by deporting the Japanese. The action proved unnecessary, cruel, and of dubious motive, since it turned out that there were interests that profited by buying up the property of the internees at ruinous prices. Fortunately, after the experience with the Japanese, the government regained its senses and left the Germans and the Italians alone. I was, however, justly alarmed. I found myself in the ridiculous situation of being privy to information so secret and important that the government kept it even from its own high military authorities and from the Congress, but I could not take a walk after sundown or go to the movies in the evening. Of course, under different regimes such contradictions might have been cured by fatal methods. . . .

I spoke to Lawrence about this situation and asked for his help. He answered me that he could do nothing; that laws were laws and must be obeyed, and dropped the subject; however many years later I saw correspondence showing that he had tried hard to protect me. I do not know why he did not tell me then.

In the meantime, I consulted with Elfriede on what to do. I took an atlas, Lo Surdo's wedding present, studied it a little, and then said: "If the government is serious about deporting us, we shall go here to await the end of the war"—and I put my finger on Santa Fe, New Mexico. "It is an isolated place, far from the coast, it has a good climate, and life should be cheap there." Within two years we were indeed there, but the reason—the creation of the Los Alamos Laboratory—was then totally unforeseeable. While the country was mobilizing, strange episodes occurred at the university. One day a gentleman arrived at the Faculty Club and asked to be introduced to several physicists and chemists. It seemed he wanted to know what was going on in some laboratories, and he struck up a conversation with me. He then followed me to my lab, always talking. With utmost politeness, I offered him a chair, oriented in such a way that he could not see anything of interest. He then asked me many questions about uranium, and I became even more suspicious. I answered him, always courteously, but being careful not to give any information not available in common books, and usually telling him the book where he could find greater details. I then came to speak about the deportation orders then in force, and since he had told me he was a government lawyer working for the Immigration Department, I advocated my case, of course without hinting at the nature of my current work. After he left, I reflected on the whole performance. The more I considered it, the more suspicious it looked to me. I concluded that the man was either a spy or an agent of counterintelligence. I therefore went to Cooksey and told him of the strange visit, urging him to report it to the FBI. In my opinion, if the man was a spy, they would catch him; if he was in counterintelligence, we would come out looking alert and diligent. In any case, there was nothing to lose. Unfortunately, Cooksey did not follow my suggestion.

About two weeks later, the same gentleman reappeared, this time in the uniform of a lieutenant colonel, and summoned the Rad Lab's scientific personnel. He told us that he had come earlier on an inspection tour in disguise to see how we kept security. On the whole he had been pleased, but he had a criticism: somebody should have reported him, because his behavior should have looked suspicious. At this, Cooksey, like the true gentleman he was, rose and said: "It is my fault, because Segrè suggested it to me." The colonel's name was John Lansdale, Jr.; he had later important assignments and testified in the Oppenheimer case.

My encounter with Colonel Lansdale had a sequel years later, when I was at Los Alamos. Riccardo Rimini was then in Uruguay, where he was practicing medicine, and I sent him a reprint of an article I had written on new chemical elements.[30] When we moved to Los Alamos, we were allowed to give our new address—Post Office Box 1663, Santa Fe, NM—to all our correspondents, and I had sent it to him. One day, at Los Alamos, I received a letter from Riccardo, saying, approximately: "I have received your reprint, which I found most interesting. I am pleased that you will be continuing your old type of work. Do not fail to keep me informed of any progress." A few days later Colonel Lansdale appeared in my office and asked: "Have you received a letter from Uruguay, saying such and such?" He obviously had a translation of the letter that had been intercepted by a censor. "Yes," I answered. "Doesn't it look strange to you?" he said. "Yes, but I can explain it to you; here is a copy of the reprint." I also told him the identity of the sender of the letter, told him that Riccardo was an Italian refugee and a known anti-Fascist, which it would be easy for him to check through the U.S. consulate in Montevideo. I learned later that Lansdale had first spoken to Oppenheimer, who told him that he could not enlighten him, and that he had better ask me himself. The matter ended there as far as I know.

In February 1942, I received a letter from Fermi hinting at the possibility of my going to Chicago to direct the plutonium work there. I think there may have been difficulties because of my "enemy alien" status, but Fermi was in the same condition and the difficulties had been overcome. However, the two cases were somewhat different: Fermi was indispensable in the full sense of the word, and I was not. It seems, however, that Seaborg had his own ambitions. A. H. Compton, chief of all the Chicago operations, had sent Norman Hilberry, his trusted agent, to Berkeley to recruit me. He arrived at 7 A.M. on March 22, 1942, and to his surprise found Seaborg waiting for him at the station. Hilberry talked with Seaborg, Kennedy, Wahl, and other chemists, but did not see me, and I did not even know he was in town. Hilberry ended by recruiting Seaborg, after having talked to Kennedy and Wahl, who told him that they would not go to Chicago if Seaborg was to be in charge there. With some surprise, I learned all this from Hilberry in 1967 on the twenty-fifth anniversary of the Chicago chain reaction.

Seaborg's departure cleared the air at Berkeley; with Kennedy the work proceeded without tension. On the other hand we found ourselves shifted to problems of less scientific interest, because many of those scientifically interesting and practically important for the making of the atomic bomb had migrated to Chicago, where Seaborg formed a powerful organization that tended to monopolize transuranic elements. It lasted several years and produced valid scientific results as well as strong feelings among his colleagues.

At Berkeley we started systematic studies on spontaneous fission of heavy nuclei, which was also of practical importance. It was expected, but it was difficult to see because it was very rare in the nuclei then known. To observe such rare events, we had to take precautions against spurious signals, considering even fairly unlikely causes. Hence we needed a particularly isolated and quiet laboratory. We first set up shop in a semi-abandoned shack belonging to the university; later in a building on a small alley in Berkeley. The work did not give important results until many months later, when we had moved to Los Alamos. Kennedy, and some chemists, Morris Perlman, Gerhart Friedländer, and Milton Kahn among them, participated in this work. Wiegand strived to perfect the electronics. On the occasion we radically improved the ionization chambers by learning how to collect electrons and not positive ions. This trick substantially increased the resolving power of the chambers.

At about that time, John Manley, a physicist I knew from my visits at Columbia University, became the secretary of a group studying the building of an atomic bomb. Gregory Breit had previously been the head of the group, but although a first-class theoretical physicist, with many accomplishments to his credit, he had weaknesses, such as an obsession with security, that seriously interfered with his performance, and he was replaced with Oppenheimer.

During the summer of 1942, a theoretical group under Oppenheimer's direction met in Berkeley to try to design a nuclear bomb. Hans Bethe, Robert Serber, Edward Teller, E. J. Konopinski, and two younger physicists, Stanley Fraenkel and Eldred Nelson, worked on this project. As they proceeded in their calculations, they needed more and more experimental data that had not been measured, and we tried to help them out as much as possible. To proceed with a concrete plan for a bomb, it was necessary to know, among other things, the fission cross section of uranium, as well as many other cross sections, as a function of neutron energy. At the time such data were few and unreliable. It was hard to obtain monoenergetic neutrons of known energy between a fraction of an eV and a couple of MeV. Some specific energies could be reached using photoneutrons. Chamberlain, Wiegand, some other students, and I used photoneutrons generated by gamma rays of Na^{24} on beryllium or deuterium. During these experiments we had a nasty accident when Chamberlain dropped a strongly radioactive solution of radiosodium. He was seriously irradiated and his blood showed sufficient alterations to require a vacation.

In the meantime, as has been described in several books,[31] the government had assumed control of the atomic bomb project through a series of often changed supervisory committees. The military took a leading role in September 1942, with the creation of the Manhattan District, headed by Brigadier General L. R. Groves. In November 1942 the need for a special lab devoted to bomb construction became irresistible. Oppenheimer was designated as its director, and he and others chose the Los Alamos site.

Oppenheimer asked me to join the new lab, to be created shortly. He became unusually friendly and invited me and Elfriede to dinner at his home high in the Berkeley Hills, a house built in an austere Spanish style, named Eagle Nest. I found him reading Petrarch's sonnets in Italian, and he fed us chicken livers and wild rice, which were excellent; we had never before savored wild rice. Concerning Petrarch's sonnets, I am afraid I did not hide my suspicion that he did not understand them. I remembered the story of a Count C. in Tivoli, who was illiterate, but always went to Mass with his breviary. "Sir, why do you read upside down?" he was asked by a lawyer. To which the count answered: "When one knows how to read, it does not make much difference which way one holds the book." I believe that during the course of the evening I told this story to Oppenheimer. Subsequent to this invitation, I received a small lemon tree from Oppenheimer as a gift. I planted it in our garden, and it was always called Oppenheimer's lemon, but unfortunately, after the war, it died.

I had no choice about going to Los Alamos. War work was a duty to the United States that I felt strongly about, and once one was asked, it was impossible to refuse.[32] In fact, whenever somebody employed in industry or at a university was asked to come to Los Alamos, a telephone call with a few appropriate words and hints sufficed to enlist the candidate, and he arrived on the spot within days. In my specific case I felt doubly obligated to help a country that had received me in particularly trying circumstances. In addition to this, the hope of being able to contribute to Hitler's undoing and to a victorious conclusion of the war appealed to me greatly.

Having agreed to go to Los Alamos, I had to think about personnel and instruments. For the latter, we would take with us some equipment from Berkeley, but we would build on the spot most of what we needed, according to the problems encountered. For the personnel, I thought of creating a group of physicists strictly connected to the chemists. Kennedy had been asked to head the chemistry division of the lab, and I was sure that we would have a smooth collaboration. It was not known who would be the leader of the physics division. Manley was a plausible candidate, most agreeable to me. For my own group I had my old students and under prevailing conditions, it was easy to enlist them. Chamberlain, Wiegand, George Farwell, and G. A. Linenberger formed the basic nucleus. The chemists Milton Kahn and John Miskel joined us at once. Kahn's wife became the secretary of our group.

On November 9, 1942, our first daughter was born. We called her Amelia Gertrude Allegra; the first two names honored her two grandmothers, the last was a wish for her life. I do not remember how we communicated the news to my parents, but we managed somehow, and my mother registered the name in a small genealogical tree she had compiled.

Chapter Seven

Los Alamos: The Fateful Mesa (1943–1946)

Smell of Piñones

Das Alte stürzt, es ändert sich die Zeit
Und neues Leben blüht aus den Ruinen.

(The old is crumbling down—
The times are changing—
And from the ruins
Blooms a fairer life.)
Schiller, *Wilhelm Tell* 4.2[1]

My work had put me at the very center of the atomic bomb project. Although I was technically an enemy alien, so were many others who were vital to the enterprise, and I found myself in a relatively important position in the extraordinary adventure that was the Los Alamos laboratory.

I have already mentioned the creation of the Manhattan District and the choice of its chief in the person of Leslie R. Groves. Son of a Protestant clergyman and military chaplain from New England, Groves had studied at West Point and later had distinguished himself in the building of the Pentagon, reaching the peacetime rank of colonel. When he was assigned to the direction of the newly created Manhattan District, he knew nothing of things atomic. He thought at first that this assignment would be the end of his career, but dutifully accepted it.

To increase his prestige, he obtained the temporary rank of brigadier and plunged into his new duties with immense vigor, courage, decision, and even intelligence. What impressed me most was the speed with which the general oriented himself in a new world totally foreign to him. He was able to deal successfully with such diverse groups as atomic scientists, unions, industrial managers, the British Mission, and assorted prima donnas. It is surprising to me how rapidly he managed to appreciate such persons as Oppenheimer, Fermi, or Wigner, so different from him in culture, outlook, language, and almost everything. Only Szilard remained beyond his grasp. Groves has written an autobiography, from which emerge not only his remarkable positive qualities but also his occasional narrow-mindedness, vanity, and prejudices.[2]

In late 1942, the heads of the Manhattan District decided to build a laboratory devoted to the development of an atomic bomb. General Groves chose J. Robert Oppenheimer to be its director, and Oppenheimer, Groves, and a few others looked for a site for the laboratory in the high plateaus of New Mexico. Some members of the party inspected several localities on horseback, with negative results. The general later joined them, and together they proceeded by car to Los Alamos, where there was a private school for boys. The original buildings were two large wooden chalets, called Fuller Lodge and the Big House, which contained the mess hall, the boys' dormitory, and the classrooms. In addition, there were several wooden or stone buildings housing the staff and various activities, as well as a corral. This handful of buildings did not intrude on the landscape; rather, they harmoniously blended into it.[3]

The mess hall in Fuller Lodge had a porch opening onto a lawn, and from it we had a spectacular view on the Sangre de Cristo Mountains, dominated by the Truchas Peak, across the Rio Grande valley. On the opposite side, behind the few other buildings of the school, there were extensive pine woods and smaller mountains that surrounded the invisible Valle Grande, where I was to spend many hours meditating and fishing. The view from Fuller Lodge could not fail to impress whoever saw it for the first time. I admired it with I. I. Rabi, and he says that my comment was that after ten years of looking at it, we would have had enough of the view. I do not know whether the story is true, but I report it because it shows what we then thought about the possible duration of our enterprise and of the war. My assessment of the war situation was greatly mistaken, and later I was surprised in reading Churchill's memoirs to find that by then he was already sure the war was won.[4] It certainly did not look that way to us simple mortals.

The view from Fuller Lodge became even more dramatic in late spring and summer at 4 P.M., when every day a big thunderstorm started illuminating the horizon toward the Sangre de Cristo Mountains, with brilliant lightning criss-crossing the sky. The storm lasted a few hours and by 8 o'clock it was over, giving way to a serene night. The phenomenon was perfectly regular: at noon big cumulus clouds started forming in a sky that early in the morning had been perfectly clear; the clouds grew and rose in the sky, until they climaxed in the afternoon's thunderstorm. These summer storms cooled the days, which otherwise would have been hot, in spite of the altitude. Thus the climate was most pleasant, with moderately cold winters that permitted skiing. In the spring there was a profusion of wild flowers, and in the fall golden aspen thickets marked the fire scars in the pine forests, inviting hiking on solitary trails. These often went through small canyons with walls of volcanic rock on which Indians of bygone times had left mysterious pictographs.

Water was scarce and occasionally slightly muddy, and the inhabitants of Los Alamos made a big fuss about it. Later we had some periods of genuine water shortage, but they lasted only briefly and were significant mostly as topics of conversation. Checking statistics, I found that the amount of water daily available per person was large compared with that in many European cities.

At Los Alamos's altitude, 7,700 feet above sea level, the atmospheric pressure is only 22.4 inches of mercury, and water boils at 198° F, substantially lengthening cooking times. Several families bought pressure cookers as a remedy; we exercised patience. Typical complaints among the residents resembled a conversation I overheard between two women at the military PX, where we did most of our shopping, who grumbled about the sacrifices imposed by war; for instance, they could not find the exact cuts of meat they wanted.

I first went to Los Alamos in March 1943, and promptly found myself attending a conference devoted to informing the scientific personnel of the future laboratory of the problems facing us and to making concrete plans for their solution. While waiting for the military to build the green wooden apartment houses that were to become our homes for the duration of the war, those attending the conference slept in the Big House in a sort of hall. There my gold Longines Chronograph, bought at the time of the League of Nations sanctions against Italy, was stolen from me during the night. The people sleeping in the dormitory were not of the kind likely to steal watches, and the Military Police maintained strict surveillance of the dormitory, where there were highly secret documents, schedules, and plans related to the atomic bomb. Nevertheless, my watch was stolen, and the MPs, whom I immediately informed, did not display much zeal in trying to track down the thief. The watch was never found, and the theft is still a mystery to me.

Oppenheimer had invited some thirty scientists to the gathering I was attending, including the future leaders of the project and consultants like Fermi and Rabi. In the conference's five sessions, Robert Serber systematically described all that was known concerning a possible bomb. E. U. Condon took notes on his lectures, which subsequently formed a report (now declassified) called "The Los Alamos Primer."[5]

Serber's lectures were followed by discussion of what to do, as well as by animated debates on the laboratory's organization. The military would have liked to put everybody in uniform, but this unhappy idea found a strenuous opposition headed by R. F. Bacher, Rabi, and others who had experience with the MIT radar laboratories, and the military gave in.

Fermi, Rossi, I, and perhaps some other Italian-speaking physicist, were lunching one day during this period at Fuller Lodge, and as usual, we slipped into Dante's language; as usual, talking loudly. General Groves was nearby, and he let us know that he did not like us speaking Hungarian (!) in public; he delicately hinted that if we wanted to speak foreign languages, we had better go into the woods.

Security originated disagreements between civilians and the military from the very beginning of the project. Military personnel were used to obeying orders from above without question and without knowing the reasons. For civilians, and especially for scientists, this was very awkward. We understood the necessity of secrecy, but we also knew that one could not develop new ideas and a new technology while enforcing strict compartmentalization of the data. The importance of information on subjects seemingly remote from the main object was obvious to the scientists but incomprehensible to the military. Furthermore, the latter were concerned with legal requirements, while the scientists were committed to the technical success of the enterprise. In short the military wanted compartmentalization of information, which was deemed catastrophic by the scientists.

General Groves wiggled out of this impasse with good sense. He had soon realized that if he wanted to make the bomb as rapidly as possible, (and in so doing make his rank permanent, or even add a star to it), he needed first-class personnel, even if they had to be aliens, even if he had to rely on Axis citizens. Who could replace a Fermi? Thus when Groves saw that the usual security rules would preclude recruiting those he wanted, he invented new rules. Each of us was to guarantee some colleague he knew well. "Guarantee" sounded good, but how? Somebody proposed an oath on the Bible, but Groves objected: "Most of them are unbelievers." An Intelligence officer then proposed an oath on personal honor, but Groves replied: "They do not have any sense of honor." "Rather," he concluded, "let them swear on their scientific reputation. It seems to me it is the only thing they care for." I thus swore on my scientific reputation guaranteeing Fermi's and somebody else's loyalty, while Bethe and Bacher, I believe, guaranteed mine. The process continued in a circle.

Other problematic aspects of security involved the handling of personal mail. We were permitted to give our address to all our correspondents, even foreign ones, and the military authorities promised that they would not censor mail. The scientists remained justifiably skeptical, and simple tests demonstrated that the military were not keeping their promise. This gave rise to almost comical scenes and to serious resentment and protests, not so much because of the censorship as because its existence was falsely denied.

The Los Alamos site was surrounded by a tall wire fence, with gates, guarded by soldiers, at which one could enter and exit at will by signing a form and showing the proper identification papers. Nevertheless, the military never suceeded in correctly keeping track of the cars entering and exiting the site. This showed up when some cars stolen from the site were found outside of the fence, without there being any signature confirming that they had exited through a gate. Physicists joked about the divergence theorem for automobiles.

Once outside Los Alamos's fence, we had to remain within a certain large perimeter, which contained the city of Santa Fe and several tourist attractions. Exiting from this perimeter required special permission. These restrictions did not seem too inconvenient to Elfriede and me.

After the programming meeting, many of the attendees returned to their usual locations to prepare apparatus and to recruit the personnel required for the new lab. Most returned to Los Alamos after a few weeks with their helpers, to remain there until the end of the war or later. In the meantime, construction on the site proceeded at full speed, and by midsummer of 1943 the lab had started operating. A cyclotron brought in from Harvard University and accelerators from the universities of Wisconsin, Illinois, and Minnesota were the first substantial bits of apparatus available. Physicists, chemists, and metallurgists, as well as auxiliary personnel, arrived daily. As soon as possible, they activated the new laboratories and shops.

At the same time, water supply and sewers, sidewalks, electricity, and everything needed for the new city grew by leaps and bounds. The residential housing consisted of several four-family apartment houses, built of wood and painted green. Bachelors lived in dormitories. The hospital that was built was staffed above all with obstetricians and pediatricians, as required by the nature of the population and the remoteness of the site.

Soon the lab had a collection of nuclear physicists that was possibly more brilliant and active than any other in the world. On average, the members of this group were young, about thirty-two years old; some who were a little older had barely passed forty and were already quite famous. Several who would become famous later were about twenty years old at the time. There were eight future Nobel prize winners (Alvarez, Bethe, Bloch, Chamberlain, Feynman, McMillan, Rabi, Segrè). Oppenheimer was thirty-nine. In these strange circumstances I again met many old friends, whose presence helped to inspire confidence in the ultimate success of the project.

During our first days at Los Alamos, Fermi and I investigated the living conditions we could expect. For instance, we washed our pants in a public launderette to see how good it was. At the Fuller Lodge cafeteria, Fermi sought to demonstrate the digestive powers of his Italian stomach by eating a clearly bad egg that had been served at breakfast. I tried to discourage him, but he insisted he could digest it without trouble. The result was that he became quite sick.

I returned to Berkeley to prepare the transfer of my group and collect the instruments needed for our assignments: measurements on spontaneous fission and on sundry nuclear data, including cross sections. We had already started most of these investigations at Berkeley.

For the proper working of a bomb, it was essential that the fissionable material be assembled in sufficient mass, and that neutrons be injected to initiate the chain reaction only after assembly. Neutrons introduced before the assembly was completed would reduce the efficiency of a bomb and make its working unpredictable. This untimely explosion was called predetonation. Now, especially in the case of plutonium, the high alpha activity could produce light-element impurities that even in very small amounts could predetonate the mass with unwanted neutrons. Extreme purification of the plutonium was the remedy. It was expected that with effort one might succeed. Unwanted neutrons could also come from spontaneous fission of the material. Nothing could be done about these neutrons. Hence the importance of assessing the presence of spontaneous fission.

This urgent job required especially clean and reliable techniques. The samples we expected were necessarily small in quantity, but at least those of plutonium would have substantial alpha activity. We observed the large ionization pulses produced by fission, and it was necessary that there should *never* be disturbances simulating such an event; no fluctuation in alpha emissions could be permitted to fool us. To avoid external electrical disturbances from the mains, we found it was necessary to power everything with batteries in a place far away from other laboratories.

For this reason, in June 1943, we got the use of the abandoned house of a forester in a canyon at Los Alamos called Pajarito (Little Bird). Seldom have I seen such a romantic and picturesque place. It was reached by jeep through trails flanked in season by great bushes of purple or yellow asters, where in full summer one could meet nice rattlesnakes. The tracks followed an open arroyo with steep, low walls, which here and there contained prehistoric Indian dwellings and glyphs with mysterious symbols. At the end of the canyon, we installed our apparatus. In one corner of our rustic laboratory, there was also a folding bed, in which one could sleep if need be.

I had commuted from Berkeley until we could obtain an apartment in Los Alamos. In June 1943, however, the whole family made the move to Los Alamos by car. Our daughter Amelia was then seven months old, and we placed her in a basket. Elfriede and I took turns driving, and Claudio, a very good boy, sat between us or in the back seat. Once in a while, the basket containing Amelia would fall to the floor from the back seat, but she did not seem very disturbed by this. We took the famous Route 66. At daybreak, when we started, the world seemed brand-new, and it was possible to believe we were its first inhabitants, but progressively the heat (without air conditioning in the car) prevailed and we slowly wilted. Troops were training for desert combat along the road, and nails from their boots gave us several flat tires.

From Santa Fe we reached Los Alamos by a dusty, winding road. First we crossed the Rio Grande by an old bridge near the pueblo of San Ildefonso, which was inhabited by Indians. From there, the road climbed to the mesa along a daring route punctuated by spectacular views. As the road climbed, the view on the Rio Grande Valley opened up until it culminated in a vast panorama, with the Sangre de Cristo Mountains as a backdrop.

At Los Alamos we were assigned an apartment on the upper floor of one of the newly built apartment houses. The other tenants included a man who used to play his trombone at night up to three o'clock.

Although we were reluctant to complain, this was too much, and we lodged a protest. It seems that trombone playing and drinking were consolations for marital problems. Our next-door neighbors, who were strict Mormons, had a son the same age as Claudio, and we started taking him with us on our Sunday jaunts, much to both boys' delight. Unfortunately, however, the Mormon parents discreetly asked us not to invite their son any more, because their religion frowned on Sunday outings.

Indian women from the pueblos of the Rio Grande valley were hired to help with the housework, and we visited the Ildefonso, Nambe, and other pueblos ourselves once in a while. During the great winter celebrations at San Ildefonso, we were invited to see the Indian ritual dances. There were no other spectators besides those from Los Alamos and some priests and nuns keeping an eye on their nominally Catholic flock.

One of the leaders of the ritual dances was Popovi, an excellent electrician working in the laboratory on our accelerators. When there were Indian religious ceremonies, he painted his face half yellow and half green and led the dance with utmost seriousness. He was the son of Maria Martinez, a celebrated ceramic artist. In 1943 one could buy Maria's black ceramics for a few tens of dollars; now they are worth at least a hundred times as much and many are in major American museums. A San Ildefonso vase even appears on a 1977 U.S. postage stamp celebrating Indian art. After the war, Popovi too became a famous artist, but lamentably, he died young, a victim of alcoholism.

From the very beginning, we found at Los Alamos old friends from California, such as R. B. Brode, H. H. Staub, Felix Bloch, Robert Serber, and Edwin McMillan; from other American states, such as John Manley, D. P. Mitchell, Donald Kerst, S. K. Allison, and Percival King; and from Europe, such as Bruno Rossi, Edward Teller, Victor Weisskopf, Hans Bethe, and others. Fermi came visiting frequently, as did John von Neumann, with whom I became friends. Joseph Kennedy, my close friend from Berkeley, had a leading position in chemistry.

Oppenheimer, in order better to direct the lab, needed the support and counsel of a certain number of trusted intimate collaborators. To them he gave the most important administrative jobs. He organized the lab into divisions and then subdivided each division into groups. I was a group leader in the physics division. Initially, the division head was R. F. Bacher, who was soon promoted to associate director. Subsequent division leaders in physics included J. H. Williams from the University of Minnesota, not a great physicist, but a person with good common sense and unusual ability in dealing with practical problems, as well as with workers, mechanics, and unqualified personnel.[6] He was followed by R. R. Wilson.

The division leaders had to put up with a certain spirit of independence and restlessness on the part of the group leaders. The division heads, often junior and scientifically inferior to their administrative subordinates, needed plenty of patience and tact to avoid awkward situations. The best they could do was to keep people happy and let them work. Goodwill, talent, and means were plentiful, and these ingredients easily produced results.

As a wise general policy, the administration gave directives and set goals, timetables, and priorities, but left ample freedom at the group level in technical choices and did not interfere in the execution of the work. For procurement, we had powerful priorities, allowing us to requisition anything we needed. D. P. Mitchell, a physics professor at Columbia University, was in charge. He knew where to find anything in the United States.

For the group leaders the problems were different. Once a goal was assigned, they had to find the way of reaching it in a limited time, and without errors that might have disastrous consequences. Furthermore there was always a problem lurking. The laboratory had one purpose only: to build a bomb as fast as possible. The physicists who worked on it often came across subjects that were scientifically most interesting, although irrelevant to the bomb; for the immediate purpose, these were a waste of time and effort. To keep balance in this predicament required a certain skill. On the one hand, discarding all irrelevant science, even if good, risked disgusting several of the best young people and might render them useless. On the other hand, it was inappropriate to devote too much time to pure science. All this produced tensions, especially among the personnel coming from the universities and used to academic freedom of choice in their work. For those coming from industry, the situation was not new, and they were prepared for it.

My group comprised about fifteen people. I was the only established scientist. The other members were either promising graduate students or fresh Ph.D.s. In addition to Chamberlain, Wiegand, and Farwell, the physics students from Berkeley were G. A. Linenberger and John Jungerman. John Miskel, R. J. Prestwood, and Milton Kahn were chemistry graduate students. Several of them had lived in the same house at Berkeley. There were also Jack Aeby and Bill Nobles, students who had been recruited into the Army and put in a "Special Engineering Detachment" (SED). They were in uniform and subject to military discipline, but worked in the lab like everyone else. All these young men received a laboratory education equal or superior to what they could have got in a first-class university; furthermore, practice was supplemented by frequents talks or short courses on subjects connected with current problems.

Oppenheimer also assigned Martin Deutsch of the Massachusetts Institute of Technology and the Pole Joseph Rotblat to my group. Deutsch was an excellent physicist and had left experiments on angular correlations in gamma decay at MIT that made him deservedly famous. Rotblat had come to Los Alamos with the British Mission, after having fled Poland at the Nazi invasion and had been separated from his wife, of whom he had not had any further news (she had, in fact, been murdered by the Nazis). Understandably, he was too upset to do strenuous work and meet deadlines. I assigned him, along with Deutsch, to study gamma rays associated with fission, an interesting line of enquiry, but not of immediate vital significance.[7]

Los Alamos was a closed society bearing some resemblance to a military garrison, but with a population not used to that type of life. Moreover, the pressure of work was immense and enhanced by the unavoidable deadlines and heavy responsibilities. No wonder the inhabitants became touchy and restless. Often they resented petty things to which they would never have paid attention under normal circumstances. Rank, housing assignments, the part of town in which one lived, social invitations, administrative assignments, everything became important, occasionally in a childish way. The fact that one willy-nilly always saw the same people added to the difficulties.

The wives, displaced from their usual surroundings, added to the problems. Without the absorbing technical work of the husbands, and unavoidably in the dark about what was going on in the laboratories, they often became depressed, quarrelsome, and gossipy. The problem became so serious that Oppenheimer consulted a psychiatrist on how to cope with it. The doctor advised him to find work to keep the women busy and to pay them so that they would have a tangible proof of their usefulness. Following this advice, many became excellent secretaries, teachers, medical technicians, clerks, librarians, and so on. For a time, Elfriede worked as a secretary in my group; later she compiled a current isotope chart, or Segrè Chart, as it was called, with my help. It was a monumental work, and after the war, it was declassified and published. More than fifty thousand copies were sold without our getting any royalties.[8]

With many children of school age, elementary schools were vital, and soon the project hired several of the best-qualified mothers to staff them, as well as a few professional teachers. They formed an excellent faculty. Other physicists' wives went to work soldering electronics apparatus, sometimes causing problems for the users, who started complaining about faulty contacts.

The magnificent surroundings of Los Alamos afforded easy and effective relaxation. Often we reached the end of the week completely exhausted and renewed our strength by going fishing on Sunday. Gasoline rationing limited our mobility, but by combining the gas coupons of several families, we could afford to go to the Valle Grande, the Jemez River, the Rio Frijoles, and several other good trout-fishing streams. For me one of the great pleasures was to wander in the cool of the morning along the meanders of the slow river that crossed the Valle Grande, looking for grasshoppers hiding under heaps of cow dung. With this bait I fished fat trout that made delicious meals. It was a prime way of recuperating from the week's hassles.

One Saturday evening I met Sir James Chadwick, the discoverer of the neutron, at a dinner. We happened to talk about fishing, and he told me that the next day he planned to go fishing on the Jemez. Since I knew the river, I asked him what bait he planned to use. He answered rather dourly, "I use only dry flies." I hinted that experience suggested that on the Jemez fish took only grasshoppers. Next day late in the afternoon I was walking along the Jemez with my limit when I saw Sir James casting, perched on a rock. Foolishly, I asked him how he was doing. "Nothing doing," he answered unsmilingly, but he had not departed from his principles. I tried to teach Fermi to fish, and it seemed to me he liked it. However, he once returned from Chicago with a lake fishing rod and reel. I told him that it was not suitable for mountain streams, but to no avail. Fermi developed a theory on how trout should bite and on how to catch them. The theory was disproved by experiment, but this did not impress him in the least. Ultimately he abandoned fishing, but not his theory.

In addition to trout, we encountered beautiful porcupines, flocks of wild turkeys, and birds of all colors, which were sometimes truly spectacular. I once even saw what looked like a sort of small leopard, although I could not identify it. Marmots, badgers, and deer ventured even among the houses.

The flora too were splendid, with columbines of all colors in the spring and mushrooms in the fall. After a while I got to know the spots where animals and plants were to be found, so that I could, for example, go mushroom hunting by car, stopping where I knew I would find them. I picked a few mushroom species that were easily identifiable and abundant, and Elfriede would make them into delightful dishes. Once we invited Mici Teller, Edward Teller's wife, to dinner and served her a delicious dish of rice with mushrooms. Highly pleased by their flavor, she asked where I had bought them. When I told her I had picked them, her face changed color, and about fifteen minutes later she said she felt sick in her stomach. Such was the power of suggestion and fear. Needless to say, nobody else had the slightest trouble.

On the rare occasions when the pressure of the work allowed it, we took some longer outings in the Sangre de Cristo mountains and to their lakes. There marmots completely ate the cork handle of my fishing rod; as an offset I caught spectacular trout, using as bait night moths that lodged in cracks between big stones. Trout took them only if they could not see the hook's leader. With that precaution, success was assured, otherwise it was a waste of time. Having discovered this trick, I challenged some of my friends and impressed them with my fishing skill.

Once, immediately after the end of the war, when Elfriede was expecting our second daughter, we went camping near a warm spring in a remote mountain spot. Next morning we explored the surroundings and found a beautiful hole of crystal clear warm water. We undressed and were enjoying the water, when some Indians who had invisibly followed us politely explained, in Spanish, that the spring was a holy place to them and invited us to decamp. On the same outing we met a newly wed young Navajo couple on horseback in their traditional attire. With the groom's permission we took pictures; the bride objected because she believed that the picture would take away her soul.

Autumn was a season of glorious yellow. The aspens and the beeches interspersed among the dark green of the conifers dropped their leaves, and their pleasant odor pervaded the usually bright, clear air along the trails. One evening I was balancing on a dead tree protruding from a small lake, fishing and admiring the colors of the fall sunset. The still, cool air announced the ending of the season and deep silence prevailed. Suddenly a loud report startled me so that I almost fell into the lake: a beaver had hit the water with his flat tail to frighten me off.

A British Mission, headed by Sir James Chadwick, joined us in 1944. It included among others my old friends Rudolf Peierls, Otto Frisch, and P. B. Moon. Later they were joined by William Penney (the future Lord Penney) and Sir Geoffrey Taylor, a first-class mathematical physicist endowed with an exceptional combination of intuition and analytical power, truly a giant when it came to problems of mechanics, hydrodynamics, and classical physics in general. Taylor had been a close friend of Rutherford's, a meteorologist, and an aviator in World War I. He came from a great scientific family, which counted among its members George Boole (1815–64), the inventor of Boolean algebra, and the surveyor Sir George Everest (1790–1866), after whom the mountain is named. Many years later he came to Berkeley as a Hitchcock Lecturer.[9]

The British mission unfortunately also included Klaus Fuchs, a German refugee who became a Russian spy. I had exchanged only a few words of introduction with him, but he passed under our window every day at noon, presumably going to lunch. Elfriede noted his sad aspect and, not knowing who he was, nicknamed him "il Poverino" (the poor soul). She was dismayed later in hearing that the "Poverino" was a spy.

We often saw the Brodes, the Tellers, and Rossi, because we were neighbors and had known each other a long time. The pianists Frisch, Weisskopf, Bloch, and Teller adorned the same neighborhood. All played well, but with differing proficiency, styles, and programs. Their music, according to Elfriede, who had a fine ear, revealed their personalities like an open book.

Common devotion to fishing and natural inclination tied me to the Swiss Hans Staub; his wife Erika and Elfriede were friends, and the Staub children were of the same age as ours. Staub came from Stanford, where he had worked with Bloch, and after the war returned to Zurich. In the evening we occasionally played poker at the Staubs' house. We played in a very amateurish unsophisticated way. Sometimes John von Neumann joined us. I do not remember that he won particularly often, but he knew the odds of every card combination and of every move.

In the fall of 1943, Niels Bohr and his son Aage arrived at Los Alamos. Shortly before, in Copenhagen, he had been warned that he was about to be arrested. He fled immediately, and after several adventures he had landed at Los Alamos. For security reasons he was given the false name Nicholas Baker and the Lab Direction ordered us to pretend that we did not recognize him, probably a useless precaution, because many knew his true identity. Bohr lived in a small house next to Fuller Lodge with his son, but traveled a lot.

Soon after Bohr's arrival, Oppenheimer convened a meeting at his own home, inviting European physicists already personally acquainted with Bohr, as well as a security officer, called in as a precaution against misunderstandings with the military authorities. Aage Bohr was also present and helped materially because he spoke more distinctly than his father and on request repeated what his father had said. Bohr gave a detailed account of conditions in Denmark and of his personal adventures. He then expanded on what he knew about the rest of Europe. It was the first time we had had direct information from an eyewitness. The atmosphere was somber; almost all those present had relatives or friends in Europe, and the news was thus personally relevant.

After reporting on events in Denmark, Bohr concluded by saying that unfortunately one had to expect things to be much worse for other countries, because the Nazis considered the Danes Aryans, of Nordic race, and treated them as such; apparently they even hoped that by befriending them they might turn them into allies. At this point, Nora Rossi, Bruno's wife, asked: "But then why don't the Danes reciprocate and collaborate with the Reich?" I still remember the expression on Bruno's face, while Bohr patiently tried to explain the reasons why the Danes did not want to collaborate with the Nazis. Nora, a granddaughter of Cesare Lombroso's, as she often emphasized, was given to very self-assured utterances on Italy. Once Fermi started contradicting her with sensible, lowbrow arguments, as was his habit. Nora did not know how to reply and tried to shut him up by saying something along the lines of: "Don't think the Nobel Prize allows you to understand these things better than I do."

During Bohr's visits I had long and frequent physics conversations with him, mainly on nuclear fission. He used to come to my laboratory asking about experimental data and then started reasoning aloud, smoking a pipe that needed constant lighting. He mumbled so badly that Aage, who accompanied him, often repeated some phrase to make it intelligible. I made every effort to understand the words Bohr was uttering; their content was obviously important and instructive, but I had great difficulty. I asked Bohr as much as possible to repeat, but there were limits to how often I could do so and I felt embarrassed at abusing his courtesy. I often remained uncertain and frustrated.

At lunch time I listened to two rather silly radio soap operas: a sentimental story about a nurse, which was broadcast exactly at the time when "il Poverino" passed under our windows, and the biblical story of Jezebel. For some reason, I don't recall why, I mentioned these programs one day to Bohr, who had just returned from a trip to Washington. He answered with great glee: "Then you will be able to give me the latest news; during my trip I lost track of Jezebel."

I was at Los Alamos on July 25, 1943, when Mussolini was toppled, and, naturally, I followed events in Italy in the newspapers. The combined ineptitude of the Italians and of the Allies, especially of the Americans, allowed the German occupation of Italy and the tragedies that followed. However, in Los Alamos, we did not know much of what was happening in the German-occupied zones. Only in June of 1944, after the Allies entered Rome, did I receive the news I coveted most. The first troups entering Rome had orders to look for Amaldi and other physicists to collect information on the German atomic project, and thus, through military intelligence, I obtained some news from home. Oppenheimer called me into his office and told me that my father was safe, but that my mother had been captured by the Nazis in one of their manhunts of October 1943. I was stunned, and Oppenheimer repeated the news to me several times, because he doubted whether I had understood. My father did not survive long; he died of natural causes in his home in Rome, at Corso Vittorio 229, on October 4, 1944, at eighty-five.

It took a long time before communications with Italy could be reestablished and I could obtain firsthand news. Only in 1947, on my first postwar visit to Italy did I learn the details, albeit even then only partially. The tragic and painful page of my parents' end is buried in the depths of my soul and must rest there.

The main and most important work of my group, the study of spontaneous fission, had been started in Berkeley by Kennedy, Seaborg, Wahl, and myself as early as 1941, but the minuscule amounts of material available and the still unsophisticated techniques used at that time allowed us to obtain only large upper limits to spontaneous fission rates, data of little practical interest. In the beginning, as we received increasingly enriched samples, we studied U^{235} with ever-increasing accuracy at Los Alamos ; the alpha activity was moderate and the efficiency for the detection of spontaneous fission could easily be calibrated using a known neutron source. Among other things, we found that the "spontaneous" fission at Los Alamos was greater than at Berkeley, which was not surprising because cosmic-ray neutrons, more abundant at a high altitude than at Berkeley, were obviously responsible for the effect. We had only to screen the chambers suitably to make it disappear.

The difficult, but important, part of the work came with the study of Pu^{239}. Once we had milligram samples, we soon recognized that Pu^{239} had a rate of spontaneous fission high enough to interfere seriously with the proposed methods of bomb assembly through predetonation. These findings, starting in April 1943 with a few counts, became firmer with better samples and longer observation times. We also checked the number of neutrons emitted per spontaneous fission. By July 1944, our results brought the Los Alamos lab to a real crisis, although the relevant information was based on a few counts only. Spontaneous fission in plutonium was so frequent that the plutonium alternative for making a bomb was excluded unless one could invent and develop a totally different assembly method. The predicament was grave indeed; it meant that about half of the total work of the project might be useless for war purposes. The statistical accuracy of our measurements was low, but I was sure that what we had seen was real and not a freak owing to malfunctioning of the apparatus. Even if we had seen only three or four events, I was prepared to guarantee that they were fissions and nothing else.

Soon we noticed another important effect. Plutonium coming from stronger neutron irradiations gave more spontaneous fissions. In the beginning we were perplexed by this finding, but soon we realized that spontaneous fission came from the isotope Pu^{240} and not from Pu^{239}. The first was produced in the pile by neutron irradiation of the second. It was thus proportional to the square of the total neutron flux used. However, high fluxes were necessary for producing enough plutonium. It was possible to envisage separating the main product, Pu^{239}, from Pu^{240}, but the prospect of separating plutonium isotopes was not alluring. We confirmed that Pu^{240} was the isotope with high spontaneous fission by irradiating Pu^{239} with neutrons and observing that the formation of Pu^{240} increased the spontaneous fission of the sample.[10]

As a consequence of this discovery, the Los Alamos project took a sharp turn. Fortunately, the implosion method invented and suggested by Seth Neddermeyer, a former pupil of C. D. Anderson's, avoided the predetonation problems connected with the slowness of the gun assembly of the bomb. Neddermeyer's invention had at first been discarded, but now it was given the highest priority and was transformed into a technically viable proposition. Von Neumann, Geoffrey Taylor, and Fermi contributed materially to its theoretical analysis. For the practical part, a new division under George B. Kistiakowsky was charged with preparing explosive lenses and assembling them. Rossi and Staub had an important part in showing that the implosion was really compressing the material and working according to plan. Through Staub I was able to follow the development of this technique.

When we irradiated Pu^{239} with neutrons to form Pu^{240}, we also formed a sizeable amount of element 95, americium. I tried to persuade Kennedy and Wahl to investigate it at Los Alamos, but they were deeply absorbed in urgent war work and did not want to subtract part of their effort from the main purpose of the lab. Seaborg, though not allowed into Los Alamos, obtained the plutonium that had been reirradiated for us and extracted americium from it. I would have liked to have been less patriotic and conscientious than Kennedy and to have pursued that investigation at Los Alamos.

In addition to regular group assignments, I once in a while received requests from Oppenheimer to look into special problems that arose suddenly and unexpectedly. For instance, once he told me to go immediately to Oak Ridge in Tennessee, to the big isotope separation plants. One of these consisted of many huge mass spectrographs, called calutrons, built under the guidance of Lawrence and the Berkeley physicists. The enriched uranium was chemically purified and converted to suitable compounds for further work. Nobody, however, had checked whether the process might somewhere accumulate enough material to start a nuclear reaction. It was no joke. With bad luck there might have been a real catastrophe. Because of compartmentalization, the local chemists had not considered that danger. Richard Dodson and I rushed to Oak Ridge to study the situation and urge instant adoption of appropriate precautions if necessary. At the same time we were under orders not to reveal secret information. Certainly, however, it was better to stretch secrecy rules than to incur a nuclear disaster. We carefully inspected the plant and found several spots that were outright dangerous, which we duly corrected.[11] When I returned to Los Alamos I had a strange surprise. I found in my suitcase, which I had jealously guarded because it contained secret documents, an alarm clock in a nicely wrapped box. How it got into my suitcase is still a mystery.

Another time Oppenheimer asked Rossi and me to figure out the effects of a nuclear explosion of a certain power and to write him a report on the subject. We were even told to consider the psychological effects on an enemy. We locked ourselves in a room and started figuring temperatures, pressures, radiation density, and so on. We had been at work for some time when we heard Fermi's voice, coming from a nearby corridor. I proposed consulting him. It turned out that he had already done a good part of the calculations and estimates on his own. We agreed well as far as we had gone. With his help we were able to finish the job in a short time; it was later pretty well substantiated by fact. As to the psychological part of the assignment, however, although we discussed it at some length, we could not reach any solid conclusion and we refused to make predictions.

Strange things happened in wartime Los Alamos. One day Oppenheimer called me and asked me to make a rather senseless experiment, which consisted in exploding a charge near a uranium salt. The outcome was easily predictable, but Oppenheimer nonetheless insisted that I perform the experiment, and I complied, obtaining the expected result. I then asked Oppenheimer why he had wasted my time in this way. Oppenheimer answered that he had known the outcome in advance too, but that President Roosevelt had personally asked that the experiment be done and that he had felt bound to comply. Possibly the president received a number of claims to inventions through irregular channels and occasionally skipped regular appraisals. To Roosevelt's credit, however, he had had a good enough nose to recognize the importance of atomic energy when it was proposed to him irregularly.

In the spring of 1945 the atomic bomb project was reaching the home stretch. The way of making the bomb had been found and the design had been fixed. Fissionable material was arriving in appreciable amounts. The purpose of the lab was changing from research and technical invention to production and testing. We thus prepared experiments that would help reveal the performance of the bomb.

A nuclear explosion is a very complex event, with mechanical, thermal, optical, chemical, and nuclear aspects. There was obviously plenty to measure; the energy released was the overall central parameter, which could be inferred in many different, independent ways. Each measurement had its particular difficulties, but one was common to all of them: the experiment could not be repeated. If something failed, there was no second chance. This was a most unusual condition for physicists and worried everybody. A second problem was that we did not know even the order of magnitude of several of the quantities to be measured; we thus required instruments or families of instruments able to cover a vast range.

In order to have at least an idea of the working conditions we could expect, we decided to set off a pile of ordinary explosive at the site of the future test and to make on it the same type of measurements we would later make on the atomic bomb. Although normal explosives could simulate only part of the effects—because, for instance, they do not emit neutrons or gamma rays—this was nevertheless better than nothing.

We were preparing this preliminary experiment when, on April 12, 1945 the laboratory's loudspeakers announced the sudden death of President Roosevelt. Everybody rushed into the corridors of the buildings; some seemed stunned, others were haggard or could not speak, and several had tears in their eyes.

The preliminary experiment was performed on May 7, 1945. While we were in the desert setting up the experiment, we received news of Hitler's suicide, of the surrender of Germany and of the end of the war in Europe. One of my reactions was: "We have been too late." For me Hitler was the personification of evil and the primary justification for the atomic bomb work. I am sure that this feeling was shared by many of my colleagues, especially the Europeans. Now that the bomb could not be used against the Nazis, doubts arose. Those doubts, even if they do not appear in official reports, were discussed in many private conversations.

The efforts to complete the bomb, however, continued unabated. The nuclear explosives, plutonium and U^{235}, were by now arriving regularly from Hanford, Washington, and Oak Ridge, Tennessee, respectively. The Los Alamos personnel converted them into metal, shaped them, and prepared the ordinary explosive lenses for the implosion, the initiators, the fuses, and everything else required for the bomb, finally delivering a complete, working weapon to the military. The military were in charge of target selection, transportation, and delivery. Of course, political and strategic decisions on the use of the bomb could, under the U.S. Constitution, be made only by the president of the United States as commander in chief.

In July many of the scientists participating in the test moved to the desert. Our group was charged with measuring prompt gamma rays emitted at the instant of the explosion (as distinguished from those due to fission products) and the total gamma radiation at several distances, as well as sundry neutron intensities. We planned and built the instruments in the lab and then tested and calibrated them in the desert, readying them for the real test.

The New Mexico desert where we were working is not completely arid; on the contrary there is appreciable precipitation, but the rain is concentrated in very few violent storms, which in a few minutes can transform a dry arroyo into a turbulent stream carrying huge amounts of water. It is not rare for somebody to lose his car or even drown in crossing an arroyo that only a few minutes earlier was dry. I could not believe it, until once Elfriede and I had a narrow escape in such a stream in our car. The desert vegetation, often curiously adapted to the dry climate, is primarily shrubs and cactus, with some grass and no trees. Animals escape the sun by going underground. Many are nocturnal.

Rattlesnakes and other reptiles, such as gila monsters, as well as spiders, scorpions, and other unusual creatures are plentiful.

The final test received the code name "Trinity," and its appointed director was Kenneth Bainbridge, who seemed to me rather disorganized. He often changed schedules, and he posted his all-important daily orders in various places, so that we did not know where to find them. He ended by placing them in the lavatories, saying that everybody would see them there. Fortunately, Oppenheimer must have sensed the problem and gave Bainbridge as second in command J. H. Williams, who was excellently suited to handle a situation requiring the laying of many miles of cables, building roads and shelters and other civil engineering jobs. Bainbridge and Williams had to coordinate a very complicated operation subject to a tight schedule, involving laborers, military personnel, contractors, truck drivers, and prima donna physicists.

We physicists lived in separate barracks, identical to those of the military personnel, and we also enjoyed their excellent food. We started working intensely at daybreak. The early morning hours were the best; as the sun rose higher in the sky, the heat became oppressive, the light blinding, and we wilted. In the evening, we fell exhausted on our cots, only to start again the next sunrise. In this way we spent several days measuring the scattering by air and ground of gamma radiations emitted by a strong radioactive source simulating a bomb. Just as I had done many years earlier at the Officers' Training School in Spoleto, I took with me a French novel by Gide; it transported me into a world totally different from the one surrounding me.

On July 14, 1945, everything was finally ready for the test. In the evening a tremendous thunderstorm broke. I had gone to sleep, but soon I was awaked by a deafening noise, whose origin I could not grasp. I got up and found that Sam Allison too had arisen. We took a powerful flashlight and went out to see what was happening. We found that a hole near our barracks had filled with water, and in it thousands of frogs were celebrating a love feast. We returned to sleep still uncertain whether the weather would allow the test, but shortly the announcement that the test would proceed at daybreak woke us up.

Oppenheimer, General Groves, and many other authorities went to previously assigned observation places. I was with Fermi, in the open, at about ten miles from the explosion. As a precaution we were lying down on the ground, and we had very dark glasses to protect our eyes. Suddenly the whole landscape was inundated by an extremely bright light, incomparably brighter than that of the normal explosive we had tested in May, and that looked, and was, much brighter than sunlight at noon. In fact, in a very small fraction of a second, that light, at our distance from the explosion, could give a worse sunburn than exposure for a whole day on a sunny seashore. At the moment of the explosion, for an instant, the thought passed my mind that maybe the atmosphere was catching fire, causing the end of the world, although I knew that that possibility had been carefully considered and ruled out.[12]

Immediately after the explosion, Fermi stood up and dropped some small pieces of paper, which instead of falling straight down were shifted when the shock wave reached us. Fermi measured the shift and pulled out a sheet of paper on which he had a calculated table of the shift as a function of the energy released. He thus obtained an instant estimate, albeit crude, of the explosion's energy.

For the rest of the day, we collected the data registered by the different instruments and prepared to return to our base at Los Alamos. This work lasted until late in the evening, when, dog tired, we departed for our mesa. After several hours of driving, we arrived at Los Alamos's guarded gate, where a sentinel stopped us. After some discussion the post commander decided to let the civilians in, although it was past midnight, the official time at which the gate was closed, but he was adamant in refusing entry to the military because they had no regular pass. I had with me a pair of young members of the SED (Special Engineering Detachment) and found it preposterous, after such a day, to prevent them from sleeping in their regular barracks. I started arguing and at the same time initiated a surreptitious march toward an emergency telephone I had spotted and that I knew connected directly to the top military commander at Los Alamos. When I came within reach, I deftly grabbed it and instantly had the commanding colonel on the line. He sounded sleepy, and I fear I had awakened him. In a few words I explained the predicament. He then summoned the commander of the guard to the phone and ordered him to let my men in. The commander of the guard, although obviously annoyed, had to obey. Next day, however, the colonel's adjutant telephoned me wanting to know the names of the SED men who had been with me the previous night. I suspected somebody wanted revenge on such small fry for the disturbance of the colonel's slumbers. I acted surprised and insisted I could not remember the incident. That closed the episode.

The bomb test had succeeded beyond expectation and the energy released was near the maximum anticipated. Although what had happened at Alamogordo was a secret, several persons had seen the light of the explosion from a great distance, and among the people who had stayed at Los Alamos at the time of the test, there was the feeling that something extraordinary had happened. At the test site it was strictly forbidden to take pictures of the event, but somebody had smuggled in cameras, and a young SED man in my group took color pictures of the explosion. On his return to Los Alamos, he developed them and they were ready before the official ones. He showed them to me, and to avoid trouble we went to Oppenheimer and gave him a copy. I believe it was immediately dispatched to President Truman. The young photographer, however, was not permitted to profit in any way.

About the time of the Trinity test, there was intensive political activity in several laboratories connected with the bomb. The Chicago scientists especially, goaded by Szilard, tried to influence the use of the bomb. Fermi was involved in high-level committees on the subject, and I gave him my own opinions, but I was not aware of what was going on behind the scenes in Washington and elsewhere. Fermi spoke to me admiringly of General George Marshall, of his speed in comprehending difficult and new problems and of his sureness of judgment in very complicated situations. Considering the times and circumstances, I can hardly see how President Truman could have acted very differently from the way he did.

We celebrated Japan's surrender by taking the day off. Elfriede, Fermi, and I drove to the Valle Grande, where we took some pictures that have been reproduced many times. In the same period, Fermi told me that he expected to become a celebrity and asked me to take pictures of him for the public. Thus we took a whole roll of thirty-six Leica pictures in my office at Los Alamos.

I have been asked innumerable times my thoughts immediately after the bomb's explosion and in the following days. I did not jot them down at the time and recollections would probably be distorted. I certainly rejoiced in the success that crowned years of heavy work, and I was relieved by the ending of the war.

I cannot remember my long-range thoughts; during the war I was mostly concerned with its progress, and with personal plans for the immediate future. More general ideas evolved with time and should always be dated in reporting them. There was optimism for the future in the sense that one trusted mankind's rationality more than was warranted. The confidence in and expectations of the League of Nations after the first world war and the United Nations after the second show how optimistic mankind can be and also how ready to bank on utopian dreams.

Many scientists, like everybody else, nurtured illusions about the farsightedness, intelligence, and reasonableness of their fellow men. Some politicians too were simpleminded, others cynically believed they could preserve technical advantages that were by nature transient. The technical people knew that the bomb could not remain an American monopoly, and I, like many others, gave the authorities that questioned me a correct estimate of the time required before other countries would have the same weapon.

With over forty years of hindsight, it seems to me that most developments were unpredictable and often depended on accidents. Politicians do not come out well as far as intelligence and farsightedness are concerned, and they often acted on erroneous information. Be this as it may the Bomb possibly had the beneficial effect of preventing major wars between the superpowers, by inspiring mutual terror.

During our stay at Los Alamos, we became U.S. citizens. We had applied for citizenship in 1939, and after the statutory period an examiner came to question us. He could not enter Los Alamos, so the examination took place in a guard post at the gate. The examiner started with Elfriede, McMillan acting as witness. The examiner asked her: "Do you believe in polygamy?" Elfriede: "No"; the examiner then turned to McMillan, who confirmed: "To the best of my knowledge, no." After several questions to Elfriede on American government and the Constitution, the examiner turned to me. By then it was about noon, and the examiner wanted to quit. He said: "I am sure that a person like you, before coming to the United States, has studied its Constitution and has informed himself in detail about it." I assented to this flattering surmise, and this ended my examination. A few weeks later we were sworn in at the superior court in Albuquerque, and the judge proclaimed us citizens of the United States. After the ceremony we returned home by a splendid detour along the Jemez. A few days later, however, a very apologetic and embarrassed judge telephoned me to say that he had to set aside the citizenship decree because he had overlooked the fact that the law prohibited naturalization decrees for a certain period before a presidential election. He begged me to take the nullification in my stride and not to raise a fuss, promising to remake me a citizen immediately after the election. Naturally, I agreed, and thus we became citizens twice. However, during the period when I believed I was a citizen thanks to the invalid decree, I had signed some patent affidavits swearing that I was a citizen. I asked the patent lawyer what to do. He told me to sit tight, and that if and when a problem arose, "We shall ask the Supreme Court."

With the end of the war, we started thinking and talking seriously about our immediate futures and long-range plans. I decided to continue my university career rather than stay permanently at Los Alamos, as I could easily have done. However, I did not have a university base, a fact that produced much uncertainty, as the next chapter will show.

Fermi was leaning toward moving permanently to Chicago, but, as usual, what interested him most were scientific projects. He sensed that the future was with particle physics and with a sardonic smile quoted the Duce's motto "Either renew yourself or perish." I had less futuristic thoughts. One expected great progress in neutron physics from the use of nuclear reactors; above all I would have liked to return to work on transuranic elements.

In the next chapter I shall tell the reasons and the events that made me return to Berkeley.

Our family was increased on November 7, 1945, by the birth of a second daughter. We named her Fausta to welcome her, and as a second name we called her Irene, as a wish for peace. My five years of Latin and Greek study finally served for something. . . .

We left Los Alamos in the middle of January 1946. Elfriede, Amelia, and Fausta flew. I went by car with Claudio, who was then about nine. On the trip I let Claudio drive for long stretches of the Arizona highways. He did very well, and I could even slumber while he drove. We briefly visited the Grand Canyon, and along the way I was amazed by the huge number of military airplanes parked in the desert wingtip to wingtip, visible proof of America's colossal industrial power.

Some of the members of my group remained at Los Alamos. Others went to St. Louis. Wiegand came with me to Berkeley to finish his studies; Deutsch returned to MIT; Chamberlain and Farwell went to Chicago to work for their doctorates under Fermi, who had accepted them on my recommendation.

Chapter Eight

Returns: Science and Struggle, Berkeley and Italy (1946–1950)

Smell of Hydrogen Sulfide, *Acque Albule*

Così con legge alterna
L'animo si governa

(Thus with an alternate law
The mind rules itself)
Giuseppe Parini, "La educazione"

The ending of the war posed me serious personal problems. I could have remained at Los Alamos, but I considered my work there an interlude required by the war. I had long ago chosen an academic career as my lifelong vocation, and I wanted to return to a good university position where I could do physics without worrying about a career and in scientifically favorable surroundings. This should have been easy, but my particular situation and my past relations with Berkeley made it difficult and unpleasant.

Many colleagues of my age had academic bases to which they could return: Bethe and Rossi at Cornell University, Staub at Stanford, Alvarez and McMillan at Berkeley. Fermi was about to move from Columbia to new institutes that the University of Chicago was creating. I did not have any certain perspective.

Berkeley was, to say the least, ambiguous, and I did not have any firm offer from other universities. People in similar situations did not even know what to ask and what reasonably to expect, so much so that to remedy this lack of information we started confidentially letting each other know about offers received, establishing a sort of stock market for physicists.

Ambassador Alberto Tarchiani (1885–1964), the first postwar Italian representative at Washington, in an official letter dated April 7, 1945, offered me reintegration into the Italian university system and return to my old Palermo chair. I answered declining the offer because by then I was established in the United States. I couched my answer in friendly, appreciative terms, as the spirit of the offer deserved. Ambassador Tarchiani also asked me for a description of the Italian contribution to the Manhattan Project, and I sent it to him within the limits of the then-prevailing secrecy rules.

I spoke to Fermi about my situation, but initially he did not propose me for a job at Chicago, probably because he loathed even the appearance of nepotism and of favoring Italians. On the other hand, he mentioned me to Professor Joyce C. Stearns, who was moving from Chicago to Washington University at St. Louis, Missouri. A. H. Compton was about to become chancellor of Washington University. Compton wanted to revitalize the sciences, and especially nuclear science. He had hired my old friend J. W. Kennedy, A. C. Wahl, and several other chemists as part of that program, and he had similar plans for physics.

In the meantime, Berkeley had shown some signs of life in the shape of a letter from Raymond Birge offering me the position of assistant professor, without tenure and with a salary suitable for a beginner. I found the offer insulting. Birge was out of touch with the realities of the situation. My only possible answer was not to answer and sit tight waiting for events to mature.

This caused me considerable annoyance. In his "History of the Physics Department," Birge remarks that part of the correspondence is missing from my personal file.[1] This missing correspondence did not show too much acumen on the side of the university. In any case, it was clear to me that unless I had some good offer from elsewhere, Berkeley would not move on its own initiative. This type of deplorable behavior is caused, I believe, by the lack of self-confidence on the part of administrators and decision makers who do not trust their own judgment. Furthermore, they want to be smart and save money for their institution by paying the faculty as little as possible, mostly with counterproductive results.

From Birge's "History," it seems that the head of the Physics Department at Washington University, A. L. Hughes, had asked for information about me already in March 1945. Finally, in August 1945, Washington University made a firm offer of an associate professorship at $5,000 a year. This offer became my baseline. Washington University was a good university, and even if it did not have Berkeley's accelerators, it was a place where one could work. R. L. Thornton, a close friend of mine, had been there before the war and had built an excellent 42-inch cyclotron; now he had been offered an important position, somewhat parallel to that of Kennedy. All told, the place was attractive. However, I thought that in the long run Berkeley would have superior facilities, and I wrote to Birge to find out whether he would improve on the St. Louis offer.

By coincidence, my letter was mailed on August 8, immediately after the explosion of the Hiroshima bomb. Birge must then have realized that speed was needed. However, he started by writing a long delaying answer dated August 11. There were also some underground maneuvers I did not suspect. Oppenheimer wrote an ambiguous letter on my behalf.[2] Lawrence too seemed to be wavering between me and others who later did not have particularly distinguished careers.

However, after the ice was broken by the St. Louis offer, Chicago came along. I went there on September 20, 1945, to inspect the situation on the spot. I had not yet answered Birge's last delaying letter, because after the first offer of an assistant professorship and Berkeley's present obvious eagerness, I believed it was better for me to let them stew in their own juice. It was also a way to let them realize the inappropriateness of their first offer. Ultimately, Chicago considerably bettered the St. Louis offer, and this put me in a quandary.

I was strongly attracted to Chicago by Fermi's presence; on the other hand there were drawbacks. Fortunately, the problem was to choose the best of two good offers. Ultimately, in order to come to a decision, I went to Berkeley to speak with the principals and to appraise the situation on the spot. I believe, however, that I always had a subconscious preference for Berkeley.

I arrived in the Bay Area by train on the night of October 1, and I remember I slept on a bench at the station because there was no way of finding a room or transportation to Berkeley. Next day I started my exploration by speaking at length with Birge, as head of the physics department, with Lawrence as director of the Radiation Laboratory, and with F. A. Jenkins as a trusted friend, conversant with the local situation. Clearly a favorable wind was blowing, and within the day I succeeded in raising Berkeley's offer to a full professorship at $6,500 a year, which I believe was the top of the regular scale.

At this point I decided to accept the offer, which fulfilled my desires and expectations. However, I somewhat delayed my final answer, because I wanted to repay Berkeley's previous dillydallying. Lawrence wisely admonished some colleagues not to be jealous that I had overtaken them. On the contrary, they should rejoice, because my promotion could only benefit them too in the future. I remembered the lesson.

When I sent my final acceptance, Lawrence wrote to me on November 2, 1945, "Needless to say I am mighty glad that you made the right decision, although I can't understand why it took you so long." I never told him why, because it would not have helped.

My appointment started effective July 1, 1945, but I began serving in the spring of 1946. At the time, ideas on relations between the Rad Lab and the university, on financial support for research, and on the possible influence of the military were still confused. I counted here on the political savvy of Lawrence, who certainly would know how to turn the tables to his advantage. I had a solid university basis and financial and instrument support from the Rad Lab. This informal arrangement left me free in my research and at the same time ensured support for my work. Lawrence generously and intelligently was willing to let me enjoy the advantages of the Rad Lab without his having direct authority over me and without paying me; I hoped to repay him by doing good work.

As a research program I wanted first to finish several studies started at Los Alamos, such as my work on spontaneous fission, and some other aspects of transuranics. Next I wanted to bring to a conclusion the unfinished work, initiated in 1940, on changing the half-life of a radioactive substance. I also wanted to investigate the chemistry of the element astatine, which we had discovered before the war. I started this work immediately with the help of Clyde Wiegand and of a chemist, Dr. R. Leininger, hired with Rad Lab money, and some graduate students I found at Berkeley.

I would have liked to continue my work in nuclear physics with a strong chemical component. I planned my operations on a small scale; four or five people. This was a serious error; I did not understand what competition awaited me. I also preferred to work in university buildings on campus, because although an excellent Maecenas, Lawrence was too demanding a boss. I found it extremely difficult, however, even with Birge's help, to get the university to fix up even a single room as an adequate lab. Buildings and Grounds worked slowly and inefficiently.

Soon I realized that Seaborg had a systematic and tightly knit net for controlling research on transuranics. Through his connections, established at the Metallurgical Laboratory at Chicago and elsewhere, he was practically the only person able to secure irradiated materials. Furthermore, he was on the declassification committees and through them could influence what was published. With his great organizing ability, he was about to create within the Rad Lab a very substantial chemistry laboratory, and Lawrence had given him strong support and a free hand. I did not have the impression that Lawrence liked him especially; to tell the truth he seemed at least equally friendly to me, but, wisely, he did not want intralaboratory squabbles, and in the end what counted with him was success and size of the enterprise.

For a while I thought that by sticking to my own specialties, spontaneous fission and radiochemical effects, I would be able to work in peace. But I soon realized that I was headed for trouble. Seaborg wanted the monopoly on transuranics, and he operated on such a large scale that I could not compete with him with a small group. Moreover, his appetites might easily extend further. A collaboration on an equal footing was also precluded. In fact, unless one was willing to accept a position subordinate to Seaborg, which I was naturally unwilling to do, all nuclear work even remotely connected with chemistry was becoming problematic; nuclear physics actually disappeared from the physics department, being transferred to the chemistry department as "nuclear chemistry."

At the same time, several physicists who had always had a dual interest in nuclear physics and in accelerators emphasized their interest in the development of the latter, which had always been Lawrence's primary interest. The abundant new financial support opened up unexpected possibilities, and an important discovery by Vladimir Veksler and Edwin McMillan, phase stability, made it possible to reach relativistic energies by a sophisticated technique and not exclusively by brute force. Also what had been learned from radar work during the war found important applications to accelerators. Berkeley physicists thus turned to new, higher-energy accelerators, which could give access to particle physics.

Higher energies might not only reveal unexpected novelties, but could also provide the key to important old problems still awaiting solutions, such as the detailed study of the nucleon-nucleon interaction. One might hope to duplicate Rutherford's feat on Coulomb forces and discover the true nuclear forces through the study of nucleon-nucleon collisions. Actually, the problem is much more complicated than we believed. From the point of view of modern "chromodynamics," which gives the forces between quarks, "nuclear forces" are a secondary phenomenon. Looked from the point of view of chromodynamics, they are similar to molecular Van der Waals forces looked at from the point of view of electrodynamics. But in the immediate postwar era, there were theoreticians who believed that observation of so-called "p waves," of angular momentum 1, in nucleon-nucleon scattering would solve all problems, or at least would be a gigantic step forward.[3]

In this long period, I was torn between physics, which I understood, and the compelling necessity of attending to business problems for which I had no inclination, and that perturbed me emotionally. My brothers, each for a different reason, made this even more painful. Angelo wrote almost daily letters that posed problems and upset me.

Marco behaved in a way that evoked mistrust and fostered serious worry.

Since the end of the war, I had felt a need to visit Italy to reestablish contact with the survivors and look after business interests.[4] This had been impossible in 1946. Not only was the U.S. government reluctant to give a passport to someone who had worked on the atomic bomb, but transportation difficulties and the devastation of Italy added further obstacles.

In the summer of 1947, I decided to go alone, leaving my family in Berkeley. This was the first of innumerable transatlantic flights. I left Berkeley on June 20 by train and stopped at Chicago to visit Fermi, at Schenectady to do some consulting for General Electric, and in New York City to visit my brother Angelo. At Chicago we talked physics, and I invited Fermi to come to Berkeley, where we needed a top theoretician, but as often occurs in these dealings, the plan came to naught. At General Electric, I found Bethe, Placzek, and Pontecorvo. In New York I saw Angelo for the first time in a long while.

He was in a mistrustful mood, and I tried to persuade him that I had no intention of taking advantage of him, but that I wanted only to come to a clear and fair settlement of our inheritance. "Angelo seems in a very nervous and sentimental state; he spent the morning vomiting as a result of the emotions of recent days," I wrote to Elfriede. Alluding to the invention of the atomic bomb, Angelo aptly said that the state of the world was as if, in ancient times, cats had by chance discovered fire.

On June 25, I boarded a DC-3 for Italy. A Catholic prelate in clerical garb was one of the passengers. He was obviously an important character, because Cardinal Spellman and several bishops came to pay their respects to him. On the plane there was also a priestling from Texas who spoke only Spanish and was terribly scared of flying. He turned to me with a demure air and said: "Mañana, Deo volente, estaremos en Madrid." Then, seeing all the bishops, he started counting: "Un obispo, dos obispos, tres obispos!" He too did not know who the dignitary was. We stopped in Newfoundland, where more *obispos* showed up. Later, in the middle of the night, we landed in Lisbon. I was looking at the sky when the prelate turned to me and said, I believe in English, "Are you looking at the stars? When I was a young man I traveled from Mecca to Jerusalem by mule, at night, and I learned then to recognize the constellations." And he started pointing them out to me. Once the ice was broken, the conversation took a surprising turn. First he enquired about myself; I told him that I was a physicist who had worked at Los Alamos, and he started speaking of things atomic, of the Russians' progress and how far advanced they were. He obviously knew what he was talking about, and when I asked him in astonishment how he had learned all this, his answer was: "There are people willing to risk their lives for the love of God." With this, the conversation ended. Who could His Eminence the cardinal be? I wondered (guessing his rank from a careful examination of his attire, including the color of his socks). The answer came on our arrival at Rome, when a solicitous customs officer shouted: "Make way, Diplomatic Passport, Eugene Cardinal Tisserand," and bore him off. So I learned that my fellow traveler was no less than the dean of the College of Cardinals. His conversation left me very perplexed as to what to do. I went to see the U.S. ambassador, but could not speak to him; instead, I met the first counsellor, Mr. Llewellyn Thompson. We had a long conversation, and he impressed me as an extremely intelligent and well-informed gentleman. (He later became ambassador to Moscow.) The information supplied by the cardinal was reported to Washington.

In Rome, I went to stay at the old family apartment at Corso Vittorio 229, where I found Ada Rimini; it was physically unchanged from the home I had left nine years earlier, but it had lost its soul, my parents. From the survivors' tales, I learned that when the Nazis and Fascists had started their manhunt in the fall of 1943, many of my Jewish relatives and friends had gone into hiding. Unfortunately, my mother was not quick enough, and she was caught by the Nazis. My father escaped, as did Ada, who lived in the same apartment. My father and some others were hidden in a papal palace under the protection of a high-ranking prelate, Monsignore Carinci. My brother Marco went into the hills behind Tivoli. A German policeman promised my father to arrange the escape of my mother after her capture and accepted a fee for doing so. When he did not succeed, he returned the check, with a psychology not too different from that of the professional assassin Sparafucile in Verdi's *Rigoletto.*

I scattered a small sample of technetium on my father's tomb at the Verano Cemetery in Rome, my tribute of love and respect as a son and a physicist. The radioactivity was minuscule, but its half-life of hundreds of thousands of years will last longer than any other monument I could offer.

As a relief from the business problems with which I was immediately confronted, I went whenever possible to swim in the sulfur baths of Acque Albule, not far from Tivoli. The sulfur of the waters eased both my physical and my psychic skin.

To make my story more understandable I shall separate the business part from the rest, starting with a summary account of family affairs, irrespective of chronological order. I hope to make it possible to follow the events that ended with the abandonment of an enterprise that had brought prosperity to the family and well-deserved credit to my father's name.

Before I went to university, I had never attended to any business. Nevertheless, I had heard daily conversations about it, mostly at meals, and I was not ignorant of what was happening around me, of the paper mill, of investments, of commercial, industrial, personnel, and legal problems. I had, however, no responsibilities in the matter; my father did everything. When I came of age, he appointed me to the board of directors of his company. I remember Father preparing the minutes of one of its board meetings and giving me a blank sheet to sign. When I said that I would like to know what he was going to write above my signature, he answered, more or less, that it was none of my business. I signed as a member of the Comintern would have done on Stalin's command (with all due respect to my father). Anyway, he owned all the shares of the company and had every legal and moral right to do as he pleased with respect to it. He had created and financed the business entirely by his efforts and had made it prosper. Nobody could complain.

After the war I was confronted by a new situation, with two brothers and partners who disagreed, made problems, and mistrusted each other. Angelo sometimes had keen perceptions and occasionally farsighted intuitions, but was mentally highly changeable, although the extent of his instability revealed itself only gradually and is clearer to me now than at that time. Marco, though intelligent, had such serious and manifest character flaws as to render him untrustworthy. Of course, this too emerged only gradually, and I understand it better now than then.

I knew that my father's intention had been to divide his estate into three equal parts, one for each child. This intent had been confirmed to me by my cousins Artom and Ravenna, both lawyers, who had visited my father during the war. It had also been confirmed to me both orally and in writing by our intimate friend Silvestro Simili (1893–1968), who was deeply involved in all our postwar affairs. He came from a distinguished Sicilian family and was a brilliant business consultant and a professor of banking at the University of Catania, where my brother Angelo had taught economic history. In due course, the two became close friends, and when Angelo left Italy, he gave to Simili a general power of attorney. During the difficult and dangerous period of racial persecution, Simili proved to be unusually imaginative and resourceful and a true and courageous friend, thus gaining the trust and friendship of my father. Simili was most amusing as a person, keenly interested in the human comedy, gifted with a very rapid and shrewd mind, and personally captivating. He ended by having an important role in the Segrè affairs.

The immediate problem facing us was to settle my father's estate according to his intention of dividing it into three parts of equal value, even if formed by different assets. My father had expressed this intention to me before I went to America in a conversation that, although without legal weight, was a clear expression of his wishes at the time. My understanding was that Father wanted to leave the mills to Marco, who was working there, real estate holdings to Angelo, who, in my father's opinion, needed a safe investment that was easy to administer, and securities to me, since I was more likely to emigrate and to be in need of liquidity.

There was a will dated May 22, 1942, during the height of the persecutions, which left everything to Marco. We all knew, however, that it did not reflect Father's real wishes. It necessarily took into account both the law requiring that shares in a corporation be registered in the name of the owner and the fact that Angelo and I were in the United States, a country at war with Italy, at the time. On May 7, 1944, when it was clear that liberation from the Fascists and Nazis was not far away, my father had made a new holographic will and entrusted it to a friend who was a well-known Roman lawyer. In it he simply stated that he wanted his estate divided according to what Italian law prescribed if he died without a will. This called for a division into equal parts among his children. I do not know exactly when Marco learned of this new will, but the lawyer who had possession of it affirmed that he had communicated it to him immediately after our father's death on October 8, 1944.

On October 16, 1944, however, Marco deposited as the legal will of my father the old one, dated 1942. He did not communicate the existence of the 1944 will either to Angelo, to me, or to Simili, Angelo's representative. The lawyer who had the 1944 will limited himself to notifying Marco of it and did not deposit it. This is most strange, but it is confirmed in writing by a letter from Angelo's lawyer.

As the racial laws in Italy were abrogated with the fall of Fascism, Marco recovered the shares of my father's paper company, the Società cartiera tiburtina, or SCT, from the various friends to whom they had been assigned for safekeeping and had them all assigned to himself. The friends to whom the shares had been assigned all proved worthy of my father's trust, and indeed had rendered a signal service.

Simili would have preferred to see the returned shares assigned to the estate or to the brothers and was worried that assignment of them to Marco alone might give him a position of strength with respect to the other brothers in any future negotiation over the division of the estate. In this delicate situation, Simili worked diligently to arrange a division, in equal shares, taking into account the operational needs of the paper mill. Ultimately, on November 20, 1944, before Simili knew of the existence of my father's 1944 will, he and Marco had an exchange of letters in which Simili acknowledged that it was my father's desire that the paper mill should go to Marco alone and Marco acknowledged that my father wanted his estate divided into parts of equal value. This could be achieved by compensation with other assets.

In 1946, about two years after my father's death, Marco revealed to Simili the existence of my father's will of 1944, which was then deposited, and Marco transferred to his brothers a certain number of shares of SCT, although retaining a controlling majority. The negotiations between Marco and Simili to reassign at least part of the shares to the other heirs were long and laborious and took place while Simili was ignorant of the 1944 will. If he had known of it, he would have had a powerful weapon at his disposal, and things would have been much simpler.

Until the end of 1945, while at Los Alamos, I had no communications with Italy, and even later I knew practically nothing of all this until I went to Italy in 1947. Simili had written to me several letters, but gave little detail. I surmise he wrote them hoping to facilitate a fair division according to my father's wishes, and trying to smooth things over. The main problem was that Marco wanted complete control of the mill and that Angelo did not trust him.

Ultimately, Simili found a compromise formula by creating two classes of shares with different voting rights. Marco obtained the operational control of SCT, but not the right to sell the business or change its capitalization. Profits would be divided into three equal shares. Marco, in addition, would receive a very high salary and a percentage of sales. I was cautioned by a friend that this last condition was objectionable, and that the bonus should be tied to profits and not to sales.

Angelo passed from great love to ferocious hatred for those he was dealing with. In the love phase, he idolized them and endowed them with almost superhuman qualities. In the hatred phase, he gravely insulted the same people. Most of them, after experiencing the love-hate cycle, did not want further dealings with him. Only Simili consistently tolerated Angelo, in part because he relished strange and paradoxical characters, in a spirit similar to that of the playwright Pirandello, his fellow Sicilian, in part out of his own pride and interest, and last, but not least, out of true friendship.

In retrospect, I believe, I may have been too patient with Angelo. He wrote me more than a thousand letters, and each of them upset my digestion or disturbed my sleep. Elfriede was rightly fed up with him, and I should have followed her advice to stop answering his letters. My patience and tolerance derived in part from a certain regard I felt for Angelo's keen intellect, and in part because in several respects I felt that I to some extent resembled him.

One of the fundamental difficulties in my Italian affairs arose from my inability to find a personal representative of my own, independently of Angelo, although it is true that we had very similar interests and that it was not unreasonable for us to be represented by the same person, namely, Simili. Simili did his utmost to avoid quarrels between the brothers, but this was beyond even his remarkable powers. He was the only person trusted by Angelo, albeit intermittently. During love periods, Simili could control Angelo completely. Simili told me repeatedly that he believed that ultimately the manifest and powerful financial interests of the parties would prevail over unhealthy mental states. He was wrong in this optimistic expectation. He had underestimated the power of uncontrollable passions.

As far as Marco was concerned, we started by offering him extremely favorable terms, leaving the management of SCT to him, with the high remuneration and percentage on sales mentioned earlier. However, no concession satisfied him; he always wanted more and, much more objectionable, he felt the need to take advantage of his partners at every opportunity. Possibly, this was his way of demonstrating his superiority over his brothers to himself.

Marco's performance as chief executive was mediocre. SCT started after the war in a miraculously favorable condition because its plant had suffered only minor damage and there were substantial accumulated reserves. We all knew that it was imperative to renew the plant, and Marco planned a new mill at Ponte Lucano, near Tivoli. He hesitated in the execution of this plan, however, and made savings in construction that in hindsight proved ill advised. He accused his brothers of obstructing his work; to them it looked as if he had never had a definite plan. He wanted only cash and a free hand. If cornered on the subject, he took offense and said that he would give all information to our representatives, not to us; that he wanted them to evaluate financial statements that were rather hard to understand. He found it undignified to give explanations, and he shrouded his pompous speeches in a smokescreen of self-serving praise. Unfortunately, I found that Marco was often far from candid.

Our trust was not enhanced when we accidentally found that he had speculated in wood pulp, buying it himself and, if the price increased, reselling it at the higher price to the factory; if the price fell, however, he delivered the wood pulp to the factory and let it bear the loss. Similarly, he bought land adjacent to the new plant, which he knew would be needed, in his own name and shortly afterward resold it to the company at considerable profit for himself. Angelo had anticipated this operation and asked Marco explicitly about it, but he got only an evasive reply. Similar deals were concluded by Marco with stock of a subsidiary corporation. I omit serious errors of judgment he made concerning the paper and pulp market; these are risks inherent in the job.

In 1953, reconstructing events, Angelo came to the conclusion that our father must have left a legal will we did not know about, and that to facilitate his wresting control of SCT, Marco had not produced it at the time of our father's death. I could scarcely believe Angelo's hypothesis, but he insisted, and I wrote to Marco on August 3, 1953, from the Brookhaven National Laboratory, where I was working, demanding a clear yes or no on the subject. Marco replied complaining about his brothers' ingratitude but evading the crucial question, saying: "I know that any consideration or proposal by me in the most favorable hypothesis is received by you with unbelief and suspicion, and thus I abstain from entering into any detail of our relations by letter." He asked, instead, that the three of us meet.

In the meantime, however, Angelo had guessed who had had my father's 1944 last and legally binding will, and the lawyer in question acknowledged having had it and having informed Marco of its contents, trusting that he would communicate it to us. Marco had, in fact, committed a serious offense, and we gave him the alternatives either of reconstituting total parity as stipulated in Father's will or of facing a criminal complaint. Confronted with this choice, Marco capitulated, and we returned to a division of SCT in equal shares. He thus lost control of the company.

I had told Angelo repeatedly that this would not help unless we had a new chief executive on hand. Angelo was in a phase of love for Simili and insisted stubbornly that he should take over the management of SCT. Simili was most reluctant to accept the job and said that he was not the right person for the day-to-day management of an industry. He insisted that he was a financier, not an industrialist. I had to return to America to my physics work and was in a quandary.

Simili took over, but he was faced with assignments he did not relish and for which he was not suited. He ended by vainly trying to bring Marco back in, hoping that he had learned his lesson, and perhaps that he could reach a modus vivendi with him.

Angelo then turned to other managers. He himself had nothing to do except to brood, but he refused to budge from the vicinity of New York, where he resided, and I ended by being repeatedly forced to rush to Italy despite the fact that I had university duties and experiments in progress at Berkeley. I therefore gave my power of attorney to my cousin Bindo Rimini, who had worked at SCT before the war. He had spent the war years in South America, but on returning to Italy at the end of the war had not resumed his old job with the firm, partly because Marco did not welcome him, and partly because by then he had other interests.

The management of SCT in postwar Italy was not an easy job; witness the fact that several major paper mills ended in bankruptcy. On the other hand, some of the smaller firms prospered. In any case, when Marco was fired in 1953, he had already badly hurt the firm. In the end, it was sold to United Paper Mills, a Finnish group that had an interest in obtaining an Italian subsidiary, which gave it access to the Common Market. The final agreement was signed in December 1959. I had gone to Stockholm to collect the Nobel Prize, and I was at the Grand Hotel putting on the pants that went with my friend Giacomo Ancona's elegant tailcoat to go to the ceremonies when Bindo called demanding that I drop everything and come to Rome to endorse it. There was no need for this, and I did not do so, but it shortened my stay in Sweden, and Elfriede, not without reason, never forgave Bindo for his ill-timed insistence.

The brothers Segrè were left with only a minority interest in SCT, and all three of us subsequently sold our remaining shares to United Paper Mills. The Finns kept the name and external appearance of the old firm. After losing a good deal of money without succeeding in revitalizing SCT, they decided to close the Tivoli mill and concentrate everything at Ponte Lucano, greatly reducing the number of workers. This produced a protracted strike, and the Finns sold out to an Italian firm, cutting their losses. The Tivoli mill never reopened. The strike ended with great losses for everybody, especially for the strikers. The available jobs were inevitably reduced to about one-third of what they had been. In view of the archaeological significance of the Tivoli mill, which occupied an important Roman site, the Italian government exercised the right of eminent domain and seized the premises.

I bitterly regret the time, psychic energy, and effort I devoted to family business after the war. With greater wisdom I could have avoided much unhappiness, financial loss, and bitterness. I wrote to Elfriede nearly every day during this stormy period, and these letters virtually form a diary of my stay in Italy. This part of my story is based on them.

"Here Marco and Family live like wealthy lords as before the war," I wrote on June 28, 1947, immediately after my arrival in Rome. "Ada too lives quite well and it is clear that it would be more appropriate for them to send relief packages to us than vice versa. They have maids, chauffeurs, etc."

Some survivors were eager to tell me their adventures, which were frequently quite harrowing; others would clam up impenetrably. I did not succeed in learning many things I would have liked to know. Sometimes I was given contradictory accounts. It was clear to me that my friends, relatives, and acquaintances were still in a state of shock. Inanimate objects, on the other hand, spoke dispassionately, but the emotion of returning to the old Tivoli house, of walking once more in the century-old olive groves, and of seeing the Villa d'Este and the paper mill again sorely tried my equanimity.

A letter dated July 5, 1947, records that "I went swimming at Acque Albule, with Bindo who brought along a girlfriend, to preserve old traditions, and we had an excellent lunch at a restaurant at Villa Adriana under one of those cool pergolas and with a Castelli wine that we miss so much in America. . . . Many things are very different from what we imagined. I have often spent the evening talking to Ada; going to bed only at midnight. . . . My impression is that here one lives 100 times better than in California and that if we were retired, we should come here at once. On the other hand it is certain that suddenly some big explosion may occur, but people do not worry too much, possibly wisely [given the political situation in Italy at that time, the coalition government could easily have turned into a communist dictatorship]. As to pleasantness of lifestyle there is no comparison with here. The beauty of the place alone would suffice. Furthermore there are here so many people we know, so many familiar faces, that one does not have that feeling of loneliness I, once in a while, mention at Berkeley. As far as work is concerned, however, things here go badly, chiefly because the professors are not paid enough to make a living ($50 per month) and there is a great exodus of the younger generation. Given the prevailing conditions, what they do is admirable, but for Wissenschaft, America is better. In any case I now think that a trip to Italy is preferable even to trout fishing, and that is saying enough."

On July 11, I gave a seminar at the Physics Institute. In talking with the physicists there, "the conclusion was that, with little money, they live here about the same as we do with our salary in America, and with plenty of money one lives 50 times better here. I believe henceforth that when I am mad at Berkeley, the talk shall not be of going to Chicago, but to Rome." On July 14, I reported on a visit with Marco to Tivoli: "The parental house is in bad shape because a bomb exploded in its vicinity. Furthermore many people who have lost their homes are squatting in it and there is no way of evicting them. I thus do not see any possibility of inhabiting it in the next years. After lunch I saw Uncle Guido [Treves] who is very well preserved (82 years old) and has not become particularly nicer than he was. At 3 P.M. Count Emo arrived by plane and I talked with him until dinner time."

From a letter dated July 19: "Among other things that popped up, there is a brand-new silver carving set that somebody gave to my parents on their wedding! There are also many books that belonged to Uncle Claudio and to my parents' home. In excavating the cultural layers deposited over almost a century, one understands why I have become a sort of living encyclopaedia, as you say, and one has truly the impression that our children grow up as barbarians. . . ."

From Limonta (near Bellagio, on Lake Como), August 5: "On Friday I saw [Carlo] Perrier at Turin. I had lunch with him and his brother at the Philharmonic Club, in a great 18th century palace with butlers in white stockings and livery, but with average food. Later I gave my talk at [Enrico] Persico's Institute and Saturday morning I left for Milan, all the time in a ferocious heat. At Milan I saw Renzo Ravenna [a cousin, formerly mayor of the city of Ferrara]. Poor fellow; the Germans killed about ten relatives of his. . . ."

From Alassio, on August 10, I wrote: "I came to Laigueglia, where Fausta [Segrè Beltrami] was supposed to be. The junket was a bad idea. Travel was a disaster because the trains are crowded beyond belief; same with the hotels. . . . From Genoa I came to Laigueglia and I have vowed not to return to the Riviera. It is beautiful, but not to my taste. . . . One has the impression of people enjoying life and without sex problems. Also boys and girls about 18 are very beautiful and make me think of 1921, at Forte dei Marmi, with J. They certainly amuse themselves and I would be glad if Claudio and his sisters could have such experiences; they are pleasant and educational even if somewhat perturbing. All told I have lived less idiotically than one would have expected, and I would like it if the children too would enjoy life. . . . I have not seen in America 18-year-old boys and girls looking as if they enjoyed life as much as here. One could say the same, however, of people in general. Here one sees elegance as before the war. . . .

"... It is remarkable how complicated the upbringing of a European is, and how many ingredients enter into it, at least in my case. Often I think that Claudio and his sisters are fed such a bland and primitive fare that they will grow up like E.O.L. and not like, God forbid, Oppenheimer who, however, I believe has not digested the food, or vital nourishment, as Dante says. (If you want to see the reference look it up in the rhyme index under 'digesto.')"

August 16, on the train to Florence: "I have rather changed my mind about Italy. . . . For instance it is impossible to send a wire from the Central Railroad Station at Rome. Here everybody behaves like a selfish pig, and they let me work like a dog, while Marco stays at Gressoney [an Alpine resort] and Angelo takes it easy in New York."

August 18: "At Florence I have seen Giuliana and Eugenio Artom [my cousin and her husband]. They live in a beautiful villa with a garden in Florentine style. I went afterwards for dinner at Marignolle where there were Marcella and Uncle Guido as well as Silvia and her husband. Marignolle's gardens and the fields are unchanged and I find again and recognize the trees I knew one by one as a child." Uncle Guido showed that he had preserved the pungency of his wit. My brother Marco had been made a *cavaliere del lavoro* [knight of work], a high decoration conferred by the Italian government on the founders of important industries or businesses. Father had been proposed for it, but his nomination came to naught because of the regime's anti-Semitic policy. After the war, he being dead, the government conferred the order on Marco, who had succeeded him in the management of SCT. "How is the knight of his father's work?" Uncle Guido inquired about Marco with a smile.

"From Florence I took the train for Padova, where I hoped to find Lorenzo [Emo]," the letter continues. "He was not there. With a shrewd move I located a cousin of his and he told me how to get here [Fanzolo], and then I arrived here on a small truck. It is one of the few times in this trip I was glad you and the children were not with me. The Lords of the Manor are not here, but will return today or tomorrow and I have made myself at home while I wait for them. . . .

"The Fanzolo villa is fantastic.[5] Unfortunately it is located in a flat, rich, and hot countryside, but the villa itself is spectacular. It was built by Palladio around 1550. It has frescoes by Paolo Veronese or pupils, and furniture, rugs, etc., all museum pieces, well kept, clean and alive. Add that it has modern plumbing, central heating and all modern conveniences. The halls, studies, and dining room are furnished with 16th century pieces, with the Emo arms; each could be the center of a modern house. Since I am here alone they serve me in a dining room in a style I knew only from my readings of Lord Fauntleroy, except that a local young girl serves the meals. In my bedroom there are 4 or 5 paintings each of which, I surmise, could provide the finances of a family. As usual, in these surroundings they look quite differently from what they would in a museum. . . . For breakfast, for the first time since I have been in Italy, I have had delicious bread and butter (naturally from the Count's estates), served on a silver tray that possibly weighed 5 kg. It is surprising there is so much stability in the world that it is feasible to accumulate and preserve a property like this one for 5 centuries. Yesterday I spent the evening reading Venetian history in a book on the Emo family; the list of their beautiful names alone is a pleasure. For instance, the brothers Emo about 1350 marry Chiara Querini, Cataruzza di Giorgio Loredan, Ginevra Corner, a daughter of Nicolò Vendramin, etc. Other names: Mabilia Venier, Belella Pisani, Elisabetta Moro, Andrianna di Angelo Badoer, Cornelia di Vincenzo Grimani; it looks like a directory of the Maggior Consiglio. There is a documented family tree up to Pantaleone Emo, who, at the Serrata del Gran Consiglio, was registered among the Venetian nobility with all his descendants. The Capodilista seem to me to be small fry compared with the Emo. All told I think Barbara [Lorenzo's Canadian wife] might be somewhat uneasy. It is a mixing of two different worlds, that might give strange results. In reading these histories it seems that the living element in such a family is the family and not the individual members, while in the modern world the opposite seems to be happening. In any case I believe that such things are easier to understand for an Italian (even a Jewish one) than for an American, the second in the whole Emo genealogy."

When Lorenzo returned, we went to Venice together. I had started feeling sick even before going to Fanzolo, and in Venice I tried to board a sleeping car train for Rome; but it was full. A one dollar bill given to the conductor made him discover that an Austrian girl had been overbooked, and with polite excuses he threw her out of her compartment and gave it to me. Such was then the power of a dollar.

My emotions in Italy, the heat, and the many bitter pills I had to swallow from Marco and from Angelo and his eccentricities were altogether more than my guts could bear and the conclusion of the journey was that I developed a duodenal ulcer. On my return to Berkeley at the beginning of September, I got the idea that the objective symptoms I had might be due to a stomach cancer. The suspicion was pessimistic, but not entirely foolish, as we unfortunately saw some years later in Fermi's case. I went to my friend Dr. Giacomo Ancona, who tried to reassure me, but nonetheless sent me to a radiologist, a man I knew well from the Rad Lab. He examined me very carefully and said that he could not see anything suspicious except a possible duodenal ulcer. I asked him how many cancers escaped him, to which he answered: "About 20 percent." My thought was then immediately, "And what if I am in the 20 percent?" Ancona then sent me to a well-known gastroenterologist, who, having studied me, concluded by asking, "What is your favorite form of relaxation?" I answered, "Trout fishing in the mountains." He then prescribed trout fishing in the mountains and advised me to forget diets, drugs, and symptoms. I followed his orders in Elfriede's company, and after some months I recovered.

I had found it necessary to organize my research group at Berkeley during my absence in such a way that I could remain informed about what was happening and not halt our work. This became easier as time went on and Chamberlain and Wiegand matured scientifically. Finally, we arranged things so that any one of us could go away for a period of up to about six months without great disruption.

Chapter Nine

Ripening Crops (1950–1954)

Smell of Ripe Wheat

Season of mists and mellow fruitfulness!
John Keats, "To Autumn"

At Berkeley after the war I found myself for the second time in my life (the first had been at Palermo) free to work according to my tastes without the need to produce a certain number of papers or results in order to survive in the profession. It is said that Verdi always complained of having to compose his operas under cruel deadlines, and that only *Othello* and *Falstaff,* the masterpieces of his old age, were composed according to his wishes, in full freedom. With all due respect to the differences, my own experience had been similar. First the Italian competitions and the compulsion to pile up printed paper, later my position as a refugee at Berkeley, and later still the exigencies of Los Alamos had forced me to work on short-range projects that could be brought to fruition rapidly or to otherwise work under great pressure. Now that I had "arrived," I could tackle longer-range projects, and I thought that the nucleon-nucleon interaction was a worthy subject of investigation. It seemed to be at the very core of the exploration of nuclear structure, the nucleus being an assembly of nucleons. Many famous physicists had devoted great efforts to its theoretical study: Werner Heisenberg, Ettore Majorana, Eugene Wigner, Igor Tamm, and others had provided deep theoretical insights into the problem, and low-energy experiments had shed light on many aspects of it, although limited to states of zero angular momentum. The higher energies now available opened up to experiment states of higher angular momentum, and they could give significant new information. To prepare myself for this experimental program, I started learning what was known about the nucleon-nucleon collision by giving a graduate course in nuclear physics with emphasis on that particular area.

The rapidly developing accelerators offered unique opportunities.[1] During the war, the great magnet planned for an ordinary cyclotron with an energy of 100 MeV, to which I had made a small contribution, served to provide the magnetic field for prototypes of mass spectrographs. After the war, it returned to an accelerator, but of a different type from that previously planned. The plans for the new machine included frequency modulation and the use of the phase-stability principle, it was to become the 184-inch Berkeley synchrocyclotron. This machine started working at the end of 1946 and had a long and glorious history. Initially, it accelerated only an internal deuteron beam to about 195 MeV. These deuterons impinging on any target produced an external neutron beam of about 100 MeV, the first we used in the neutron-proton collision investigation started in 1947.

These neutrons caused a funny episode. A hasty investigation of the angular distribution of the neutron beam, performed as soon as the accelerator started functioning, indicated a forward distribution, as expected, but with a minimum at deflection zero. This was reported in a colloquium attended by Oppenheimer, and he immediately gave a learned theoretical explanation of the phenomenon. I listened to it and then said that it was better to check if by chance there was not a lead brick just in front of the target, projecting a shadow. Immediately after the colloquium somebody rushed to check my hypothesis. It was correct.

In the summer of 1948, Owen Chamberlain, who, as previously noted, after having served in my group at Los Alamos, had gone to Chicago and obtained his Ph.D. under Fermi, returned to Berkeley as an instructor and rejoined my group. Our detailed study of the nucleon-nucleon collision was greatly enhanced at the end of that year by the availability of an external proton beam. For many years the measurements we did between 1948 and 1955 remained authoritative.[2]

Many people collaborated in the experimental side of the enterprise including H. F. York, Chamberlain, Clyde Wiegand, and Tom Ypsilantis.[3] The contribution of all these colleagues, postdoctoral fellows, and graduate students was essential; without it, it would have been impossible to carry on the enterprise. Measuring not only absolute collision cross sections, but also polarizations, correlations, and similar beauties, we collected a wealth of experimental data. The next step was to derive from them the phase shifts of the partial s, p, d, etc., waves. For this analysis, Ypsilantis, Henry Stapp, and Nicholas Metropolis used the Los Alamos computer.[4]

In the end, the problem turned out to be less fundamental than either we or the theoreticians believed in the 1940s. At the time, it was believed that the nucleon-nucleon force was a primary natural force mediated by pions according to Hideki Yukawa's ideas. Progress has changed the outlook. We now believe that the true fundamental forces are those between quarks, described by chromodynamics. Nevertheless for nuclear physics in a strict sense, the nucleon-nucleon force is still of capital importance, independently of its origin. Even with this reservation, however, ours was an important piece of work of durable value, or at least as durable as such things are in present-day physics.

In addition to research I regularly taught one or two courses, usually one to upper-division undergraduates and one to graduate students. Over the years I greatly varied the subjects. Often I taught nuclear physics and quantum mechanics, trying to make the subject simple while at the same time presenting significant problems. It seems that the reactions to my courses were of two kinds: one group of students were very appreciative, others not so much. I believe my lectures were not as polished as others, but they had a degree of freshness that made them appealing. I believe one saw I was not reciting a textbook, but rather telling from experience as a practicing scientist.

Seminars were an important component of my teaching. I often held them with some other professor, such as Chamberlain, with an attendance of about ten students. Students prepared to explain some good review article, often from *Annual Reviews of Nuclear Science,* to the audience. I tried to pay attention and to learn something new. I also demanded that the speaker understood what he was saying. It was surprising how often formulae would occur where the speaker did not really know the meaning of the symbols he was using, or diagrams where the speaker could not explain what was plotted. Unfortunately, I had a tendency to fall asleep, especially when I did not understand. At a slightly higher level I tried to elicit an explanation of the various "It is easily shown"s that adorn the scientific literature. All told, I hope the courses gave not only specific information on technical points, but also an education in scientific attitude.

I well remembered how once as a young assistant professor I had cut a bad figure on one of the rare occasions when Corbino attended a seminar. He had asked for a subtle, though necessary, explanation in a question of adiabatic demagnetization, and I obviously demonstrated that I did not know what I was talking about.

In the same vein, a few years before his death, in a conversation in which I complained about the many subjects that are supposedly "well known" but in fact are just the opposite, Fermi suggested I make a note of any such questions I came across—such as validity conditions for Born's approximation, subtle questions of B and H in magnetism, signs in the energy expressions in thermodynamics, innumerable questions related to phases in quantum mechanics, and so on—and that when he retired, he would write a book giving all the explanations. Unfortunately, this did not come to pass. It would have been the best-seller in physics. Of course, there are many other physicists who could write such a book. I hope one of them will oblige, and write it with Fermi's clarity and simplicity.

In that period I also started to busy myself with physics literature, in which I had a long-standing interest. Immediately after the war, knowledge acquired while working on military projects was classified and could be used only by those with appropriate clearances. Much of it, however, had only little and indirect military value, although it was important for scientific and technological progress. Moreover, the very existence of secret reports produced awkward results, because, while cleared personnel knew that certain things had been done or could be done by methods and techniques developed during the war, they could not use or teach what they knew without violating military secrecy. The government had started a big declassification process, but by its nature it was a slow operation, very bureaucratic and occasionally even subject to manipulation.

I then thought that it might be a worthy endeavor to compile a big treatise on the model of the prewar German *Handbuch der Physik* edited by Hans J. W. Geiger and Karl Scheel, but limited for organizational and time reasons to nuclear physics. It could not be written by any one person, because each chapter would require the expertise of one or more specialists with firsthand information. I assembled a group of experts and directed the compilation, in three fat volumes, of *Experimental Nuclear Physics,* which was very successful (it was even translated into Russian). The enterprise took longer than I had anticipated, however; the first volume appeared in 1953 and the last in 1959.[5]

In the same spirit of helping disseminate scientific information, I also started to work for *Annual Reviews of Nuclear Science.* The series was initiated by others, but I soon became one of its mainstays, and in 1952 its editor. I kept the job until 1977, when I retired because I thought younger forces should carry on.[6] The work was practically unpaid, but it helped to keep me current, and I believe it has not been the smallest service I have performed for the profession.

I remained in Berkeley during all of 1948. Fermi came there to teach the summer session, after which we went to Los Alamos for a few weeks together. He drove all the way from California, because he did not trust anybody else. In 1949 the problems of SCT required my presence in Italy, and I returned there with Elfriede and children. I would have liked to spend a good part of the time at Tivoli, but Elfriede demurred. I never fathomed the cause of Elfriede's dislike for Tivoli. A possible explanation, although strenuously denied by her, may have been the contrast between her unhappy childhood at Ostrowo, and later at Breslau, and my own at Tivoli.

In September 1949, I attended an international physics conference at Basel, where for the first time since before the war, we were once again able to meet our old European colleagues and friends and report at least in part on the great novelties developed during the war. I spoke there of our work on nucleon-nucleon collisions.[7] The great Wolfgang Pauli was in the audience; he listened shaking his head from left to right and simultaneously shifting his heavy body up and down. After my speech, I was going away with the Swiss physical chemist Egon Bretscher, who had been at Los Alamos, when Pauli accosted me and said: "I have never heard a worse report than yours." I did not answer; what could I say? But Pauli turned to Bretscher and added: "I stand corrected; when you spoke [and he mentioned the occasion] it was even worse." Whereupon he departed. Another physicist, who knew Pauli well and had witnessed the performance, smiled at me and said: "Don't listen to him. Your speech must have been quite interesting to him because he oscillated all the time, which means that he was listening carefully." I knew Pauli well enough to know that there was no reason to worry about his remarks.

In an excursion following the Basel meeting, we went to the Cosmic Ray Lab at Testa Grigia near the Theodule Pass under the Matterhorn. I revisited the Breuil basin where I had spent memorable days twenty years earlier. The Breuil was already spoiled by new hotels and ski lifts, which had transformed that Alpine gem into a vulgar commercial resort. At the Cosmic Ray Lab, work was proceeding chiefly under the leadership of Gilberto Bernardini. Several of the Italian physicists were fed up with the difficult working conditions that prevailed there, however, and shortly thereafter, Bernardini, Gian Carlo Wick, and other friends of mine packed up and came to work in the United States, at least for a few years.

Back in Berkeley, Robert Brode, Francis Jenkins, and I kept our eyes open for promising young recruits to the staff of the physics department. This was one of the reasons for attending the annual meetings of the American Physical Society in Washington, D.C., which functioned as a placement market. Interviewing possible candidates in 1948, I recalled the days when I had attended similar meetings looking for work, particularly my difficult time in 1940. The tables had now turned. The situation in 1948 was radically different; before the war there had been few jobs, whereas now able candidates could always choose among several excellent positions. Birge, though head of the department, hardly took the initiative in the selection of young recruits, but the University of California was in a phase of expansion and qualitative improvement, which facilitated the recruitment process. If one could find an outstanding prospect, we could count on the administration being able to find the means to make an attractive offer. Furthermore, for people interested in nuclear or particle physics, the accelerators and Rad Lab facilities were a powerful attraction. We thus were able to hire many future celebrities. Unfortunately, several left because, especially to theoreticians, the eastern United States offered strong competition by reason inter alia of its superior scope for communication and exchange of ideas. Harvard enticed away several of our best young professors, among them Steven Weinberg, Sheldon Glashow, and Michael Tinkham.

In 1949 the Regents of the University of California, its supreme authority, came up with the idea of requiring a loyalty oath of the faculty, demanding among other things that they declare that they were not members of the Communist Party. The wording of the oath was relatively harmless, but it was not harmless to demand it of professors while exempting all other state employees. The controversy expanded into bitter arguments, and in my opinion the Regents showed that they were not up to their task. They seemed to me more concerned with their prestige and with asserting their paramount authority than with the welfare of the university. The governor of California, Earl Warren, sided with the moderate faction among the Regents; the extremists were led by J. F. Neylan, a former Hearst lawyer noted for his extreme political opinions. This radical faction prevailed, and those who did not swear the loyalty oath were dismissed.[8]

In physics this resulted in serious losses, among them Geoffrey Chew, Wolfgang Panofsky, Marvin Goldberger, Gian Carlo Wick, and Robert Serber. For theory, it was a body blow; for experiment, something a bit less. Luckily all these men found excellent positions and were not forced to make severe personal sacrifices, except for the inconvenience of moving. None of them were communists, but they refused on principle to take a discriminatory oath.

Lawrence took a hard line, following Regent Neylan and those who demanded the oath. He did not appreciate the nature of the objections of the non-signers; to him they seemed byzantine quibbles. Grotesque episodes ensued. One involved the special pass required at the time for admission to the Rad Lab. Lawrence summoned Wick to his office and, in the presence of Alvarez, curtly asked Wick whether it was true that he had not signed the oath. Wick, taken by surprise, gave an equally curt affirmation. Looking straight into his eyes, Lawrence responded: "Then you can no longer work at the Radiation Lab." Wick, in turn, coldly offered to return his pass to the security officer, and Lawrence said, "Please do." Wick gave Lawrence the pass, and that ended the interview. Under the apparent coldness of the words, feelings ran high. It was clear to Wick that Lawrence was acting illegally, because he had no authority for withdrawing the pass, and that he was in a vulnerable position. Alvarez at once understood the implications and ramifications of Lawrence's action. He must have cautioned and calmed him down, and after a few days Alvarez called Wick on the phone and asked him to forget the whole incident. Ernest, he said, sometimes acts in a fit of emotion; please do not take what he said seriously. Wick got back his pass.

I personally sympathized with the non-signers, but I was not incensed by the controversy. My friends Jenkins and Brode, who shared my opinions on the subject, insisted that the oath and the connected excitement were transient lunacies of a type that had occurred many times in the history of the United States, and that they would pass. In any case, I signed the oath, although I thought such oaths meaningless.

I calculated that I had sworn my allegiance to king, Mussolini, party, constitutions, and institutions at least fifteen times, and I even remembered a pronouncement by Pope Pius XI, elicited by a Fascist oath, explicitly stating that under certain circumstances one could take such oaths with mental reservations that made them void.[9] I dug the papal document out in the library and translated it, and some colleagues to whom I had sent it posted it in Los Alamos, which administratively depended on the Regents of the University of California. At Berkeley it circulated less openly.

With all this brewing, it was not clear how the situation would evolve, and I thought it advisable to distance myself a little from the University of California and to wait the turn of events, prepared to go elsewhere if worst came to worst. In the meantime the non-signers had started a legal action against the Regents, which ultimately went to the supreme court of the state of California. The court ruled that the oath was unconstitutional, forced the Regents to abolish it, to reinstate the professors who had resigned or been fired, and to pay them damages. All this, however, was still in the future.

The unpleasant atmosphere created by the oath deteriorated further, for me, when Bruno Pontecorvo unexpectedly defected to the USSR. He was vacationing in Italy from his permanent post in England, when about October 20, 1950, he traveled to Finland with his family and vanished. He left no traces, but it was reasonably supposed that he had gone to Russia. I do not know the reason for his flight. One can formulate several hypotheses, few of them flattering to him. In 1940, escaping from the Nazis, Pontecorvo had sought asylum and found hospitality in the United States, thereby indirectly incurring a moral obligation toward the country. I had no information about his disappearance other than what I read in the newspapers. Although I had been Pontecorvo's friend since his student days in Rome, had helped him obtain a job at Tulsa in 1940, and had seen him at the Basel conference, I certainly was not privy to his secrets.

After Pontecorvo's flight, several officers of the U.S. government questioned me, but only as a witness, and with full acknowledgment that I had not had anything to do with his disappearance. On the other hand, G. M. Giannini, who had an interest in our neutron patents and dealt with other parties about them, telephoned to me as though he wanted to implicate me in Pontecorvo's actions, and Luis Alvarez attacked me, as I recorded in a notebook on October 24, 1950:

> Alvarez enters my office. O. Chamberlain present. Asks about Pontecorvo. Then says it is improper to ask for compensation for the Fermi patent because we came to this country and the shelter received was to us worth more than a million dollars. We are guests here and we should be glad to be able to repay in part the USA for the privilege of citizenship. I answer I did not think that citizenships were for sale. That the law fixed such things. He said that to bring a suit (for patent compensation) was like settling a quarrel by fisticuffs in a bar. I answered that a US court will not be flattered by the comparison. He concluded that I should let him know when Pontecorvo writes me from Russia.

I spoke to Birge, Thornton, and Brode about Alvarez's outburst, and all three advised to keep my cool, since nobody suspected me. On Giacomo Ancona's suggestion I also spoke to a former California supreme court justice who was his close friend. He listened intently and then advised me to do absolutely nothing and wait for the storm to blow over. Pontecorvo's flight also made trouble for Serge Scherbatskoy, his old boss at Tulsa, who was the son of a czarist general.

Much bigger and more ominous things were brewing at Berkeley, however, especially after the explosion of the first Russian atomic bomb (August 29, 1949). Lawrence grew increasingly concerned about the United States's atomic weapons, which seemed to him inadequate, and sought to improve them with various technical initiatives, which ended in serious failures.[10] I had heard of some of these plans, which seemed ill advised to me, and I decided, feeling duty bound, to speak to Lawrence on the subject, trying to give him my reasons and my numerical estimates as well I could. He reacted with great vehemence, accusing me of being unpatriotic, lazy, selfish, and God knows what more. I was not surprised and did not lose my temper. I must also add that Lawrence, apart from a few insults, did not do anything against me, left me the use of the machines, and continued his support as before.

In the meantime, Lawrence and Teller were striking an alliance for the building of the hydrogen bomb. Teller had been obsessed with it since the time of Los Alamos, and this had given rise to serious conflicts with Oppenheimer, Bethe, and others. Fermi, whose scientific prestige was paramount, was against the hydrogen bomb, but felt it incumbent upon him to inform the government about the technical situation, although he advised against building such a weapon.[11] After the development of the Soviet atomic bomb, Teller redoubled his efforts and entered an alliance with Lawrence, the Air Force, and other scientists and politicians to push the development of the hydrogen bomb.

I personally did not want to get involved in a military project that I regarded as being of doubtful usefulness. In any case the pressures exerted on me were moderate and easy to resist. Not only did I not take part in the struggles, machinations, and intrigues of this period, but I did not even know of them. I was not among Lawrence's confidants. We both knew the chasm between our ideas and it was not worth his while for Lawrence to try to convert me to his views, especially since I was unimportant in science policy. I, on my side, neither knew how to approach Lawrence nor had any hope of changing his ideas. On the other hand, I had several conversations with Teller, whom I had known well since my time at the Physics Institute in Rome. I soon realized, however, that he was dominated by irresistible passions much stronger than even his powerful rational intellect.

Students and postdoctoral fellows were in a much more difficult position than I. For them to say no to Lawrence might seriously threaten their careers, and Lawrence also had a technico-scientific ascendancy over them that he lacked over me. Moreover, although presented chiefly as a patriotic duty, war work was not devoid of financial attractions. I tried to give to my students and younger colleagues my candid appraisal of the situation as I saw it, but I always concluded by insisting that the decision must be theirs alone. All these struggles poisoned the scientific atmosphere and part of the scientific community. Too often the political authorities were deprived of cool, informed technical advice; too often they heard only confirmation of their own wishful thinking. Battles strangely similar to those of that period still persist and are a serious threat to humanity.

At that time I had frequent offers of university positions, and, considering the turmoil generated at Berkeley by the loyalty oath controversy, I decided to go to the University of Illinois at Urbana to wait things out. The chairman of the physics department and local czar at Urbana was W. F. Loomis, an able and agreeable person. He was entirely dedicated to the improvement of his department, which was then undergoing expansion, and was striving to enrich an already excellent faculty. Gilberto Bernardini had moved to the University of Illinois from Columbia, and Urbana was relatively close to Chicago, where I could visit Fermi. Bernardini and I had been close friends since our student days, and we had pleasant exchanges in and outside science, including memorable parties playing *scopone* (an Italian card game). Donald Kerst's betatron seemed to be an interesting accelerator, although it failed to completely fulfill my hopes. All this was needed to compensate for the climate and surroundings, which could not compare with those of California.

Before going to Urbana, however, I was able to spend a few months in Europe and in 1951 we went to Paris, where I renewed my acquaintance with the Joliot-Curies. I had received a Fulbright fellowship for Italy and had been invited to give the Accademia dei Lincei's Donegani Lectures. (This was quite an honor because they had been given only once before, by Fermi, in 1949.) The lectures I gave in Rome and Milan in April 1951 were later published in a booklet by the Lincei.[12]

The president of the Accademia dei Lincei, Professor Francesco Giordani, a physico-chemist, told me that the Academy would like to elect me a member or honor me in some other way. Being an American citizen, however, I could become only a foreign member, and they would have preferred to have me as a national member, with voting rights. I said that I did not mind becoming a foreign member, which happened in 1958, such being the speed at which the Academy moves. In the meantime, I was awarded the prestigious Cannizzaro Medal (1955).

On the occasion of the Donegani Lectures I also met Dr. Luigi Morandi, then vice president of the Montecatini Company, the biggest Italian chemical company, the steel magnates Falck, the publishers Mondadori, the newspaper publisher I. Montanelli, and several other Italian personalities, with whom I remained in contact.

At the end of 1951, International Business Machines Corporation offered me the directorship of a scientific laboratory it proposed to establish, which would be connected with Columbia University in New York and devoted to fundamental research without any short-term commercial application. The offer was alluring and merited serious consideration. The university connection, important for me, would have been preserved by a chair at Columbia. On the other hand, I would hardly be able to continue my nuclear work, which was too remote from IBM's interests. The salary IBM offered was about twice what I was earning at Berkeley.

"Dr. Segrè, where do you come from?" Thomas Watson, Jr., the president of IBM, asked me in the course of our discussions.

"From Berkeley, California," I replied.

"Strange, I have been to California many times, but I have never met a person who talks like you," Watson said.

We both started laughing, and I explained that I was Italian. I also went to see IBM's great research laboratory at Poughkeepsie, a trip of which I remember best the red flame of the autumn foliage against a deep blue sky, a natural spectacle I still vividly recall after so many years. During my negotiations with IBM, I started formulating some plans for the future lab and suggesting some people it would be desirable to hire. Among them were Erwin Hahn and Richard L. Garwin. I had heard of the first at the University of Illinois and of the second from Fermi, who had praised him to me in unusual terms. The two were ultimately hired by IBM, and Garwin is still there in a senior position, while Hahn has become a Berkeley professor.

To further complicate my deliberations, I also received another attractive offer, from the Brookhaven National Laboratory. While I was trying to make up my mind, importuned by these *fastidi grassi* (fat troubles), as my father would have called them, the Berkeley situation was improving. The Regents were getting a tough and deserved lesson from the California supreme court, and at the same time the Board's composition was changing, as the most extreme members were being replaced at the end of their terms by more moderate ones. In the end, having considered my options, I decided to return to Berkeley.

Not that everything there was easy. I had to wobble between the physics department and the Rad Lab. I say wobble, because on the one hand I needed the accelerators, and my work was supported by the Rad Lab, but on the other hand I had difficulties with Lawrence. What had happened in my early years at Berkeley had left its scars, and besides I did not like the little I knew of Lawrence's military activities. I grant that he was friendly and generous in giving me access to the machines, but I felt that if I had entered the circle of his close friends, I would have lost my freedom. I thus avoided going on his payroll even during the summer, while almost everybody else did. I wanted to make sure that he could not consider me as his employee. I also rather childishly avoided appearing in group photos taken on the inauguration of accelerators and similar occasions.

It was not easy for me to speak to Lawrence, as it had not been easy many years earlier to speak to Corbino, and possibly Lawrence too was uncomfortable with me. The differences between our scientific outlooks, cultural backgrounds, and political ideas were obstacles difficult to surmount. I clearly saw that Lawrence's relations with Alvarez (then, because later they changed appreciably), McMillan, and even Panofsky were different from what they were with me; they ran on a much smoother and easier basis. R. L. Thornton and Don Cooksey, two true gentlemen and close friends of both Lawrence's and mine, when necessary acted as intermediaries. It was easier for them to speak with both parties. I do not want to appear ungrateful to Lawrence, and certainly I owe him much, as I said in my Nobel speech, but the truth is that I rather avoided him. I note now that I have always called him "Lawrence" here; everybody in the lab called him "Ernest," but it took me many years and considerable effort to do the same.

Like everyone involved, Lawrence recognized that my group was doing good physics, and it did not cost him much; as a consequence my requests were always promptly satisfied, and more than once Lawrence personally intervened to assign me time I needed on the accelerators. In hindsight, I believe I should have made greater efforts to pierce the crust that separated us, and perhaps we would both have profited from it, although I am afraid a close friendship would have been unlikely.

It is not easy for me to describe my scientific work. The titles of my publications speak to physicists, but without explanations, my work is not easily understandable to lay people. I must anyway emphasize that science was by far my main occupation, and that it absorbed my time and energies. Even when I was away from the lab, I thought of physics, and I generated many ideas in places not usually connected with work. Mountain outings had always also been scientific occasions.

The general thrust of my work, from the very beginning, was to explore various more or less recondite consequences of modern theories, or to measure things thought to be important. My aims were not spectacularly inventive, like those of people who look for unexpected new phenomena, and neither were they based on development of new techniques that made new regions accessible to experimentation, although accelerators, developed by others, were mostly essential to my work. My strong points were a good knowledge of physics and a certain imagination, which enabled me to see things not immediately apparent to everybody. For many years the techniques I used were very simple, almost rudimentary, and I spoke of doing "physics without apparatus." Later, mostly thanks to Chamberlain and Wiegand, we refined known techniques and applied them with a critical eye, avoiding errors and obtaining results that at the time were the best available.

Physics strategy changed very much after the war. Many experiments are rather obvious, and the problem is only to perform them at the highest possible energy and with a clean technique. At Urbana I used to call such experiments "battleship experiments," implying a parallel between physics and naval warfare. I compared the function of Admiral Nelson with that of an admiral at the beginning of World War II. Nelson had to guess where the enemy would be, divine the weather, and maneuver his fleet accordingly, whereas a modern admiral who had guns that shot farther than his opponent's could simply hit him without danger of retaliation. Today in physics it is often possible to make discoveries simply by having more powerful apparatus—not that this counts for little, because to create such tools and to know how to use them is not given to everybody. Of course, it is also essential to know which problems are important and promise solution—the more so as the investments in time, money, and effort involved in each experiment have grown immensely. The time when a Faraday could perform several significant experiments in a single day is as long gone in physics as that of Nelson in naval warfare.

Berkeley after the war was especially suited for "battleship experiments" because its accelerators excelled both in energy and in the quality of their performance. Working conditions were very different from prewar times; with more money available it was possible to build adequate detectors, and it was easier to get access to the accelerators, because all their time was not taken up with development work or medical experiments, as often happened before the war.

Among my endeavors in "physics without apparatus," I count, in chronological order: my work on forbidden lines, and in particular on quadrupole radiation;[13] that on "swollen atoms" (now Rydberg states);[14] the finding of the new chemical elements technetium, astatine, and plutonium;[15] the chemical separation of nuclear isomers;[16] and the changing of the radioactive decay constant by chemical action.[17] On the other hand, my investigations into nucleon-nucleon collisions and antiproton work were typical "battleship experiments."[18] I have omitted the neutron work performed with Fermi and the work on molecular beams in Otto Stern's lab.[19] In the neutron work, Fermi's contribution was paramount; in the molecular beam work performed in the general framework of Stern's research, I devised the experimental trick that made that specific experiment possible, but Rabi better understood its significance.

I want to add a word on how accelerator time was assigned. The apportionment was an administrative decision, taken after consultation with user groups, which defended their requests at scheduled meetings. I found that the key for obtaining what I wanted was to show that we used the time effectively. This in turn depended upon going to the machine with well-prepared experiments and achieving interesting results. It was also very important to secure the enthusiastic cooperation of the technical staff who operated the machines. This we obtained by telling them what we were doing, why, our hopes, and the reasons for our operational requirements. This resulted in an intelligent, diligent, and generous cooperation that greatly helped the work, for which I am still very grateful.

In recent times, physics has evolved in the direction of increasing specialization and technical complication. Apparatus now costs sums unthinkable only thirty years ago; research teams have expanded to hundreds of people, the part played by computers has become preeminent and so has engineering collaboration. Physics is certainly unrecognizably different today from what it was when I started work. It requires personal qualities quite different from those once required. The chief of a team must often be an organizer and a charismatic type more than a thinker. What I have said applies especially to particle physics, but the situation in other specialties is evolving in the same direction. Chemistry has undergone a similar sea change. The whole process is connected to applications, military, industrial, and of other kinds. It is not clear where all this will lead.

In 1953 the old questions relative to nuclear power patents came to a head. I had an interest in two patents: the Rome neutron patent and the one pertaining to plutonium. The latter, however, had not yet been granted; it was still only an application.

The Fermi group applied for the neutron patent directly after the discovery of slow neutrons. As soon as he heard of our findings, Corbino recognized their potential practical importance and urged us to file for patents. We applied with the help of the attorney Laboccetta, and the Italian patent N. 324 458 was granted on October 26, 1935. It concerns a method of producing radioactive substances by neutron collisions and in particular covers the increase of efficiency obtainable by slowing the neutrons with multiple elastic collisions. Because slow neutrons are central to nuclear power production, the patent is basic to the nuclear industry. It is basic also for military applications that use both slow and fast neutrons. The Italian patent was later extended to other countries, including the United States. The inventors were

Fermi, Amaldi, Pontecorvo, Rasetti, and Segrè. We agreed however to share profits with Oscar D'Agostino and G. C. Trabacchi as well, in equal shares.

The increasingly precarious European situation prompted us from the very beginning, in 1935, to transfer our rights to an American company, in the hope that the patent might better escape a possible European catastrophe. For this reason, we entered into an agreement with G. M. Giannini, a young businessman we knew, who had emigrated to the United States. Giannini took title to the patent, sharing in the profits as one of the other partners. Patent expenses were paid by Philips of Eindhoven, which also had a share in the profits. Fermi and I tried to interest some of the big U.S. corporations, such as General Electric, but without success, although Fermi personally tried to illustrate the potential of the field to their technical bigwigs. Their reaction contrasted with the vision of Corbino and of Philips.[20]

After the discovery of fission, when nuclear energy development started in earnest, the neutron patent became obviously fundamental to all applications and hence of considerable value. This was well understood by Fermi and by myself, who were involved in secret work. On the other hand during the war it was neither possible nor desirable to raise questions about compensation for the patent. At the end of the war, there was a long period of uncertainty while Congress debated the Atomic Energy Act, which among other subjects, was expected to regulate the whole question of patent rights and compensation for their expropriation by the government.

Fermi, as the chief inventor, had the paramount voice in all decisions. After the war, in the transition period, the U.S. government bargained doggedly with Fermi through his patent lawyers. He would readily have renounced his own rights, as he had done for other most important inventions relative to the pile, but he felt an obligation to protect the rights of the Italian inventors. After all, the invention had been made in Italy, ten years earlier, and the patent had been granted long before the U.S. government had any interest in the matter. The shenanigans used by the lawyers to obstruct and minimize the "just compensation" mandated by the law ended by disgruntling Fermi to the extent that he declined reappointment to important government advisory boards on which he served (of course without compensation). I believe that the zealous government lawyers, in finding possible conflicts of interest and other technicalities, did grave damage to their client. While the negotiations proceeded slowly and laboriously, Pontecorvo's flight in October 1950 further complicated matters. Ultimately, compensation for the neutron patent was fixed at $400,000, which became a sort of standard for the most important patents. After expenses, each inventor received about $20,000.

The history of the plutonium patent is different. On request of government agencies, Kennedy, Seaborg, Wahl, and I filed a patent application on plutonium. The application covered work done by us, on our own initiative, before any government participation. If the government, ex post facto, had not asked us to file the application, we might not have done so. Later, however, the government changed its position and wanted to obtain our rights free of charge. There were years of maneuvering and negotiations on the subject, first with the Manhattan District and later with the AEC.

After the war, one did not know how the law would treat patents or patent application rights whose content was classified secret and that were of public interest, as in our case. The McMahon Act of 1946 required the government to expropriate these rights, paying "just compensation." A special committee was appointed to determine "just compensation," but it had hardly any guidelines to go by. The uncertainty extended over several years.

At one point, the University of California made claims, and these interacted even with the loyalty oath controversy. The majority of the Regents were determined to assert their authority over the faculty in every possible way, and patent rights entered, although remotely, into this picture. Lawrence, a strenuous defender of this majority, took an adversary position against the inventors. Among other things, he said he feared the effect that compensation for our patents would produce on the morale of the Rad Lab. Seaborg negotiated with great ability on behalf of the inventors; he succeeded in placating everybody concerned and very effectively protected our interests.

The AEC paid $400,000, the same as for the neutron patent. Since there were no legal expenses, each inventor received $100,000, which in those times was an appreciable sum. When the inventors settled with the goverment, R. G. Sproul, president of the University of California, wrote us a nice letter thanking us for the monies he expected we would turn over to the university. He must have been disappointed when we did not give him anything. Later, when my dear friend Joseph Kennedy, much to our grief, fell ill with stomach cancer, he wrote me a letter thanking me for insisting that the compensation to the inventors had been fully earned, and that we should spend it as we liked. The sum he had kept gave him a certain peace of mind for the future of his family.

I think the cases of the two inventions were very different. When the government entered the picture, the neutron patent had already been granted and had international validity; it was much more than an application. Moreover, it had been granted to inventors who had nothing to do with the U.S. government. I think we were vastly underpaid for it. The treatment we received as inventors from the U.S. government reflects the mindset of lawyers and bureaucrats, who believed that by squeezing the inventors as much as possible, they were properly serving the government, and who also hoped to acquire merit. They may have saved a few dollars, but how much did they lose in the advice a person like Fermi could have given the government? And what about the goodwill of many others?

The British have done better; I do not believe they have been lavish with money, but they conferred knighthoods and even life peerages on men like John Cockcroft, Rudolf Peierls, James Chadwick, William Penney, and Patrick Blackett. It is an inexpensive form of compensation, but it gives satisfaction to many. I believe that the pettiness, the jealousy, and the inclination to litigation prevailing in a democracy such as the United States are in the long run sources of weakness.

In 1953 I went to a Gordon Conference in Laconia, New Hampshire, which gave me the opportunity of visiting part of New England, and I spent the rest of the summer at the Brookhaven National Laboratory, where I found Fermi and Chamberlain. Riccardo Rimini came to New

York too, accompanying one of his wealthy Uruguayan patients, and we spent many hours together. In the same period I first met the brilliant physicist Oreste Piccioni, whose ideas on impulse approximation with virtual particles and similar subjects I was interested to hear. At Brookhaven I again worked on astatine;[21] I also did some chemical experiments, which I have never published.

Meanwhile, Clyde Wiegand and some of my students continued our experiments at Berkeley. Among the students was Tom Ypsilantis, who had studied chemistry, but had recently come to me because he wanted to change to physics. I soon recognized his human qualities as well as his uncommon scientific ability. During my absence, Tom and Clyde succeeded in polarizing the proton beam of the synchrocyclotron by collision. The method was not new; it had been theoretically predicted and experimentally demonstrated at Rochester, New York,[22] but Ypsilantis succeeded in obtaining superior results and started the exploitation of polarized protons, opening up new possibilities to the study of nucleon-nucleon collisions. The success obtained and Ypsilantis's spirit of initiative impressed me, and I proposed a faculty appointment for him. He was one of the most promising young physicists at Berkeley, where he continued to do brilliant work for several years. Unfortunately, his bright flame did not last long; some demon, still unidentified by me, attacked him. He lost some of his drive and later sought to resign his Berkeley post. His friends on the faculty, myself among them, tried to persuade him to reconsider the decision and ask for a year's leave of absence instead. The following year, however, he insisted on resigning. Thereafter he held several positions, but he ended by achieving less than his great potential had seemed to promise. Ultimately, he moved to CERN in Europe and elsewhere. He is an exceptionally agreeable and gifted person and one whom I sincerely love.

In February 1954, during a visit at Chicago, I talked at length about our group's polarization work with Fermi, my last serious scientific conversation with him. Fermi developed the formulae at the blackboard while I took notes, and he subsequently wrote a paper on the subject.[23]

Chapter Ten

Triumphs and Tragedies (1954–1982)

Odor of Laurel and Cypress

Ehret die Frauen! Sie flechten und weben
Himmlische Rosen ins irdische Leben.

Honor to Woman! To her it is given
To garden the earth with the roses of heaven.
Schiller, "Würde der Frauen"[1]

I had been invited more than once to lecture in Brazil, in part through G. C. Lattes, who had helped detect the first pions formed by the Berkeley synchrocyclotron. In 1954 the time seemed ripe for a visit to Rio de Janeiro. Our children were too young either to take along on such a trip or to leave alone, so I went to Brazil by myself that July while Elfriede stayed behind with them in Berkeley. When the Jenkinses heard of our predicament, however, they offered to take care of the children for a while, giving us yet another reason to be grateful to them, and Elfriede was able to join me in August.

Brazil fascinated me. I am hard put to describe, let alone explain, my feelings toward my exotic, but at the same time almost familiar, new surroundings. The novel tropical beauty, nostalgic reminders of the colonial period, my affection for Dom Pedro and his well-ordered empire, and the behavior of the people formed a mix most agreeable to me. We also liked the food, the exotic fruits, and the many kinds of bananas. Furthermore, we found excellent company: Georg von Hevesy, with whom we often spent the morning walking on the Copacabana beach; G. P. Thomson, who was lecturing at the same institution as I was, and Lattes himself. Almirante Alvaro, chief of the Conselho de pesquisas físicas, entertained us at the Bosque Tijuca, where he planted a tree in my honor and recited poems by Camoëns.

Guido Beck, an Austrian physicist whom I knew from my time in Rome, helped immensely in guiding me in the new strange world of Latin America. Among other things, he found a way of getting us a visa for Peru, which we did not know we needed. When we asked for it, much to our surprise, Elfriede was refused and declared "peligrosa a la seguridad nacional del Peru." It was because she had been born at Ostrowo, which in the meantime had become part of Poland and thus was behind the Iron Curtain. Our friend ultimately succeeded in obtaining the visa through personal intervention, but not without a few comic scenes. A Peruvian professor helped me materially, and I ended up becoming an honorary professor of San Marcos University in Lima.

From Brazil we went to Uruguay, where we stayed with my dear cousin Riccardo Rimini, and then to Argentina, which was under the Perón dictatorship. The slogan "Perón cumple y Evita dignifica" (Perón delivers and Evita dignifies) was everywhere, evoking somewhat cynical comments, which we naturally kept to ourselves. In Argentina we visited several Italian émigrés, some physicists, some not, whom we had known before our own emigration.

The last leg of our trip took us to Peru, a country we found extremely attractive owing both to its peculiar natural beauty and to its Indian culture. We were wise enough to allot sufficient time for sightseeing to allow us to gain a real impression of this beautiful world, so different from anything we had seen before. We flew from Lima to Cuzco in an unpressurized airplane. We wore oxygen masks but I took mine off in opening a window to snap a picture of the Andes. I barely made it back to my seat, where I fainted briefly, while Elfriede put the mask back on my face. We remained at Cuzco for a few days to acclimatize ourselves to the altitude and see the Inca monuments, then descended to Macchu Picchu, where we spent the night. Next day we climbed Vaina Picchu by the very steep stairway hewn from the rock by the Indians, under a tangle of orchids.

On October 8, 1954, shortly after our return, while I was resuming my regular routine, I had a telephone call from Chicago. The caller was Sam Allison, and from his tone of voice, I realized at once that he had very bad news. From his almost incoherent words, I gathered that Fermi had been operated on shortly before, and that the surgeon had found an incurable stomach cancer. I was stunned. When I had seen Fermi in February, I had noticed that he looked a little tired, but it did not cross my mind that there was anything to worry about. During the summer, Fermi had gone to Italy and we to South America; we had not been in touch. In Italy, he had begun to feel seriously ill, and as soon as he returned to Chicago, he went to Billings Hospital. The first doctor who saw him, an intern, did not make the correct diagnosis, but the chief surgeon shortly thereafter performed an exploratory operation and found a hopeless situation.

As soon as possible after hearing the news, I caught a plane to Chicago. I found Fermi at Billings Hospital, fed by a tube that ran directly into his stomach. The patient was measuring the flow of the fluid by counting the drops, using a stopwatch, as though performing a physiology experiment. He was perfectly aware of his condition and started talking about how many months or weeks he might survive, and what he would do in the short span still allotted him. He asked me to summon Edward Teller to see him, adding with a slightly ironical smile, "What nobler deed for a dying man than to try to save a soul?" Fermi thought that Teller's behavior in connection with the hydrogen bomb and in the Oppenheimer hearings had been reprehensible—among other things, it had split the scientific community into factions—and he wanted to make him realize this. "The best thing Teller can do now is to shut up and disappear from the public eye for a long time, in the hope that people may forget him," he added. Needless to say, as soon as I got back to Berkeley, I relayed the summons to Teller, who subsequently went to visit Fermi. Teller has given his own report of this visit.[2]

Fermi then spoke pessimistically about the world's future. Atomic bombs were making possible the destruction of civilization. All it would take for them to be used was for a madman to come to power in a great nation. Since this happened every few centuries, he reckoned that civilization might, with luck, last roughly that long. He said, too, that if he lived long enough and had the strength to do so, his last service to science would be to write down his lectures on nuclear physics, which were preserved only in the form of notes taken by students.[3] This was, in fact, his last effort at scientific writing. In a lighter vein, he told me that he had been blessed by a Catholic priest, a Protestant pastor, and a rabbi. At different times the three had entered his room and demurely and politely asked permission to bless him. He had given it. "It pleased them and it did not harm me," he added.

We spent several hours talking about various subjects. Among other things, Fermi observed that since his wife, Laura, had just finished her book *Atoms in the Family,* his death would come at the right moment for promoting it, and that he hoped the literary success he anticipated for it might help her overcome the difficult times she faced.

At the end of the afternoon I left. When I got out of the hospital, I felt ill; the emotional upheaval produced in me by the visit was too much for my constitution. I could scarcely stand, and I remember going into the first bar I came across to fortify myself with a cognac, something exceedingly rare, perhaps even unique, in my life. I returned to Berkeley gravely upset, and as soon as possible I went back to Chicago. I found Fermi much worse and in a more somber mode. He spoke of his sufferings and of other subjects I will omit. We talked until late in the evening. During the night, I was awakened by a phone call announcing that Fermi had died. It was November 29, 1954.

I stayed in Chicago for the memorial service at the University of Chicago. Searching for an appropriate text, the university chaplain proposed several that did not seem right. Finally, he suggested St. Francis's "Cantico delle creature" ("The Song of Brother Sun and All His Creatures"). It seemed to fit the occasion, and he used it.

Fermi's unexpected and premature death shook me deeply. Even now Fermi often appears in my dreams. In their grief, his former pupils and friends sought an appropriate memorial. Ultimately, the idea of publishing his collected papers crystallized. The Accademia dei Lincei and the University of Chicago Press undertook to do so and appointed me chairman of the editorial committee.[4]

In 1955 we moved from Berkeley to Lafayette, a suburban community about ten miles east of Berkeley, behind the coastal hills. Our new house was on a dead-end road on a hill; the address was 36 Crest Road. It had been part of a large estate, had a beautiful view, and was in a most attractive natural setting.

During my lifetime I came to love three homes in particular. The one in Tivoli, the Treves villa at Marignolle, and our house on Crest Road in Lafayette. Naturally, I remember the others—229 Corso Vittorio at Rome, my apartment in Palermo, 1617 Spruce Steet at Berkeley—but I do not have a special attachment to them, and I do not dream of them at night.

I loved the Tivoli house where I was born, because I spent my childhood there. Conflict with Marco separated me from it, but much more serious has been the deterioration of the surroundings. The Tivoli of my childhood does not exist anymore, and even if the walls of the house are still there, all the rest is gone: landscape, roads, gardens, neighborhood.

Marignolle was never my home except during summer visits. The place struggles to survive the changes in the Treves family. Those of my generation have almost become slaves of the villa, while those of the next will be hard put to maintain its spirit. Irresistible forces have transformed the agriculture, the society, and even the face of Tuscany.

Geology drove me from 36 Crest Road, the house I loved best in the United States. When I bought it, I did not realize the serious geological problems affecting the location, nor the importance of a lower lot, which was also for sale at the time, to its stability. I never thought that somebody might want to build on that lot, because it was manifestly foolhardy to do so. The surroundings of the place were also different from what they are now. There was no Highway 24, and the area now occupied by a church was a beautiful meadow with a few old oak trees. The church in due course destroyed the beautiful setting to create parking lots and make money, which convinced me that it was an enemy of God and His works. Furthermore, in 1956 somebody bought the lot below my house mentioned above, and without my becoming aware of it, obtained a building permit. When I saw the new owner of the lot excavating the slope in an obviously dangerous way, I warned him and took photographs of the terrain, but I did not start a legal action, which I would possibly have lost. During the winter of 1958, heavy rains caused a slide on the excavated slope.

This neighbor, in 1960, sued us because the slide he had provoked had damaged his house! We countersued him, and he lost, but the damages awarded to us paid only part of the cost of a retaining wall we had to build to stabilize the slope; nor did they compensate for the depreciation of our property, not to mention the time lost and the anguish caused to us. In practice we never succeeded in repairing the damage satisfactorily. Ultimately, in 1978, we sold the house for much less than it would have fetched without its history.

I grasped too late the type of house I would like; furthermore, my ideal home has contradictory elements in it and hence may not exist. I like a country setting, but I never really enjoyed gardening; I like a well-finished house, but I am not a handyman. All told, Adalbert von Chamisso's poem "Schloss Boncourt," mourning the demolition of his childhood home, perhaps best reflects my feelings:

Ich traüm' als Kind mich zurücke
Und schüttle mein greises Haupt;
Wie sucht ihr mich heim, ihr Bilder,
Die lang ich vergessen geglaubt!

Hoch ragt aus schatt'gen Gehegen
Ein schimmerndes Schloß hervor;
Ich kenne die Türme, die Zinnen,
Die steinerne Brücke, das Tor.

Es schauen wom Wappenschilde
Die Löwen so traulich mich an,

Ich grüsse die alten Bekannten
Und eile den Burghof hinan.
. . . .
So stehst du, o Schloß meiner Väter,
Mir treu and fest in dem Sinn,
Und bist von der Erde verschwunden,
Der Pflug nun über dich führt.[5]

Back to physics! The problem of the existence of antiparticles arose in 1928 with Dirac's relativistic theory of the electron. This theory gave solutions that corresponded to a then-unknown stable particle of the same mass and spin, but opposite charge and magnetic moment, in other words a positive electron. This particle was called the antiparticle of the electron, or positron. At the time of Dirac's prediction, the positron was unknown, and its absence was considered a serious flaw in Dirac's theory. Dirac, as a last resort, tried to identify the positive electron with the proton, but this proved untenable.

Things changed radically with C. D. Anderson's discovery of the positron in cosmic rays in 1932. (The prediction of the positron is one of the triumphs of Dirac's theory.) The notion of antiparticles was generalized into the postulate that every particle has its own antiparticle. In the case of neutral particles, particle and antiparticle may coincide.

The extension of Dirac's theory predicting antiprotons was very plausible, but not certain. Furthermore, most physicists were surprised when, around 1931, Otto Stern measured the magnetic moment of the proton and found it to be very different from the naive theoretical expectation based on a literal extrapolation of Dirac's theory. This result suggested caution in generalizing from Dirac's theory. Even in 1955 at least one distinguished physicist had bet money against the existence of antinucleons. For many years, experimental physicists had looked for antiprotons in cosmic rays, with inconclusive results. Among others, Bruno Rossi and his collaborators, using a cloud chamber, and Edoardo Amaldi and collaborators, using photographic emulsions, had observed particles in cosmic rays that may have been antiprotons. Their observations were not, however, sufficient to establish the particle.

In planning the bevatron, Lawrence and the Rad Lab physicists had consciously chosen as a goal an energy of 6 GeV, slightly above the threshold for the formation of nucleon-antinucleon pairs from a proton colliding with a nucleon at rest. In 1955 the bevatron reached this design energy and thus afforded the opportunity of proving the existence of the antiproton unequivocally, and we wanted to settle the question once and for all.

Several Berkeley groups started the hunt. My group had for some time studied the problem and prepared for it. I decided to attack the problem in two ways. One was based on the determination of the charge and mass of the particle. The other concentrated on the observation of the phenomena attendant on the annihilation of a stopping antiproton. The stopping antiproton and a proton of the target should mutually annihilate each other, and the rest mass of the two particles should transform itself in one of many possible ways into other particles such as pions. These would leave tracks in a photographic emulsion and the annihilation would thus become evident.

For the first attack, Chamberlain, Wiegand, Ypsilantis, and I designed and built a mass spectrograph with several technically new features. For the second attack, Gerson Goldhaber, who was then in my group, exposed photographic emulsions in a beam enriched in antiprotons by our apparatus. Many other people were involved in the enterprise, and we had agreements on how to publish the results and give appropriate credit to everyone. The proper working of the bevatron under Edward Lofgren was of paramount importance. We were in competition with physicists of other groups trying to detect antiprotons at the same time, but this did not prevent frequent mutual help.

We started the run on August 25, 1955, and after a few days of tuning up, we began observing antiproton signals. We based the identification on measurement of the velocity, momentum, and charge of a particle. The signals for velocity were oscilloscope traces recording the passage of a particle through a velocity-selecting Cerenkov detector, corroborated by a measurement of the same particle's time of flight between two detectors. The trajectory followed by the particles gave their momentum and the sign of their charge. Velocity and momentum determined the mass of the particles and this, combined with the sign of their charge, identified them as antiprotons. We also checked, among other things, that protons below the threshold energy did not produce our signals.

We detected about one antiproton for every few hundred thousand other particles crossing our apparatus, and the good signals arrived with a frequency of a few per hour. Naturally there was considerable enthusiasm in the laboratory, and many people came to see our progress. So as to be able to work undisturbed, we wrote up a bulletin of our results on a blackboard. In the meantime, we had to think about writing a paper and there were also delicate questions of the order of names and formulation of the text to consider. We decided to write a letter to the *Physical Review* and an article for *Nature*,[6] to which I had written every time I had something important to say, or at least something I thought was important. We listed the authors in alphabetical order, as we had done in most of our many common papers before. Some original pieces of the apparatus, such as the Cerenkov velocity selector, were later described in greater detail by Chamberlain and Wiegand.[7]

I had no doubt that *antiproton* was the right name for the new particle. Lawrence preferred *negative proton*, but he did not insist. The mass-spectrograph experiment concluded on October 1, 1955, having proved the existence of the antiproton, and soon thereafter the emulsion work confirmed it.[8]

At that time the physicist Oreste Piccioni wrote a scathing letter to Lawrence accusing us, and me in particular, of several misdeeds. Lawrence looked into Piccioni's accusations and dismissed them. Piccioni had made some good suggestions during the planning of the experiment, and these were duly and repeatedly acknowledged in publication. This was his pretext for starting a legal action against Chamberlain and me eighteen years later, in 1972, in which he maintained that we had stolen his ideas. The complaint went all the way up to the U.S. Supreme Court, but all the courts, from the Alameda superior court to the Supreme Court, refused to hear the case, because the statute of limitations had run its course.[9]

One day when I was complaining about the aggravation caused to me by Piccioni, a famous physicist who was an old colleague of his kept exclaiming, "Poor Oreste! Poor Oreste!" I resented this and demanded, "Why poor Oreste and not poor Emilio?" To which he promptly answered: "No; poor Oreste and not poor Emilio, because Oreste is crazy, and you are not!"

At the time of the antiproton experiment, Amaldi and his wife Ginestra were at our home in Lafayette as our guests. He and I established a collaboration for the study of photographic emulsions exposed at Berkeley, taking advantage of the numerous well-trained scanners available in Rome. When Amaldi returned to Italy, some Italian newspapers wrote inappropriate comments and tried to ascribe to him a part he had not played. This misreporting could have had unpleasant consequences, but Amaldi set things straight and we kept calm. The experiment was widely acclaimed and soon we, and Lawrence, started receiving numerous compliments on it. Lawrence politely answered those addressed to him with a form letter saying that he had passed the congratulations to Chamberlain, Segrè, Wiegand, and Ypsilantis, the people directly involved. Shortly afterward, another group in the lab, including Piccioni, observed the antineutron, obtaining it by charge exchange from the antiproton.[10]

Once we had discovered the antiproton, we obviously wanted to know the properties of our particle and build on our initial success. Some "battleship experiments," such as cross-section measurements, were possible, and we performed them, but our antiproton source was weak and it was soon surpassed by other accelerators that were coming on stream. With our means we could not do much more than what we had already achieved. However, in collaboration with Wilson Powell's group, which had a propane bubble chamber, we obtained some good pictures of antineutrons obtained from antiprotons by charge exchange,[11] and using photographic emulsions we started developing statistical information on the annihilation process. In the meantime, Alvarez's group had developed the hydrogen bubble chamber, and I proposed a collaboration, but he felt we did not have an adequate contribution to offer and demurred. Soon the Alvarez group, using their hydrogen bubble chamber, started obtaining capital results. Bogdan

Maglic, a Yugoslavian postdoctoral fellow, pioneered in detecting the first resonance between annihilation pions, and this was the curtain raiser for a whole series of brilliant investigations.[12] By now (1986) there are accelerators forming beams of antiprotons and using proton-antiproton collisions in great storage rings. This is a measure of the pace of progress in particle physics.

Theoreticians had speculated on the rho and omega mesons, and members of my group tried to see them experimentally with a big new instrument, planned and developed chiefly by Ypsilantis and Wiegand, which we called the "fly eye" because it contained many scintillators that formed a sort of big compound eye. The technique was state of the art, and this work contributed, in a small way, to the discovery of the rho meson.[13]

In 1955 the discovery of the antiproton reopened the possibility of my winning a Nobel Prize. After the war I had started thinking that my work on the new chemical elements and on radiochemistry might bring me that distinction. I saw Seaborg's efforts at getting it on similar grounds, but I did not know how to stake my claim. I hoped that the Nobel Committee would somehow split the award. A poll among members of the Chicago section of the American Chemical Society in 1947 had chosen me as one of the ten best radiochemists in the United States.[14] Lawrence, too, as I found out many years later, considered me a good candidate.[15] "Contrari ai voti poi furo i successi" (Events turned out contrary to hopes; Ariosto, *Orlando furioso* 1.9.5); while I was at a cocktail party at Donald Kerst's house in Urbana in October 1951, I heard that the Nobel Prize for chemistry had been given to McMillan and Seaborg "for their discoveries in the chemistry of transuranium elements." I was deeply disappointed.

During the summer of 1954, I met Hevesy in Brazil. We were friends and I could speak freely to him. Thanks to his Swedish connections, he knew many of the secrets of the Nobel Committee, and he told me that I had not been specifically nominated in the year 1951, which had automatically eliminated me. He advised me to try to interest Fermi. I did not do so because I knew perfectly well that Fermi could not be influenced in matters such as competitions and awards.

However, a few years later, after Fermi's death, his widow, Laura, asked me to look at her husband's papers before she gave them to the Regenstein Library at the University of Chicago. In so doing, I found out, to my surprise, that both Fermi and James Franck had proposed me repeatedly for the Nobel Prize in chemistry. I saw also that Fermi had proposed, in physics, Maria Mayer, Hans Jensen, and Wolfgang Panofsky. His spontaneous proposal deeply moved me, for the same reasons that had prevented me from asking for his support. Nomination by him was, for me, almost as important as getting the prize. Much later I had the opportunity to tell Mayer, Jensen, and Panofsky that they had been nominated by Fermi, and all three had the same reaction. Of them, Mayer and Jensen had had the prize. Panofsky had not.

The discovery of the antiproton had some unpleasant consequences for the structure of my group and for relations between its members. Owen and Clyde, who were charter members of the group, developed most of the electronic detectors and counters of different kinds. Gerson Goldhaber, the group's expert on photographic emulsions, was recruited by me in the early 1950s at Columbia University, where he had studied with Gilberto Bernardini.

After the discovery of the antiproton and connected publicity, the moods of Owen and of Clyde separately darkened. Owen wanted to be more independent than he already was, which was hardly possible. He wanted to have his own group, but our group was so small that I felt further splitting would seriously impair its efficiency. Owen was then invited to go to Harvard, where he spent a period as a Loeb Professor; on his return, he started a small separate group. Clyde, too, wanted to go it alone, and above all to work independently of me and of Owen. Perhaps he wanted to show his personal prowess, although his ability was widely recognized, above all by me and by his other colleagues in the group. It is possible that even Ypsilantis had similar wishes, but being younger, at the beginning of his career, and of a sunny disposition, he was less affected.

In my opinion, the strength of our group came from the combination of different talents. Nobody could dominate by his obvious and disproportionate superiority, as had been the case with the various groups led by Fermi. As things were, I firmly believed that fragmentation or dissolution of our group would damage us all and impair our scientific output. No one else among us had Owen's critical mind, Clyde's technical ability, Tom's enthusiasm and optimism, and so on. Nor did I think that my contribution was as negligible as it perhaps then appeared to Owen and Clyde. One element of discomfort was the fact that both had been my students and co-workers for over fifteen years; the problems to some extent resembled those that arise between fathers and sons.

I thought that for me the best course was to give broad autonomy to the younger members of the team and try to aid their personal initiatives as much as possible. Some of these initiatives went well, some were less successful. Our group was too small to compete with the much larger groups then entering the field of particle physics, but to enlarge it greatly did not suit my modus operandi.[16]

A few years after we received the Nobel Prize and Owen seceded from the group, he changed his mind and, to my great joy, rejoined us. I expected that, being fifteen years younger than I was, he would in time succeed me as head of the group, with Ypsilantis as second in command. This happened for Owen, but unfortunately Tom left Berkeley before he was offered the opportunity.

Of the experiments we did after the discovery of the antiproton, I have already mentioned the one on the rho meson. Others, such as the pion beta-decay experiment,[17] were successful, but took much time, above all because the authors, in our tradition, properly insisted on measurements of superior quality. Much later Clyde Wiegand continued excellent experiments on mesic atoms on a small scale with a few students.[18]

A few months after the antiproton work, in the spring of 1956, I unexpectedly received a telegram from the secretary of the Soviet Academy of Sciences inviting me to an international science conference soon to be held at Moscow. A few hours later, a similar telegram reached Owen Chamberlain. Lawrence, whom I consulted, objected to our going, mainly for political reasons. I thought otherwise and decided to accept the invitation, but I had to maneuver a little to avoid a direct clash with Lawrence. Our invitation was perhaps the first to come from the Soviet Union, and it arrived at a time when scientists, especially those who had been at Los Alamos, were considered privy to "atomic secrets," and when very few Americans had visited the Soviet Union. After a few days, invitations also arrived for McMillan, Alvarez, Panofsky, and others, so that it became difficult for Lawrence to thwart so many people eager to go.

The trip to the Soviet Union lasted about six weeks, and besides Moscow and Leningrad, we also went to Armenia. It was some time after the famous Khrushchev speech revealing Stalin's crimes. The Russians did not know its text, but large excerpts of it had appeared in the Western press, and our hosts asked us about it. My impressions of Russia are too superficial to be of value. We were obviously favored guests—witness our advantages, such as tickets to superb ballet shows, our priority in visiting the Kremlin, and similar privileges. To everybody's surprise, including mine, I even succeeded in obtaining payment from the Bureau of Foreign Translations for the translation of the first volume of *Experimental Nuclear Physics,* which I had edited. When I asked for royalties or compensation, adding that I would appreciate payment in U.S. dollars, the Russian bureaucrats were nonplussed, and answered that they would ask their superiors, and that I should return in a couple of days. When I returned, the answer was that the superiors had to ask still higher authorities and that I should return in a couple of days. I doubted anything would come of it. However, after three days, when I returned to inquire, I was told that the request had been granted and that I would receive the money in New York. I could scarcely believe my ears.

On this trip I saw Bruno Pontecorvo again for the first time since his defection. He was so little Russified that at the conference the Russians present told him to speak English, because they had difficulties in understanding his Russian. Some of his colleagues treated him as a "Party zealot."

The person that impressed me most among the scientists I met was Igor Tamm, who subsequently shared the Nobel Prize for physics in 1958. I immediately liked this cultivated and refined gentleman's warm personality. He was also obviously a courageous person who, although he dearly loved his country, did not hesitate to help it by saying what he thought true and fair.[19] L. D. Landau seemed to me very arrogant; he reminded me of Oppenheimer, although of greater ability as a physicist. I also saw Peter Kapitza, whom I had known at Cambridge in 1934, as well as his son, whom I had then seen in a cradle, but who now looked like his father twenty-two years earlier. Among other able physicists, I met the Alikanian brothers, Pavel Cherenkov, J. A. Smorodinsky, D. D. Ivanenko, I. P. Nikotin, Nikolai Bogoliubov, and Vladimir Veksler for the first time.[20] The small fry were scared to mention even the most innocent subjects. When I asked a young chemist working on technetium what he was doing, he answered evasively and referred me to his superiors; when I pressed him to say what he had in a test tube he was holding, he said disconcertedly that he did not know! Furthermore, in the laboratories, I noted doors sealed with wax seals, as if there were great secrets behind them.

Later we flew to Armenia; during the flight we passed over the Turkish-Russian border, where one could see an abundance of military airfields. I asked for permission to take pictures, and the guide who accompanied us assented freely. I was surprised, but took the photos. Years later I came to think this may have been foolhardy on my part.

In Armenia we climbed to a high altitude observatory. A snowstorm trapped us inside the observatory, and the Armenians, feeling at home, started freely expressing thoughts that at sea level and among Russians would have been dangerous. I also saw ancient churches and monasteries, in one of which an old priest took me aside and bitterly complained to me in French about the negligence of the authorities, who did not provide the necessary funds for the preservation of the monuments of the past.

I returned twice to Russia, the last time for the centenary of Mendeleyev's table of the elements in 1969. In 1957, some of the Russian scientists we met returned our visit by coming to Berkeley, and I invited them to our home. On the way there, as luck would have it, we had a blowout, the only time this had happened to me in thirty years. The Russians were amused by this failure of American technology and chuckled freely, but fortunately a colleague of mine, the physicist Herbert Steiner, was in the car. As a student, he had worked in a gas station and he showed the Soviet visitors the speed with which one changed a tire in the United States. The guests were impressed. Unfortunately, the Russians permitted by their authorities to visit us were few, always the same, and often not those we were most eager to see.

In 1957 Tsung-Dao Lee and Chen Ning Yang proposed the nonconservation of parity in weak interactions. Very crudely, the nonconservation of parity means the following: if one performs an arbitrary experiment—for instance, if one observes the disturbance of a magnetic needle by an electric current—and one looks at the experiment or at its image in a perfect mirror, and there is no way of telling which is which, parity is conserved; on the other hand, if it is possible to tell apart object and image, parity is not conserved. In all experiments performed up to 1957, parity seemed to be conserved.

In the same year, the tau and theta meson decays (now they are both called K mesons) showed a peculiarity. The particles have the same lifetime and the same mass but decay in final states of different parity. Lee and Yang suggested that they were one particle with two different modes of decay. There are many examples of dual decay, but the difficulty in this specific case was that the decay to two states of different parity necessitates a parity change in the decay. This had never been seen in electromagnetic or strong decays but had not been ruled out experimentally in decays by weak interaction.

Lee and Yang pointed this out. Chien-Shiung Wu and her colleagues at the National Bureau of Standards showed, in a case of beta decay, that the Lee-Yang hypothesis was correct: parity was not conserved. Within a few days this surprising result was extended to muon decay by Richard Garwin, Leon Lederman, and Gabriel Weinrich at Columbia University, and it turned out that Valentin Telegdi at the University of Chicago had previously had indications of the same phenomenon. The sensational discovery removed an old and firmly established prejudice and opened new horizons to the theory of weak interactions. Everybody rushed to work on the subject, with an eagerness reminding me of that following the discovery of fission in 1939, or of high-temperature superconductivity in recent years.[21] Fermi may have had some thoughts on the subject; he had occasionally cryptically remarked to me that nobody had ever inverted space (like a glove), transforming a left hand into a right one, but he left no written document of what he had in mind. I was deeply interested in the discovery of parity nonconservation and tried to read and understand the new papers on the subject that flooded the literature. Less agreeably, from a narrow and selfish point of view, I realized at once that the new discovery postponed the possibility of my winning the Nobel Prize; I was sure Lee and Yang would have priority. Tough luck, but there was nothing I could do.

In October came the announcement of the awarding of the Nobel Prize in physics to Lee and Yang. Nobody was surprised; it was an almost perfect opportunity to follow literally the wishes of Alfred Nobel as expressed in his will. I was curious whether the sages of Stockholm would also include C. S. Wu, but they did not. Many years later I was pleased when she won the important Wolf Prize.

In July 1958 I went to Geneva for an international scientific conference at CERN. At the same time there was a disarmament conference between United States and the USSR. Lawrence and Panofsky were among the American experts; Igor Tamm, whom I had come to know in Moscow, among the Russians. I met Tamm on the street and, knowing that we both liked hiking, I suggested we hike Mount Saleve, in France. He answered that he could not because he did not have his passport. Foolishly thinking that he had simply left it at his hotel, I said, "Let us go and fetch it." Tamm then explained to me that he did not have his passport because on his arrival the Soviet consul had impounded it. I blushed at my lack of tact in asking, and I am still amazed at a country that would take away the passport of one of its important delegates at an international conference, and that of a man of Tamm's stature.

I saw Lawrence only in passing. While he was in Geneva he had a serious recurrence of a colitis that had long afflicted him. He returned to Berkeley and went to Stanford University Hospital. I was worried by his condition and looked in a medical manual for information about his illness. I found that it is an insidious disease with acute periods alternating with remissions. The patient used to this cycle may delay an operation too long, until it becomes dangerous. I had read this when I heard the sad announcement of Lawrence's death on August 27, 1958, in circumstances similar to those described in the manual.

Lawrence was an intense, impulsive, optimistic, and very active individual, more a doer than a thinker, and a born leader of men. He was full of contradictions, which made him unpredictable. His personality was fundamentally generous and magnanimous, but he could occasionally be petty. His optimism and enthusiasm, basic ingredients to his success, led him sometimes beyond where he should have gone as a scientist. He enjoyed life to the full and drew great satisfaction from his scientific successes and those of his associates, but he also pursued childish ambitions of consorting with rich and powerful people. His political activity, the dark side of his life, is scarcely known to me. In his youth he started as a liberal in the midwestern tradition of Robert La Follette, as one might expect given his family origins; but he ended as a reactionary. I personally am grateful to him for the help and the opportunities he gave me.

At the beginning of 1957, Seaborg told me that he thought Lawrence's nomination would be indispensable for the awarding of the Nobel Prize to anyone working in the Rad Lab. I said that I would not speak to Lawrence on this subject, but that if he, Seaborg, would do it, I would be grateful. A few weeks later, Lawrence's secretary, without a word, showed me, on Lawrence's orders, a letter from Stockholm acknowledging receipt of my nomination by Lawrence. No word on the subject passed between us.

Also in 1958, I was given the Hofmann medal of the German Chemical Society. At first I was uncertain whether or not to accept it, given recent German history, but I decided, I believe correctly, to do so. It was a high distinction, and I liked being recognized by chemists. I went to Wiesbaden at the end of September 1958 to receive the medal at the meeting of the Gesellschaft Deutscher Naturforscher und Ärzte. It is a big affair, with scientists of all specialties and also philosophers. I spoke, in German, on a subject I was interested in—that is, on systems similar to atoms, but constituted of particles different from electrons and ordinary nuclei, something on the borderline between spectroscopy and chemistry.[22] At Wiesbaden I found Otto Hahn, with whom I renewed an old friendship. We sat next to each other in the front row at a conference addressed by the philosopher Karl Jaspers, who had attracted a huge crowd. The speaker was rather theatrical and, I thought, tried to look like the old Goethe, but I did not have the impression he said much. Hahn had fallen asleep, but at a certain point Jaspers started attacking science and scientists. Exactly at that moment, Hahn woke up, turned to me, and said: "By *scientists,* he means you and me."

The death of Lawrence necessitated the appointment of a new director for the Radiation Laboratory. The obvious choice was Edwin McMillan. He belonged to Lawrence's old guard, he was an eminent physicist who had greatly contributed to the laboratory's success, he was a distinguished accelerator's expert, and he was well liked by most of the personnel. He lacked Lawrence's charisma, but that could hardly be duplicated.

Under McMillan, the laboratory changed its name, becoming the Lawrence Berkeley Laboratory (LBL), while the laboratory in Livermore became the Livermore Lawrence Laboratory (LLL). The administration became less capricious than it had been under the creator of the lab, but also less enthusiastic and more bureaucratic. McMillan's assignment was tough. The unavoidable comparisons with his predecessor and the implacable personal hostility of Alvarez, who antagonized him constantly, added to the difficulties.

I spent a good part of 1958 in Rome as a Guggenheim Fellow. In applying for the fellowship, I gave Franco Rasetti as one of my references; a couple of weeks later, I received a letter from the Foundation asking me for a letter of recommendation for Rasetti, who had also applied and had given my name as a reference. The accident was comic, but also embarrassing, and I wrote to the Foundation explaining our predicament and innocence of collusion in the matter. We both received fellowships, and later I served for many years as a consultant to the Foundation.

I devoted the time of my fellowship to the preparation of the two volumes of Fermi's collected papers. Since he had died at such an unfortunately early age, there still were many witnesses who were able to write valid historical introductions to individual papers. I took responsibility for the organization of this work. I felt an obligation, having been close to Fermi in Italy as well as in America, and I remembered, without wanting to make ridiculous comparisons, that Maxwell had edited Cavendish's collected papers and Marie Curie those of her husband. The job required considerable time and effort even with the help of other members of the editorial committee; Amaldi, Anderson, Persico, Rasetti, and Wattenberg carried a substantial part of the load, as did several outsiders. The biographical introduction I wrote for Fermi's collected papers served me as the basis for his biography, which I published ten years later.[23]

On January 19, 1959, the University of Palermo gave me an honorary degree, which I highly appreciated. On the occasion I visited Palermo for the first time since the war. The city had greatly deteriorated and the elegant surroundings of our house at the beginning of Viale della Libertà had badly decayed. I saw again the Istituto fisico in Via Archirafi and several of my old Sicilian friends. After our stay in Palermo, we toured Sicily once more and visited Syracuse, which we did not yet know.

A little later, in February, I had to go to Scandinavia. I was invited to visit Bohr's Institute in Copenhagen, and to Stockholm and Oslo to lecture for Nordita, an association of Scandinavian universities. I was thus able to see Bohr again, as well as the two Siegbahns, father and son,[24] Oskar Klein and other colleagues. I had a friendly reception everywhere, and I suspected that I had not been invited solely to show me the wintery attractions of Scandinavia.

On my way back I stopped at Hamburg, where I lectured on February

16, 1959, on the invitation of W. Jentschke, a physicist I had befriended in Urbana in 1952 and who later became director of CERN. I spoke at the old Stern Institute, where I had worked about thirty years earlier. In the audience were some professors who had been Stern's assistants; I knew they had become zealous Nazis under Hitler and avoided them.

By chance I had read a newspaper advertisement placed by a detective who specialized in locating people. Out of curiosity, I wrote to him a few days before my arrival in Hamburg, asking him to find my old girlfriend, "I." All I had to go by were her maiden name and her address in the 1930s, but for a very modest fee the detective supplied me with her married name, address, and telephone number, as well as a description of her husband, his profession, and their financial and family situation. Immediately after the detective left my Hamburg hotel, I called her number. She answered herself, and I did not have the impression she was excessively surprised. We made an appointment to see each other the next day, and when we met we spoke about some of the past and little of the present. She did not believe in the reality of the Nazi crimes, which, in an intelligent person, astounded me. The denial can be only explained by the terrible difficulty of facing the facts. She had two daughters, whom I did not see. A few years later they sent me a printed announcement of the death of their mother.

On my return to Italy, in March 1959, I attended an award ceremony for old employees of SCT, my father's paper mill, and I was asked to confer the medals. I deeply appreciated being chosen for this as the representative of the family.

When I returned to Berkeley, the date for the announcement of the Nobel prizes was approaching and some Swedish journalists called me from New York asking for biographical details. I was high in the balloting of the awarding committee, they said. That year, contrary to all precedents, the Nobel committee had leaked information to the press about a week before the final vote. Thus they kept me, and other hopefuls who had been named, on tenterhooks for a week. Finally, on October 26, I heard the announcement on the radio, and shortly thereafter I received an official telegram. Needless to say, before the announcement I did not know if and how a prize given for the antiproton would be divided between Chamberlain, myself, Wiegand, and Ypsilantis, since the paper reporting the discovery had been signed by all four of us in alphabetical order. The Swedish Academy of Sciences decided to award the prize to Chamberlain and myself. Chamberlain was at Harvard at the time, and he telephoned me to plan the speeches we would give at Stockholm. I left the choice to him, and he asked to speak about the technique followed in revealing the antiproton, leaving to me the consequences of the discovery. I willingly agreed.

The trip to Stockholm was more or less the same as that of all other laureates, very interesting and satisfactory. The three children accompanied us. Amelia had caught poison oak a few days earlier, and her face was badly swollen. I told her that if she did not scratch herself, she would most likely recover before the time of the ceremonies. So it was. I always admired the willpower of the little lady.

Following tradition, I gave a short speech of thanks on behalf of all the laureates at the royal banquet that celebrated the awarding of the prizes.[25] As I have observed elsewhere, I had borrowed my friend Ancona's tailcoat for the occasion. I had also obtained a white waistcoat that had been used by several Berkeley Nobel Prize winners from McMillan, who asked me to sign it. One of its wearers even told the king of Sweden the story, saying: "Sir, please look carefully at this waistcoat. You have seen it repeatedly." At the royal dinner I had a most interesting conversation with the king, who knew Italy very well indeed.

Later in the evening, at the students' dance, I had to give a second speech. Here is what I said:

> Students, Ladies and Gentlemen:
>
> Although we have a poet in our midst [Salvatore Quasimodo], who would be far more eloquent than I can be, I have been chosen to answer your gracious and heartfelt greeting, and I will do my best.
>
> We Nobel laureates, although we work in widely diverse fields, share at least one thing in common: we spend a good part of our life teaching and working with students and young people like you, the new generation on which the future depends. Usually we are before you to discuss our special fields of interest. Tonight we may well speak to you in broader terms.

> It has almost become a custom to tell animal fables on this occasion. Two years ago perhaps you heard a wise Oriental one from my friends Lee and Yang. I do not know the origin of the one I am going to tell you. Perhaps it is Swedish, and so you may already have heard it. The person who taught it to me was an old Quaker lady from Pennsylvania [Lorenzo Emo's grandmother].
>
> Two frogs were leaping and frolicking in a meadow when they spied a strange object. Being curious, they decided to investigate it, and the way frogs investigate things is by jumping into them.
>
> In this particular object they found themselves very much at home, because it was a pail full of milk. For a time they had a splendid time swimming about. Then they felt tired and began to seek solid ground, because, as you know, frogs cannot live indefinitely in a liquid.
>
> Much to their consternation they found that there was no island in this pond of milk. Panic-stricken, they tried to jump out of the pail, but the walls were too high and too slick and they fell back. Again they jumped and fell back, and then again and again. The situation became more and more desperate.
>
> At last one of the frogs gave up. The walls were far too high, the surfaces far too smooth to climb up, he reasoned. Clearly there was no hope. He fell back and drowned.
>
> The other frog, perhaps a little less intelligent, but far more stubborn and persistent, continued jumping. Over and over he leaped up and fell back. He was at the point of complete exhaustion and nearly resigned to joining his fellow.
>
> And then he felt something firm and hard under his legs. A little island of butter was forming. With a few more jumps, he churned an island that was big enough so that he could rest and then jump out of the pail, and so he was saved.
>
> I leave the moral to you, but it must be a powerful one because I still remember the old Quaker lady of Pennsylvania telling me the story in 1940, during the darkest days of the war.

I always liked the story, which I fancied reminiscent of my own experience. On this occasion my audience bestowed on me the "Order of the always smiling and jumping little frog."

I had time to see Oskar Klein and Lise Meitner (by then rather aged) in Stockholm. I also went to Uppsala to Kai Siegbahn's institute to give a lecture, but I had to hurry to Rome for SCT business.

I have always regretted that neither my parents, my uncle Claudio,

Corbino, nor Fermi were able to see me getting the prize. My parents' satisfaction would very likely have exceeded my own. I can hardly imagine that of Uncle Claudio, who gave me a pair of gold cufflinks simply because I had received a superior grade in a mechanics exam. By now I had all the public recognition I could hope for; self-esteem is something else. I believe I never got a swollen head. That there is no honor that can affect my accomplishments is a hard fact I have always kept in mind.

The reason for the prestige of the Nobel Prize for physics is that, all told, it has been given well. This does not mean that there have not been some lucky mediocrities who have received it and some eminent deserving scientists who have been passed over. For the former, it was a stroke of luck; for the latter, apart from worldly disappointment, it is unimportant. Persons such as Einstein, Planck, Rutherford, and Bohr have given the award its prestige. If one or more of them had not received it, the loss would have been entirely to the prestige of the prize, not to them. Considering all the laureates, one can divide them into three groups: one group has given prestige to the prize, one has been exalted by it, and one has more or less broken even.

In the nominations for Nobel Prizes or other important awards, the decision is easy when there are truly extraordinary candidates, but even there, the diversity of the fields in physics may make some choices difficult, and I have found that there have been some glaring omissions, such as that of G. E. Uhlenbeck and S. A. Goudsmit, who discovered the electron spin.

The monetary value of the prize was initially very substantial, corresponding to about fifteen times the yearly salary of a distinguished professor. In 1959 it amounted to $21,184 for each of us, and my net annual salary at the time was about $13,000. Of course, the prize also provides many less tangible advantages: invitations, prestige among one's colleagues, the chance to be on various committees, numerous opportunities to serve as an ornamental plant, and even some minor monetary advantages. At Berkeley, in recent years, one is even given a private parking place on campus!

There are also drawbacks: one automatically becomes an oracle on every topic, and one is subject to distractions from work and difficulties with jealous colleagues or collaborators. Above all, sensible persons, which Nobel Prize winners usually are, know that what counts is their work.

"Emilio, you could take all your work and exchange it for one paper of Dirac's and you would gain substantially in the trade," Fermi once said to me. I knew this to be true, of course, but I answered: "I agree, but you could likewise trade yours for one of Einstein's and come out ahead." After a short pause, Fermi assented. I know of scientists who cannot resign themselves to being inferior to contemporaries, with dire consequences for their personalities and happiness.

Finally, being a celebrity may give rise to amusing episodes. For example, at the time of the discovery of the antiproton, I happened to read an article in the *New Yorker* in which Salvador Dali said he had abandoned Freud, and that his "father" was now Heisenberg and his credo the uncertainty principle. In particular, he had painted or was about to paint an antiprotonic madonna. I then wrote to him and sent him some beautiful pictures of antiproton stars in photographic emulsions, saying that I was curious as to how he visualized antiprotons. He did not answer. Shortly after receiving the Nobel Prize, I was in New York at Robert Serber's house at a cocktail party with several physicist friends. During the party George Placzek and I modeled a mink cape I had bought for Elfriede, with everybody laughing merrily. In this joking mood I said I would look up Dali and see him before leaving New York. Everybody laughed at the idea, but the next day I found Dali's address and telephoned him at the hotel where he lived, explaining who I was and reminding him of the photos I had sent to him.

He was most friendly and invited me to come to see him at 8:30 that evening. Given the time, I thought this was after dinner, and I ate before going to the appointment. I called him from the lobby of his hotel, and he came down very shortly. I had started having doubts about what would happen next and thought that perhaps he might want to amaze me in some way or other. I therefore hid behind a column from where I would be able to see him immediately, but he would have to look for me for a few seconds at least. I counted on this interval to prepare myself in case he had something up his sleeve. Indeed, he arrived with his moustache stiffly pointing upward, a thin cane, and strange attire that seemed a caricature. I looked at him from my hiding place for a fraction of a minute and then greeted him in the most natural way, without showing the slightest surprise. "What language shall we speak?" he asked. "Any," I replied, as if I knew them all. By now we were competing in one-upmanship. We settled on French, mixed with much English. It turned out that Dali, who was accompanied by his attractive and interesting wife Gala (whose history I did not know), intended to invite me to dinner. I did not say I had already eaten but ate once more, lightly. After a while, it must have been apparent that the competition in one-upmanship was a draw, and the game subsided.

Dali explained to me that he was truly interested in modern physics and that he had read several articles in the *Scientific American*, which was obvious from the way he spoke of quanta, the uncertainty principle, antimatter, and so on. All these ideas had, however, suffered a sea change in his mind that I could not grasp, but that was obviously sincere and interesting. His paintings of soft watches bending and dripping as if they were cheese had hidden physical and psychic meaning for him. He explained to me that the madonna he had painted was "antiprotonic" because only the annihilation of matter could give sufficient force to propel a woman to heaven. As he spoke I became convinced that he had a way of seeing the world different from that of a scientist, certainly more subjective, but also valid in its own terms. We then passed to artistic technique. He said he painted many hours a day, slowly and with extreme care. "If you look at my paintings with a magnifying glass, you will discover many things, because I often paint using a magnifying glass, and details are almost invisible without it." I hoped that he might give me a drawing in exchange for what I had sent him, but instead he sent me a book of reproductions of his work with a dedication. I regretted not having read the book before our meeting; it also contained the whole history of his wife Gala. She had listened to our lively conversation, speaking only rarely, mostly helping us when we searched for words, but I saw she was his constant model and obviously in many ways an inspiration.

In March 1960 I gave the Faculty Research Lecture at Berkeley, for which I had been selected before receiving the Nobel Prize. This lecture is a high local honor conferred on members of the Berkeley faculty by their colleagues.[26] Following the lecture, I was on the selection committee for ten years or so, and there, contrary to what had happened on other Berkeley committees, I saw a certain spirit of partisanship based on disciplines. Ultimately, on Alvarez's suggestion, the Berkeley Academic Senate resolved to appoint two faculty research lecturers yearly, one for the humanities and one for science.

The year 1960 was saddened by several tragedies. On April 23, Cornelius Bakker perished in an airplane accident. We had been friends since 1930, when we collaborated in Zeeman's laboratory. After the war Bakker had visited me in Berkeley, and we had even conducted a small investigation together. He had later become the director general of CERN, and I had visited him at Geneva. Amaldi and Bernardini thought of nominating me as his successor at CERN. I was somewhat surprised, but I agreed to stand for the appointment. The CERN directorship would have given me a new activity that at my age and at that phase of my career appealed to me. However, the nomination was received coolly and a campaign for another person was immediately started; I then withdrew my candidature.

A few months later an unexpected blow struck us. Our dear friend Francis Jenkins, who had helped us so much in difficult times, and to whom we were very close, fell incurably ill and died. During the summer, Elfriede and the children had gone on a tour of New Mexico, revisiting Los Alamos. I did not leave Berkeley because I knew of Jenkins's condition and wanted to be on hand in case I could in any way be of help. I wrote a deeply felt obituary and spoke at his memorial service. In 1967 his wife, Henriette, also died, to our deep sorrow.

In the autumn of 1960, the Rockefeller Foundation invited me to go to Nigeria for that country's independence proclamation, scheduled for October 1. It was a unique opportunity, and Elfriede and I stayed there for about three weeks as guests of Ibadan University. It was my first trip to Africa, and I knew very little about it. I bought light clothes suited to the climate and a pith helmet of the kind once usual in Africa, but I was told on arrival to suppress this headgear at once, as it was considered a symbol of colonialism. After our first official dinner, with everybody formally dressed in black tie, I was surprised when the ladies retired and the men went to urinate on the host's lawn.

One evening while touring the country, we visited the Oba of Benin, who lived in a big mud palace, together with his court, which included many wives and about fifty children. The anthropologist M. J. Herskovits, who was also in the party, asked the Oba many questions on justice and law in Benin, and how he reconciled them with British law and the religious commandments of the several prevailing faiths. The Oba had been educated in England, spoke the language well, wrote with a modern fountain pen, and had inscribed photographs of several members of the English Royal Family. He complained about his children, who all wanted to go to Eton, an expense he could not afford. As he spoke, some of these children showed up stark naked in the reception hall, only to be promptly dismissed.

After a while the Oba politely hinted that he had enough of the anthropologist's questions and turned to me, saying that he knew I was a physics professor and hoped I could help him to clear up the confusion he felt when they told him the earth was spherical, while old traditions said it was flat and so it appeared to him. Furthermore, he could not understand how the sun disappeared every day from a certain part of the horizon and reappeared on the opposite side the next morning. I tried as best as I could to explain these mysteries to him, and he then passed to the moon. How far was it? Would the Americans or the Russians arrive there first? On the moon's distance I gave him some information, but I refused to predict who would arrive there first. On my return to Berkeley I bought an illustrated astronomy book for young people for the Oba. I did not want to offend him, however, so I sent it to him with a letter to the effect that I thought it might interest some of the children I had seen at his court.

At the proclamation of independence, the man slated to become the first prime minister of independent Nigeria, Sir Abubakar Tafawa Balewa, made a speech that impressed me greatly for his realism, and equilibrium. He pointed out grave problems facing the new nation and some of the necessary remedies. Within a few months he was murdered.

In my academic career, I long avoided administrative work. This was easy to do, but I nevertheless ended up shouldering many responsibilities in this field too, especially in later years. As a rule I have accepted assignments if asked to by the university, the government, or some other public institution, but I never strove to join boards, directorates, committees, and so on. Since there are always many people eager to serve in such offices, those who do not show a keen interest in the jobs do not get them.

However, at the University of California at Berkeley I have ended by serving on most of the committees of the Academic Senate. Some were a waste of time, but I remember the Budget Committee, on which I served from 1961 to 1965, as a superior school and as an occasion for really helping. This committee practically controls all academic promotions and appointments. It was then composed of five members indirectly elected by the whole faculty. I was impressed by the dedication and fairness displayed by its members in working on assignments that were often quite delicate. I did not see examples of partisanship or of sloppy procedures, and all the members labored assiduously at their job. It is justly said that this committee has been the source of the quality of the Berkeley campus.

I have occasionally been summoned to Washington by the National Science Foundation or the National Academy of Sciences, or invited to serve on panels of NASA and similar organizations, and I have been a trustee, advisor, member of visiting committees, and so on for several universities. I always conscientiously did my homework and said what I thought. I did not have the impression that I was particularly popular as a member of committees. Maybe I spoke too much and not diplomatically enough. In the Rad Lab, I had few administrative functions, except in the direction of my own group. Our views were too divergent and our personalities too different to make me useful to Lawrence in the direction of the lab. Indirectly, I carried some weight, but not by reason of any official position.

I was influential in the physics department because the chairmen, on their own, consulted with and listened to me. I was insistently offered the chairmanship, but I did not accept it, because at the time I would have had to sacrifice too much of my scientific work. Later I served as chairman for two years, from 1965 to 1967.

I have also been on selection committees for fellowships. Generally there were excellent candidates, and as regards the first places there was only the predicament of choosing among them. The less worthy were also pretty obvious. The problems arose at the dividing line.

Letters of recommendation always play an important role. On the many occasions when I was asked to write them, I always tried to help the addressee by giving him as much information as I could, good and bad, and trying to put myself in his shoes. I believe one of the results was that my letters carried weight. I once recommended an excellent former student, Fred Noel Spiess, who had done his Ph.D. with me. Lawrence read my letter and reinforced it with one of his own, in which he added the comment: "As you know, Segrè is rather conservative in his statements about people, and his letter, I would say, is a strong recommendation" (which certainly was my intention).

The recipient of a letter of recommendation must know the author to understand the import of the letter. It is said that Einstein was so nice to everybody that his letters ended up being discounted. Fermi was stingy with praise, but precise in his statements. Once he wrote that Richard Garwin had been one of the very best of his students. "One says this of everybody," commented the personnel director of the Rad Lab, to whom the letter was addressed. "But if Fermi says it, it is most important," I replied.

In February 1962, Arthur H. Compton came to Berkeley as a visitor, together with his wife. Besides being a great physicist, Compton was also an appealing human being. He had acted courageously and nobly during the witch-hunts of Senator Joseph McCarthy of Wisconsin, but he had been very cautious with respect to refugee scientists and foreigners in general, seeming worried above all that they would take positions away from Americans. Now, after the great successes achieved with atomic energy, after his close acquaintance with Fermi, and so much water under the bridge, he was a changed man. We invited him to dinner, and I went to listen to his speeches, which were in the nature of sermons on science and religion. Unfortunately, while still at Berkeley, Compton fell ill and died, without being able to complete his announced series of lectures, although he had given most of them. The Regents refused to pay the honorarium agreed upon, because Compton had not fulfilled his contract. Somebody asked me to fill in and give one or two lectures so that the widow might be paid. I willingly complied.

In 1963 the Accademia dei Lincei asked me for suggestions for the Donegani Lectures, named for the founder of Montecatini, the largest Italian chemical company. I thought that it might be more worthwhile to establish a summer school rather than continue the series of formal, polished orations that the lectures had become. I wanted them to teach something new and practical for the benefit of young Italian chemists and technologists. I discussed this possibility with my friend Dr. Luigi Morandi, vice president of Montecatini.

I suggested that materials science, which was not much studied in Italy, but was flourishing in the United States, might be an appropriate field. The idea was to study the technological properties of materials scientifically, using all the tools of modern physics. Ceramics, modern metallic alloys, semiconductors, and plastics were suitable subjects. Much of the technology had been known for a long time, but in recent years it had been transformed, passing from an empirical to a scientific stage. The fruits of the marriage between science and technical practice are abundant and pervade modern industry.[27]

I outlined a program for a summer school. Although I am not an expert on the subject, I knew to whom to turn for help and advice. When it came to the practical organization, Morandi leaned heavily on his right-hand man, Umberto Colombo, to whom he introduced me, and Colombo and I soon became good friends. Colombo's scientific training was mainly in geochemistry, which he taught at the University of Genoa, but his interests extended very broadly to economics, applied science, futurology, and science policy. Eventually he became one of the foremost European science administrators. Indirectly I helped him to buy a large country estate near Florence by introducing him to a friend of Emo's who was a real estate broker. I admired the promptness with which Colombo closed the deal; he needed hardly more than a couple of hours.

After 1946 I often taught courses in nuclear physics, as well as in what was then known of particle physics. For this purpose I used Bethe's famous articles, a book by Rasetti based on Fermi's lectures, and Fermi's own remarkable notes. None of these was up to date and wholly suitable for the students, however, and in 1958 I decided to write my own textbook on nuclear and particle physics. The spirit and the aim of this project are given in the preface to *Nuclei and Particles*, which took five years of intermittent work for its completion.[28] The result was a thick volume, which has been translated into several languages and required a second edition in 1978.

The history of science has always interested me. When I was a boy, my parents gave to me books on the subject by Gaston Tissandier, which were long my favorite reading. Later I read René Vallery-Radot's *Vie de Pasteur*, which was one of my mother's favorite books. As an active scientist, I also subsequently read books on the history of physics, chemistry, and mathematics. I knew that in our times physics was making colossal strides, and that we were probably living in an exceptional age. I had saved some documents on the subject, especially with regard to the work done in Rome. Unfortunately, the bulk of these papers were lost in a parcel I sent from Italy to Berkeley that went down with the ocean liner *Andrea Doria.* I never kept a diary, and I regret not having done so, at least during certain periods of my life.

I had professional contact with history of science for the first time in the early 1950s while serving on a committee charged with revitalizing the philosophy department at Berkeley. Among several alternatives, we proposed hiring Thomas S. Kuhn, who later became a noted historian of science and proved to be a happy choice. When the great project of collecting the sources of quantum theory was started, I helped Kuhn in interviewing Georg von Hevesy and Otto Stern. It was not an easy assignment, and it revealed to me the problems of human memory and of oral history. My own interview in 1967 did not entirely satisfy me when I read the transcript. It seems that it is very difficult to convey one's deep thoughts and feelings in an interview.

Around 1960 I started giving occasional historical lectures. Thus I was Sarton Lecturer at the Tenth International Congress of History of Science at Ithaca, N.Y., where I spoke of the consequences of the discovery of the neutron.[29] Later at Berkeley I prepared a set of lectures on twentieth-century physics, which I repeated many times in different formats. I called these lectures my accordion because I used them like a minstrel, reciting them at various places; moreover, I could extend or contract them according to need, like an accordion. This was the origin of my book *From X-Rays to Quarks*, which appeared in 1976 and has since been translated, to my knowledge, into Italian, French, German, Greek, Japanese, Spanish, Portuguese, and Hebrew.

In 1962 we received a telegram from President John F. Kennedy inviting us to a White House dinner on April 29. He had invited all American Nobel Prize winners and some other notables, such as the poet Robert Frost and the science advisor George Kistiakowsky, whom I knew from Los Alamos. The president and the first lady arrived in an official procession with silver trumpets and alarums. Kennedy greeted everybody and made a nice little speech, saying among other things that there had not been such a concentration of intellects at the White House since the time when Jefferson dined there alone. He then made some appropriate remarks on science and politics, expressing his appreciation for science and intellectual prowess, and finally we had an excellent dinner. I was sitting next to Mrs. Kistiakowsky, as indicated by a card marking her place, and I tried to show off by saying that I remembered her from when she had got married at Los Alamos. (Kistiakowsky had divorced his first wife at Los Alamos and married his secretary.) "Your memory is at fault; that one was number two; I am number three," she answered coolly.

For a long time I had wanted to go around the globe, and in 1962 I had enough leisure to do so. I contacted the State Department, which often sponsored trips by Nobel Prize winners, offering to give a series of lectures. This proposal was accepted. I was completely free to say whatever I wanted, without any strings attached, except to mention that the U.S. Information Service was sponsoring my speech. The government would buy my ticket, but not Elfriede's, and would pay me a per diem rate for the days on which I lectured. The government auditors were very strict. I still remember a Chinese accountant at a U.S. embassy seriously challenging the amount of a taxi fare from my hotel to a university because it was ten cents higher than the return trip! Our journey lasted from September 17, 1962, to January 27, 1963, but we spent the last month in Italy. We visited Japan, Formosa (now Taiwan), Cambodia, Thailand, India, Ceylon (now Sri Lanka), Nepal, Pakistan, Iran, and Israel. The children were by then old enough to stay home by themselves and have a good time of it.

After having been a few days in Cambodia without reading newspapers, Elfriede and I returned to Bangkok for a lecture. An embassy official met our plane and took me directly to the lecture hall. On the way he told me that the news about Cuba was somewhat better. I did not have any inkling of the Cuban missile crisis, then raging; in the car the official gave me some of the news, and I was flabbergasted. I clearly remember the strange sensation of giving a lecture like an automaton, practically without knowing what I was saying, with my mind turned to what I had heard in the car and concern for the children in Lafayette.

We were also forced to cut short our planned stay in Ceylon because India had gone to war with China and was about to requisition all planes. We were told that if we did not leave at once we would be stuck in Ceylon indefinitely! In India I visited Homi Jehangir Bhabha and several other physicist friends.

I liked the little I saw of Nepal immensely—it reminded me of the Abruzzi in Italy on a much enlarged scale.

On our way to Italy we stopped in Israel. It was my first visit, and we were coming from Third World countries. The last we had visited was Iran, which, in spite of its impressive art and magnificent monuments was still a Third World country. The contrast was striking and fostered my appreciation of Western civilization.

In many countries I had seen the misuse of funds given by the United States; in Israel one had the impression that the money had been well spent and that each dollar had been transformed into something useful.

I hoped to be able to spend some time with Giulio Racah, now president of Jerusalem University, but he had to leave suddenly and unexpectedly for the United States on university business. However, I met another theoretical physicist, Yuval Ne'eman, who tried to persuade me of the importance of the SU3 group in particle physics. I was skeptical, but on my return home I studied the subject and acknowledged my error. Ne'eman asked me to give a talk at Tel Aviv University, of which I subsequently became a "governor" and from which I received an honorary degree in 1972.

My scientific activity was diminishing, both because of my age and also because of the internal situation of my group, already described. In addition to giving my regular courses, I went to the laboratory every day and talked to students and co-workers, following their progress, offering criticism and suggestions, calculating results, or formulating some simple theory, but without personally executing the experiments. Several of them were concerned with mesic atoms of various kinds. The relevant publications were properly authored by Wiegand, who was the principal investigator. Chamberlain at that time was mainly concerned with polarized hydrogen targets and with developing a technique based on a method introduced by Carson D. Jeffries.

The scientific profession belongs to youth; any perusal of scientists' biographies bears testimony to this. It would be too long and out of place here to give the reasons for this fact, which in any case is not too mysterious. Experimentalists last a little longer than theoreticians, but even for them scientific productivity declines at a time when other abilities are still well preserved.

December 16, 1965, marked a fateful turning point for the Lawrence Berkeley Laboratory. On that day the U.S. government decided on the location of a powerful new accelerator that would serve the whole nation, to the planning of which the Rad Lab had devoted years of study. Seaborg was by then chairman of the Atomic Energy Commission, which should have helped the candidacy of Berkeley, but in all probability the final choice was made by President Lyndon Johnson on the basis of political considerations. The selection favored Illinois and originated the present Fermilab, in the vicinity of Chicago.

The choice was a severe blow to Berkeley, which thus lost a prime reason for its preeminence in physics. It was also a jump in the centralization of high-energy experimental physics, which was now concentrated at three national laboratories: Fermilab, SLAC at Stanford, and Brookhaven, Long Island. The universities formed users' groups that would prepare and analyze experiments carried on in the three major labs. This evolution was unavoidable because of the cost and complexity of the gigantic facilities required, but nonetheless it was traumatic and painful, especially to Berkeley.

From 1965 to 1968 I was one of the trustees of Fermilab (I do not know who suggested the name) and one of the proposers of R. R. Wilson as its first director. I attended its official inauguration in 1974 with Laura Fermi.

In 1966 Elfriede and I paid another visit to South America, which had unexpected long-range consequences, because at Montevideo, in Uruguay, we met Rosa Mines, a friend of the Riminis. She came to their home to meet us and to see whether she could get an affidavit from us to help her immigrate to the United States. I was somewhat reluctant to give it, but Elfriede was favorably impressed by her, and by the strong endorsement she got from Riccardo, and gave her her own personal affidavit of support. Perhaps Elfriede recalled the period when she had left Germany and her own first difficult steps in Italy. Thus, a few months later, Rosa arrived in California, and after a hard beginning, she found a satisfactory position in a bank. She continued to visit Elfriede, but not frequently.

I thus come to 1970, my sixty-fifth year, a year full of events, some happy, some tragic.

I have not written here about the life of my children. It belongs to them and they are entitled to their privacy. I want however to make a comment. Some of their decisions left me perplexed, and although I am sure I followed my conscience and my best judgment, helped by my greater experience, and in their sole interest, I would prefer not to have influenced their decisions. I believe parents should educate their children in a broad sense, but afterwards it is better to leave them their independence. It is not always easy to follow this policy, and sometimes, as for instance in historical families, it is subject to exceptions. In our case, we followed it, for better or for worse.

In 1967, our son Claudio married Elizabeth Bregman, a student of French he had met through Amelia; they now have three children, Gino, Francesca, and Joel, and live in Austin, Texas, where Claudio teaches history at the university.

Early in 1970, Amelia told us that she planned to marry Joseph Terkel, an Israeli fellow student of animal behavior at Rutgers University. That January she and Joseph came to Lafayette, where they built a canopy of leaves with their own hands for a Jewish wedding, which took place in the presence of a few close friends. Joseph very rapidly endeared himself to Elfriede and me. The Terkels have two children, Amir and Vivian. He is professor of zoology at Tel Aviv University, and Amelia is the curator of Tel Aviv Safari Zoo.

I always greatly loved nature and the outdoors, from the mountains of my youth, to fishing and mushroom-hunting in my old age. One's way of enjoying nature obviously changes with age and physical strength. In the 1960s several raft trips on rivers such as the Rogue in Oregon and the Salmon in Montana introduced us to some aspects of the American wilderness of rare loveliness and romantic appeal.

At the beginning of July 1970, we went with Amelia and her husband and Fausta and her friends on a canoeing trip in the Trinity Alps. We first climbed through a long valley and arrived at a place called the "Stone House," where a colossal granite block created a delightful camping spot. During the expedition we had the usual minor adventures, crossing of a ford with high water, bears scrounging among the food, other nocturnal animals prowling the camp, and so on. We enjoyed the encampment, climbed lateral valleys, and visited high glacial lakes with excellent fishing. We even used a small boat that Joseph had carried up there. All told we had an excellent time and hoped to be able to repeat the trip.

Later that year, Elfriede and I went to Italy, where I had promised to attend a Donegani school at Lake Como. Everything went well, and after the school session we went on to Florence, where we were considering settling on my retirement, due in a couple of years. Our plan was to visit the Teramo side of the Gran Sasso, the highest mountain in the Apennines, and then go on to Rome. On October 15, we visited the poet Leopardi's house at Recanati, and that evening we went to Teramo. During the night Elfriede died in her sleep.

I still shiver in writing these lines.

Four or five intimate friends came from Rome, and we buried Elfriede at Teramo in a peaceful cemetery in view of the Gran Sasso. I returned to Rome in a sad state. My son-in-law Joseph later told me: "You must remember you are like a man that has suffered a major amputation; a leg cut off at the groin, or some other terrible trauma." I cannot report the horrors of that tragic period.

When I returned to Lafayette all the children came home, and they were an incomparable solace. All in their grief turned inward to the family, and if Elfriede could have seen them she would have been proud of the fruits of her labors in raising them. A letter they wrote to friends answering messages of sympathy shows their feelings:

> There are many ways in which we think of our mother, but most of all we seem to remember her working in the garden, rooting or transplanting, watering or trimming, or simply worrying over a faltering plant. All around the house we saw, season after season, the fruits of her gardening: the roses, the African violets, the philodendrons, the fuchsias, the little fig tree, the lavender, the tomato patch, and all the myriads of flowers and shrubs that flourished in every corner. At times we wondered how she could keep track of all her charges. We asked ourselves if even she knew how far her garden extended.
>
> Now that she is gone we wonder even more. We children are scattered, yet when we are at home, we have to laugh a bit sometimes when we realize just how far her garden did extend. We have to laugh

when we realize how firmly Mamma rooted us in her ways: a way to set the table, a way to bake the panettone, a way to show hospitality. And when we scatter to our own homes again, we wonder at how much of Mamma's gardening we carry with us. We ask ourselves if we will ever see the full extent of her garden.

Claudio, Amelia, Fausta.

At the time, Amelia lived in Los Angeles, and Fausta at Santa Barbara, so that I could go to them for weekends. Claudio was in Texas, too far for short visits. My physician labored to bring me back to normal, both physically and mentally. When the children left, my old friend and colleague Carl Helmholz and his wife Betty, who lived near me in Lafayette, invited me to sleep at their house. They offered not only tactful and compassionate hospitality but also precious company. I want to remember these good deeds, which deserve my deepest gratitude.

I realized myself, and from talking to others who had had similar losses, that work was the best way to get out of the abyss into which I had fallen. But how and on what? I was not in a state to embark on original scientific endeavor. Both my Japanese gardener and one of my colleagues suggested that I go to some senior citizens' center to find company.

While I was trying to find something to bring me back to life, I received unexpected help. Ugo Fano phoned me from the University of Chicago inviting me on behalf of the university to give a series of Fermi lectures on a subject of my choice. At first I said, truthfully, that I did not feel up to such an enterprise, but both he and Laura Fermi insisted I accept. I decided to give a series of historical lectures on modern physics similar to those I had given at Berkeley in 1968. Laura Fermi put me up in Chicago, where I stayed for several weeks. I was given an office at the university, and I slowly started writing down my lectures, an activity that kept me busy and gave me a written text I could later use for a book.

Among the letters of sympathy I received on Elfriede's death was one from Rosa Mines. I have already told how we first met. I shall not tell here how, in a couple of years, we passed from the letter to marrying, which occurred on February 12, 1972.

The marriage of two people, one sixty-seven years old, one much younger, with very different life experiences, has aspects I shall not go into. The analysis of such a relationship would require more psychological insight than I can muster; this at least is Rosa's authoritative opinion. Suffice it to say that it is a little like mixing spring and autumn, with their storms and periods of good weather, until a new equilibrium is formed.

Four months after our wedding, in June 1972, I reached the compulsory retirement age at Berkeley. There was the usual ceremony, with a dinner and speeches. "Rosa has shown me that sometimes even a tree hit by lightning can resprout new life, in late fall as in spring," I said in concluding my thanks.

My direct experimental activity had ceased, but not my teaching. Now, however, I taught history of physics rather than physics itself. The University of California recalled me to service repeatedly, and I had the impression that my historical courses were popular and successful. The number and quality of students attending them, in spite of their not being required courses, were most encouraging.

However, I also wanted to go back to Italy with Rosa, and to leave Lafayette, where there were too many ties to the past. For years Italian friends and colleagues had been suggesting I reenter Italian academic life. I appreciated the offers, but declined them because the working facilities in Italy were not comparable to those in America. Furthermore, I could forget neither the treatment my family had suffered at the hands of the Fascist government nor my debt of gratitude to the United States, a country that had allowed me to rebuild my civil and scientific life. By now my family had become American, and Americans treated me as one of them and did not discriminate against me in any appreciable way.

This, of course, did not detract from the deep gratitude I felt toward all those who had helped my parents in the "time of iniquity" (to use an appropriate term coined by Ugo Amaldi, Edoardo's father) or my appreciation of the behavior of innumerable Italians that clearly set them apart from their despicable government.

I would have liked to teach in Italy again, and, especially in the first postwar years, I tried to help the reconstruction of the country by fostering scientific relations, student exchanges, and outright collaborations. However, I would not assume either heavier responsibilities or employment as a professor, which might, I thought, jeopardize my U.S. citizenship. I discussed these problems with Edoardo Amaldi, with Francesco Giordani, president of the Accademia dei Lincei, with Giordani's successor, Beniamino Segre, and even with Giovanni Leone, then president of the Italian Republic. They proposed several alternative solutions, centering on appointments at the Accademia dei Lincei.

In November 1972, as a first step, I gave a course on the history of modern physics at the Lincei in Rome for six months. It was a time of student unrest, and being far from the university was an advantage. Whoever came to my lectures did so out of interest and not to make trouble. Besides the students, many old friends and acquaintances came to listen. I had an office and lectured in an old mansion near the Villa Farnesina, a jewel of Renaissance art famous for its Raphael frescoes. The buildings were surrounded by a formal Italian garden. Occasionally I dreamt of donning a cardinal's hat and walking in the gardens with some artist of the time of the Farnese. In the Lincei library I found several rarities I had never seen, such as Wilhelm Röntgen's original papers and the *Transactions of the Connecticut Academy* containing Josiah Gibbs's papers. Life in Trastevere, on the north bank of the Tiber, was pleasant, but access to the rest of the city was not easy because of Rome's notorious traffic; not even the common remedy of simply walking helped very much.

My stay also had another purpose. It was the first time Rosa had been to Europe, and I was eager to introduce her to Italy and to my friends there. We traveled in the vicinity of Rome, in Tuscany, and in the northern provinces. Everywhere old friends received us with great warmth, and Rosa was able to meet people I had often spoken of to her and to experience Italian behavior, so different from the North American and Latin American ways she knew.

Italian law had recently changed, and it had become possible to appoint foreign citizens to university chairs. There were no vacant physics chairs in Rome, and in the end Parliament passed a law creating a chair for me ad hominem, which would be abolished on my retirement. Unfortunately, overcoming all these obstacles took time, and I was appointed only one year before reaching the Italian retirement age.

When I returned there in November 1974 to teach nuclear physics, the University of Rome was in bad shape, torn to pieces by unruly students and by politicized professors. In spite of my best efforts, I was unable to make any appreciable improvement even in the restricted field of the physics department. Most of my colleagues were interested in quite other things, which had little to do with what I thought were the purposes of a university, and I did not find any scientific project I could be of much help to. The best scientific work was being done at CERN in Geneva or at the Frascati Laboratory, and they certainly did not need my help. Among the Rome faculty there were several old friends, with whom I enjoyed consorting again, and my colleagues and students were always friendly and respectful to me, in contrast to their behavior toward excellent persons such as Edoardo Amaldi, Giorgio Salvini, and Giorgio Careri.

Rosa learned to speak Italian perfectly, and we both made several new close friends. One of them was Cesare Tumedei, one of the best civil lawyers of his time—a "prince of the forum," as we say in Italian. Tumedei represented Donegani, the founder of Montecatini, and was on the boards of numerous large companies. He was seventy-nine and in excellent physical and mental condition. Every Sunday, he organized excursions, by car and on foot, leaving at 8 A.M. from his villa on the Via Cassia. A small caravan of cars followed his. His chauffeur left us at the starting place of the hike and later picked us up at the other end. The walks usually lasted several hours and passed places of splendid natural and artistic beauty. The group changed from week to week, but there were usually about a dozen of us. Among others, all of them noteworthy people, and many of them quite old, I recall Prince Schwarzenberg and his wife; Signor Biamonti, a lawyer, the son of a friend of my father's; Dr. Bignami, the son of a famous researcher on malaria; and Monsignore Sticker, prefect of the Vatican Library, who became a cardinal in 1985. Most of them, notably Tumedei, were brilliant speakers and always had interesting stories to tell. The conversation was always as stimulating as the places we visited. One Sunday we went to Campo Catino, where I had skied forty years earlier, when one could only reach it by a horse trail; now it was accessible by car. From there we descended to the Trisulti Abbey, following a steep, picturesque path through cool woods embellished with orchids and other magnificent wildflowers. Meeting shepherds and enjoying freshly made ricotta was another pleasure along the way on these excursions. The chauffeur's intuition was extraordinary, and even when we emerged at a different place, he was there waiting for us.

Every now and then I went to the sulfur baths near Tivoli. It became a kind of pilgrimage back in time. I would take the bus near the train station in Rome and ride along the length of the Via Tiburtina, changed by time beyond recognition, a road I had traveled by tram before World War I and by car before World War II. Now, jammed with traffic, it had lost a great part of its beauty, but here and there one could still recognize remnants of its ancient nobility. Arriving at the baths, I enjoyed using the card that identified me as a Tiburtino and qualified me for reduced admittance. Sometimes the cashier would recognize me and make some comment. The baths had not changed much since I used to visit them with my teacher, Signorina Maggini, almost three-quarters of a century before. After my swim, I would buy a sandwich roll with the same filling and smell they had always had. From the baths I could see Tivoli and distinguish our home and the cypresses of the Villa d'Este, as well as the roofs of the paper mill. I would soon be surrounded by friendly ghosts from my childhood and youth, and they often rejoiced in meeting and talking to me again. The strong hydrogen sulfide smell was disagreeable to many, but not to me; it was the scent of many memories.

When I reached the Italian retirement age, I returned to Lafayette. Rather than telling the plain (and uninteresting) truth, the Italian newspapers tried to see in my leaving a protest against the conditions I had found in Rome. I did not like this misrepresentation of my feelings, and I tried to correct it, but to little avail.

Nevertheless, while I am sure that it is possible to acquire a scientific education of the same quality as in any Western country in Italy, the average level is low, and the degradation of Italian university degrees seems permanent. After two suicidal internecine wars, no part of Europe is what it used to be, but Italy's decay is especially vivid to me. As far as living conditions go, Italy is still a very pleasant country, but the United States seems livelier than Europe, although this is not to say that it shows much wisdom or greater civilization. I often think of myself (with all due respect) as Einstein did, who ended by feeling himself to be a "world citizen."

• • •

For many years I had wanted to tour Tuscany on foot, because I realized that, given the size of the towns, the many beautiful and interesting things to be seen there could not be properly appreciated from a car. In 1981, at the age of seventy-six, I finally fulfilled that wish in the company of my old friend Giuseppe Occhialini. He had a property in Tuscany, as original as its owner, where he planted different trees in honor of deceased friends. One of the nicest was in memory of Lorenzo Emo, so dear to both of us.

We had decided to use only the railway or public transportation on our trip, and above all our legs. On foot we would be able to follow narrow paths and even trails across the fields, thus avoiding the main roads, which actually destroy the local charm. We visited Montepulciano, Pienza and its surroundings, Siena, and fascinating, out-of-the-way places, always on pleasant hikes, talking of interesting subjects and staying at small, modest hotels with a homelike atmosphere and food. One day, we stopped and asked some old women who were sitting in the sun in front of their home, sewing, to show us the road to a castle. They gave us the information, adding on their own that the place was quite far and that it would be imprudent at our age to try to reach it. We told them not to worry and started walking, reaching the castle after about an hour. When we returned, we were hoping to be able to brag to the old women, but they were gone.

Talking to Occhialini was fascinating. His ideas were usually highly original and interesting. For example, he was against teaching so much of Dante in the public schools because he inspired seditious and vindictive thoughts. In physics, he spoke of events in which he had participated. Certainly his life and his way of doing physics were very different from what I knew. In physics he stressed technique above all; life was romantic, full of adventures and women.

• • •

When writing *From Falling Bodies to Radio Waves* in 1979, I devoted a chapter to Faraday, and being in London I went to visit the Royal Institution, his home and laboratory. I was much impressed by how well preserved and tastefully displayed everything pertaining to Faraday was, as though in a still active and living laboratory. The personality of Faraday pervaded the place; one could see his desk and home furniture in the apartment where he lived, his lecture room and apparatus. One almost expected to find the owner of the house performing his experiments in some corner. In no other place is a scientific presence so vividly felt as there.

Early in my career, my Dutch mentor Pieter Zeeman had described the Royal Institution to me and told me in detail from his own experience of the ceremonies accompanying a lecture there. Until the time comes for him to speak, the lecturer is locked up in a little room to prevent him from taking flight in panic, as happened once in 1846 with Charles Wheatstone. The lecture starts at 8 p.m., only a few seconds' delay being tolerated, and lasts for an hour, give or take a minute. No notes are allowed, and one must perform at least one experiment. At the end of the lecture, the speaker is restored with a glass of whiskey.

In May of 1982, I myself came to lecture at the Royal Institution, repeating to the letter the ritual followed by Zeeman in 1906, when I was one year old.

• • •

The discovery of technetium 99 earned me a certain notoriety among specialists in nuclear medicine, and not unfairly, since that isotope is their bread and butter. As a consequence, I had been invited on various occasions to speak at meetings on nuclear medicine, and I have become an honorary member of associations like the Society of Nuclear Medicine and the American College of Nuclear Physicians. The latter held a meeting at the end of January 1982 in Tucson, Arizona, and I had been invited to give a talk. I gladly accepted, and Rosa and I amused ourselves comparing the luxurious lifestyle of these doctors with our own.

Once the meeting had finished, we went to see Tucson, especially the Sonora Desert Museum, where I greatly admired wonderful scorpions and live tarantulas and the excellent reproduction of the desert's environment. After exploring a bit more of Arizona, I thought that this would be the opportunity to release myself from my old vow to see the Canyon de Chelly, so highly praised by Rasetti back in 1929, which I tried in vain to reach in 1936.

The weather was threatening, and in the deserts of northern Arizona there is always a chance of getting stuck in a heavy snowstorm, but we set off in our small rented car for Sedona, which reminded me of Cortina d'Ampezzo, although less delicately beautiful. After spending the night there, we moved north, in the direction of Highway 66, with which I had become familiar in 1936. We arrived in Flagstaff with great uncertainty, both in the weather and in my mind. Finally, I decided to head east, toward Holbrook. About halfway there, we found a long, straight road going north, toward Hopi territory, and much to Rosa's surprise, I turned onto it without hesitation; we were on our way to the Indian reservations.

The skies were gloomy, with dark clouds, and the desert had a strange glow, with magnificent and unusual views. We arrived at dusk at Second Mesa, where we found the pleasant surprise of a comfortable motel, run by the Hopis and with mainly Hopi clientele. The food was either Hopi or American, and we decided to try the Hopi. The next morning the skies were heavy with snow, and the roads started getting white, but I was determined to see the Canyon de Chelly. I took the wheel, as Rosa was afraid of skidding on the snow-covered road. (I remembered that when I had been there in 1936 there were no roads.) Along the way we picked up an Indian mother and daughter who were almost frozen; we left them at Chinle, where they wanted to buy a birthday cake for the man in their family. We checked in at the local motel and then headed for Canyon de Chelly. We received some information, unwillingly given, and finally found the road on the south side of the canyon, the only one allowed to white people who are not accompanied by an Indian guide. We started descending the picturesque trail. Looking down, one could see some agricultural land at the bottom, several hogans, and a few goats. When we arrived there, in order to get closer to the magnificent Indian ruins visible from a distance, one had to cross a small but icy creek. I did not hesitate in removing my shoes and socks; I rolled up my pants and got into the water. Rosa, after some hesitation, followed my example. The ruins of most of the communal dwellings were, as usual, leaning on and protected by huge stone walls. At that time of the year, there were no visitors, and one could wander freely, admiring the petrographs. The place was made still more poetic by the sound of the creek and its black poplars, whose light wood is used for many Indian sculptures.

It was a wonderful way to celebrate my seventy-seventh birthday.

A Few Words from Rosa

Many of Emilio's colleagues told me that they were very eager to read his memoirs. This was also the reason he wanted this book published posthumously. "I meet many of the protagonists on campus or at meetings, and they don't remember the facts the way they happened, but the way they would like them to have happened," he said. Emilio kept an enormous number of notes, notebooks, documents, and letters, and we spent months verifying details and dates.

Around that time, I found myself sitting next to Luis Alvarez at a dinner party, and he told me that he was writing his memoirs. "Do you spend so much time going through your filing cabinets reading old letters?" I asked him. "Oh, no," he said; "I just write down what I remember." Well, memory is likely to play tricks after so many years, and a lot of wishful thinking replaces the true facts. For all his previous books, Emilio researched tirelessly in his obsession with accuracy and truth; his memoirs were treated no differently. He had this deep sense of duty toward history; when he read it, he wanted the real facts, and when he wrote it, he tried to provide the reader with them according to his own expectations.

Writing his memoirs came rather easily to Emilio: there was so much to tell. As time went by, he trimmed a bit of the first manuscript, eliminating a few names and explicit details regarding the weakness, and I can add, wickedness, of relatives and colleagues. Having reached eighty made him more tolerant and forgiving, but not toward all; some wounds still hurt.

What was a major problem for him was when to end this book. At what stage of his life—after receiving the Nobel Prize, after retirement, when? He talked this over with several trusted friends, both in the United States and in Italy. One advised him to conclude it at the point where he returned to teach in Italy. Another told him to end the book philosophizing about the progress of science and the good things nuclear research offered the world in energy, medicine, communication, and so on. A third suggested a final chapter on Emilio's ideas regarding the future of science and how he visualized the world of his grandchildren. But nothing was decided.

For the reader who wonders what Emilio did after retirement, I'll try to give an idea of his life during his last years. At Berkeley, he was recalled to teach, usually during winter quarters, and he chose history of science rather than physics. The series of lectures he gave over the years on this subject eventually became a very successful book, *From X-Rays to Quarks,* and a few years later a second volume, *From Falling Bodies to Radio Waves,* appeared. Both were translated into several languages, and we did the proofreading of most of these editions. There was a yearly six-to-eight-week trip to Europe, organized so that it would coincide with the meetings or conferences he wished to attend, and this included the annual Board of Governors' Meeting at Tel Aviv University, which gave him the opportunity to visit his daughter Amelia and her family. He was a sought-after speaker, with a strong voice, who knew how to adapt his subject to the audience and keep them interested and awake. On our last trip to Europe, in June 1988, he gave the opening address at the congress of the Union of Producers and Distributors of Electric Energy (UNIPEDE) in Sorrento.

He always was up to date as regards the developments in his field and in science in general; he read new books and re-read old ones, he subscribed to publications, and attended weekly seminars and discussions; and when there was a subject he did not understand clearly, he did not hesitate to attend a class, sometimes given by one of his ex-students, or to ask "the younger generation" to explain the mysteries of the latest discovery to him. He also made a point of reading at least two books a year in German and in French, "just not to forget the language," and he was extremely proud of the fact that when German television interviewed him in 1985 for an hour-long documentary, he could do it in German without difficulty. At seventy-nine he learned how to use a personal computer, and this book was his first project on it. He subscribed to the *Wall Street Journal* and other financial publications and followed his investments almost daily; his knowledge of tax law was well above average.

You have just read the story of a very interesting and busy life. You know how it started and you may wonder how it ended. Emilio was well, physically and mentally, until his very last minute. At the time of his death there was a furor going on about "cold fusion." He phoned the scientists who announced the discovery in Salt Lake City, talked things over with them, and was mailed their papers, then decided that "it was not true." "When we made our discovery in Rome," he said, "within a week other laboratories over the world could replicate it with the same results. This thing won't fly."

In January 1989 Emilio delivered a brilliant address on the discovery of fission at the Annual Joint Meeting of the American Physical Society in San Francisco. He looked forward to attending the conference commemorating "50 years with Nuclear Fission" in Washington, D.C., on April 25–28. He and Glenn Seaborg were general co-chairmen of the event, and both had worked for months in the organization of it. "It is a good way to close a scientific career," he told me. Two weeks before, though, during a fund-raising dinner at UC Berkeley, he suddenly did not feel well, and I decided to drive him to the hospital. Although he did not suffer a heart attack, his electrocardiogram showed irregularities and his doctor decided to keep him in intensive care for a couple of days. Once home, he returned to his normal routine, reading, writing, making telephone calls, but the doctor did not permit him to travel to Washington, and he was extremely disappointed but accepted it. At that time, his old computer broke down and a friend sent him a used IBM computer he didn't need anymore. Now we had to convert all our diskettes to the new system, and we looked forward to the weekend, when Emilio's 20-year-old grandson Gino, who was studying at Berkeley and was familiar with the IBM PC, was to come for a visit and help us in this process. On Friday we drove to the doctor for a follow-up appointment and after that to the university, where Emilio picked up some books at the library, talked to colleagues, and so on; then we returned home with Gino. Saturday, April 22, 1989, seemed like one of Emilio's happiest days. He was so excited with the speed and all the potential of his new computer, he looked like a child in a toy store. After lunch, the men took a nap, and then continued their work. Gino had to be back on campus by 5 P.M., and we drove him to the train station; on the way back, Emilio suggested that we go for a walk. We selected an easy trail, he changed from his slippers to his tennis shoes, and we started walking slowly and talking about the events of the day. It was cool, but sunny, and we were the only people on the trail. After about five minutes Emilio calmly said, "Wait a minute" and stopped; he turned toward me and put his hands on my shoulders. I thought he was going to give me a kiss, as he used to do when he felt happy (and nobody was looking). Instead, his weight started pulling me down. The life that started in Tivoli eighty-four years before had ended.

Notes

Preface

1. Sir Rudolf Ernst Peierls, *Bird of Passage: Recollections of a Physicist* (Princeton: Princeton University Press, 1985).

Chapter 1

1. On the Villa d'Este, see D. R. Coffin, *The Villa d'Este at Tivoli* (Princeton: Princeton University Press, 1960), and C. Lamb, *Die Villa d'Este in Tivoli* (Munich: Prestel Verlag, 1968).

2. In 1986, in the Brera Library in Milan, I found a copy of the issue of *Scienza per tutti* dated March 1, 1914, which contains the articles mentioned.

3. See *Il collegio Ghislieri* (Milan: Alfieri e Lacroix, 1967), esp. p. 415. On Claudio Segrè, see L. Maddalena, "Claudio Segrè," *Boll. soc. geologica italiana* 47 (1928); A. Stella, "Commemorazione del socio Claudio Segrè," *Rend. acc. naz. Lincei (scienze fisiche)* 10 (1927): xi. On Gino Segrè, see G. Grosso, "Gino Segrè," *Temi emiliana* 19 (Milan, 1942); A. Candian et al., "Si respira nell'altitudine," *Temi* (Milan, 1962); Provincia di Torino, *Consegna del medaglione in memoria di G. Segrè* (Turin, 1963); V. Arangio-Ruiz, "Commemorazione del Socio Gino Segrè," *Rend. acc. naz. Lincei (scienze morali)* 2 (1947): 607; E. Betti, "Gino Segrè," *Riv. italiana sc. giuridiche* 18 (1942): 200, 302.

4. I myself have frequently been confused with the mathematician Beniamino Segre (whose name is written without the accent), and even with his famous cousin Corrado. Once, knowing that I was coming to visit him, Fabio Ferrari, a theoretician who studied with Geoffrey Chew, and who subsequently became rector of the University of Trento, made his young son read an article about me in an Italian encyclopedia; in it, under my name, there was a portrait of Beniamino Segre, and as soon as he saw me,

the little boy declared that I was not the man for whom he had prepared himself.

On another occasion, I placed a person-to-person call to Elfriede, who was in different city. The operator had some difficulty in understanding the name, but suddenly he said: "Ah, Segrè, like the physicist?" I felt flattered at being recognized and said: "Exactly. Actually, I *am* the physicist." Operator: "That is not possible, because that one is dead." I insisted: "I can assure you that I am the physicist and that I am alive." Operator: "That cannot be true because I have read your name on a street in my neighborhood, near Piazza or Via Fermi, and streets are named only after people who are dead." When I checked later, sure enough, near Piazza Fermi, there was a Via Corrado Segre.

5. My late cousin Silvia Treves Vidale wrote a privately circulated monograph on the Treves family. The Florence synagogue designed by my grandfather Marco Treves, and by M. Falcini and V. Micheli, is depicted on a 1987 Israeli postage stamp.

6. On Guido Treves, see A. Sapori, *Compagnie e mercanti di Firenze antica* (Florence, 1965); G. Devoto, *Civiltà di parole* (Florence: Vallecchi, 1969), vol. 2.

7. Among the paintings of Ettore Roesler Franz (1845–1907) is a series of watercolors titled *Roma sparita* (Vanished Rome), now at the Museo di Roma in Rome; many of them have been reproduced as postcards. Franz sold his watercolors to tourists as souvenirs, in the tradition of the Venetian Canaletto. I bought two such paintings at auctions in San Francisco.

8. Gaston Tissandier, *Le ricreazioni scientifiche* (Milan: Treves, 1897).

9. Adolphe Ganot, *Trattato elementare di fisica* (Milan: Pagnoni, 1863).

10. Ganot-Maneuvrier, *Traite élémentaire de physique*, 25th ed. (Paris: Hachette, 1913).

Chapter 2

1. The translation is from Ernesto Grillo, ed. and trans., I sepolcri *in Italian and English* (London and Glasgow: Blackie and Son, 1928).

2. For Enzo Sereni, see R. Bondy, *The Emissary: A Life of Enzo Sereni* (Boston: Little, Brown, 1977), as well as D. V. Segre, *Memories of a Fortunate Jew* (Bethesda, Md.: Adler & Adler, 1987); the latter author's impressions are not too different from mine. Enzo's brother Emilio became a prominent member of the Italian Communist Party; see *Enciclopedia europea,* s.v. "Sereni, Emilio" (Milan: Garzanti, 1980).

3. Sir Richard Glazebrook, *Light: An Elementary Textbook . . . for Colleges and Schools* (Cambridge: Cambridge University Press, 1894); Sir Robert S. Ball, *The Elements of Astronomy* (New York: Longmans, 1891); James Clerk

Maxwell, *Theory of Heat* (New York: Appleton, 1875); Fritz Reiche, *The Quantum Theory* (New York: Dutton, [1922?]).

4. *Albert Einstein, the Human Side: New Glimpses from His Archives,* ed. Helen Dukas and Banesh Hoffmann (Princeton: Princeton University Press, 1979), p. 32.

5. See *B. J. Moyer: A True Humanist* (pamphlet in memoriam, Eugene: University of Oregon, 1972).

Chapter 3

1. Walther Nernst, *Theoretische Chemie vom Standpunkte der Avogadroschen Regel und der Thermodynamik.* I read a French translation of this work, which I found among the books of my brother Angelo.

2. The course was later published as T. Levi-Civita and U. Amaldi, *Compendio di meccanica razionale* (Bologna: Zanichelli, 1928).

3. Much has been written on Majorana, often pure fiction. For a serious life, see Edoardo Amaldi, "E. Majorana: Man and Scientist," in *Strong and Weak Interactions*, ed. A. Zichichi (New York: Academic Press, 1966). See also E. Segrè, *Storia contemporanea,* vol. 19 (Bologna, 1988), p. 107.

4. Enrico Fermi "Sui principi della teoria dei quanti" (On the principles of quantum theory). FP22. The general reference for all scientific publications by Fermi is Enrico Fermi, *Collected Papers*, ed. E. Segrè et al. (Chicago: University of Chicago Press, 1961). Single papers are cited by the notation FP, followed by the number in *Collected Papers*. For biographical data, see E. Segrè, *Enrico Fermi, Physicist* (Chicago: University of Chicago Press, 1970), and *Enrico Fermi, fisico* (Bologna: Zanichelli, 1971, 1987), as well as Laura Fermi, *Atoms in the Family: My Life with Enrico Fermi* (1954; New York: Tomash Publishers / American Institute of Physics, 1987).

5. On Franco Rasetti, see T. Nason, "A Man for All Sciences," *Johns Hopkins Magazine* 17 (1966): 12.

6. On the Volta Conference at Como, see *Atti del congresso int. dei fisici (11–20 sett. 1927 Como- Pavia- Roma) pubblicati a cura del comitato per le onoranze ad A. Volta nel primo centenario della morte* (Bologna: Zanichelli, 1928).

7. Enrico Fermi and Franco Rasetti, "Una misura del rapporto h/k per mezzo della dispersione anomala del tallio" (A measurement of the ratio h/k using anomalous dispersion in thallium). FP40b.

8. FP2: 673.

9. Laura Fermi vividly describes the company in *Atoms in the Family.*

10. On O. M. Corbino, see *Conferenze e discorsi di O. M. Corbino* (Rome: Edizioni Enzo Pinci, 1937), and Epicarmo Corbino, *Racconto di una vita*

(Naples: Edizioni scientifiche italiane, 1972), as well as *Gli atti del convegno ai Lincei nel cinquantenario della morte*, the proceedings of a conference at Lincei on the fiftieth anniversary of Corbino's death.

11. Enrico Fermi "Un metodo statistico per la determinazione di alcune proprietà dell'atomo" (A statistical method for the determination of some atomic properties). FP43.

12. On Edoardo Amaldi, see E. Segrè, "Italian Physics in Amaldi's Time," and Edoardo Amaldi, "The Years of Reconstruction," in *Perspectives of Fundamental Physics: Proceedings of the Conference Held at the University of Rome, 7–9 September 1978, Dedicated to Edoardo Amaldi*, ed. Carlo Schaerf (New York: Harwood Academic Publishers, 1979).

13. E. Segrè and Edoardo Amaldi, "Sulla dispersione anomala del mercurio e del litio" (On anomalous dispersion in mercury and in lithium), *Rend. Lincei*, ser. 6, 7 (1928): 407–9.

14. See, on this, Laura Fermi, *Atoms in the Family*, pp. 69–73.

Chapter 4

1. See chapter 3, n. 13.

2. Edoardo Amaldi and E. Segrè, "Sulla teoria dell'effetto Raman" (Quantum theory of the Raman effect), *Rend. Lincei*, ser. 6, 9 (1929): 407–9.

3. E. Segrè, "Sulla dispersione anomala negli spettri di bande" (Anomalous dispersion in band spectra), *Rend. Lincei*, ser. 6, 10 (1929): 590–94. See also *Nuovo cimento* 4 (1930): 144–47.

4. See FP1: 444, 445.

5. F. Rasetti, "Raman Effekt und Struktur der Molekeln und Kristalle," in Peter J. W. Debye, *Leipziger Vorträge*, 1931 (Leipzig: S. Hirzel, 1931).

6. E. Segrè, "Evidence for Quadrupole Radiation," *Nature* 126 (1930): 882.

7. E. Segrè, "Über den Zeemaneffekt von Quadrupollinien" (On the Zeeman effect of quadrupole lines), *Zs. f. Physik* 66 (1930): 827–29.

8. On Zeeman, see the obituary by Lord Rayleigh in *Obituary Notices of Fellows of the Royal Society of London* 4 (1944): 591.

9. On Cornelius J. Bakker (1904–60), see Armin Hermann et al., *History of CERN*, vol. 1 (Amsterdam and New York: North-Holland Physics Pub., 1987), pp. 127, 158.

10. See E. Segrè, "Righe di quadrupolo negli spettri di raggi X" (Quadrupole lines in X-ray spectra), *Rend. Lincei*, ser. 6, 14 (1931): 501–5.

11. Arnold Sommerfeld, *Atombau und Spektrallinien: Wellenmechanischer Ergänzungsband*, vol. 2 (Braunschweig: F. Vieweg, 1939).

12. See E. Segrè, "Otto Stern," in Nat. Ac. of Sciences, *Biographical*

Memoirs, vol. 43 (Washington, D.C.: National Academy of Sciences, 1973), p. 215.

13. Otto Frisch has written an autobiography entitled *What Little I Remember* (Cambridge and New York: Cambridge University Press, 1979).

14. Friedrich von Schiller, "Sprüche des Konfucius." One of Niels Bohr's favorite quotations, "Nur die Fülle führt zur Klarheit / Und im Abgrund wohnt die Wahrheit" (Naught but fulness makes us wise,— / Buried deep, truth ever lies!) comes from the same poem. The translation is from *Poems of Schiller* (New York: John D. Williams, n.d.).

15. Adrienne Thomas's antiwar novel *Die Katrin wird Soldat* (Berlin: Ullstein Verlag, 1930; Frankfurt am Main: Fischer Taschenbuch, 1987).

16. Enrico Fermi and E. Segrè, "Sulla teoria delle strutture iperfini" (On the theory of hyperfine structure), *Reale acc. d'Italia, memorie cl. scienze fisiche* 4 (1933): 131–58, summarized in "Zur Theorie der Hyperfeinstruktur," *Zs. f. Physik* 82 (1933): 729–49, and in *Nuovo cimento* 7 (1934). FP75b.

17. E. Segrè and C. J. Bakker, "Zeeman Effect of a Forbidden Line," *Nature* 128 (1931): 1076.

18. E. Segrè, "Un metodo per l'osservazione dell'effetto Zeeman quadratico" (A method for observing the quadratic Zeeman effect), *Ricerca scientifica* 4, no. 2 (1933): 531, and "Effetto Zeeman quadratico nella serie principale del sodio" (Quadratic Zeeman effect in the principal sodium series), *Nuovo cimento* 5 (1934): 304–8.

19. Edoardo Amaldi and E. Segrè, "Einige spektroskopische Eigenschaften hochangeregter Atome" (Some spectroscopic properties of highly excited atoms), in *Zeeman Verhandelingen* (The Hague: Martinus Nijhoff, 1935), pp. 8–17.

20. R. Frisch and E. Segrè, "Über die Einstellung der Richtungsquantelung II" (On the dynamics of space quantization), *Zs. f. Physik* 80 (1933): 610–16; also R. Frisch and E. Segrè, "Ricerche sulla quantizzazione spaziale" (Investigation of spatial quantization), *Nuovo cimento* 2 (1933): 78–91; and see I. I. Rabi, "On the Process of Space Quantization," *Phys. Rev.* 49 (1936): 324.

21. Ernest Rutherford, James Chadwick, and C. D. Ellis, *Radiations from Radioactive Substances* (Cambridge: Cambridge University Press, 1930).

22. E. Amaldi, O. D'Agostino, E. Fermi, F. Rasetti, and E. Segrè, "Radioattività 'beta' provocata da bombardamento di neutroni, III" (Beta radioactivity produced by neutron bombardment, III), *La ricerca scientifica* (henceforth cited as *RS*) 5, no. 1 (1934): 452–53 [FP86a]; E. Amaldi, O. D'Agostino, E. Fermi, F. Rasetti, and E. Segrè, "Radioattività provocata da bombardamento di neutroni, IV" (Radioactivity produced by neutron bombardment, IV), *RS* 5, no. 1 (1934): 652–53 [FP87a]; E. Amaldi, O. D'Agostino, E. Fermi, F. Rasetti, and E. Segrè, "Radioattività provocata

da bombardamento di neutroni, V" (Radioactivity produced by neutron bombardment, V), *RS* 5, no. 2 (1934): 21–22 [FP88a]; E. Amaldi, O. D'Agostino, and E. Segrè, "Radioattività provocata da bombardamento di neutroni, VI" (Radioactivity produced by neutron bombardment, VI), *RS* 5, no. 2 (1934): 38–82; E. Amaldi, O. D'Agostino, E. Fermi, B. Pontecorvo, F. Rasetti, and E. Segrè, "Radioattività provocata da bombardamento di neutroni, VII" (Radioactivity produced by neutron bombardment, VII), *RS* 5, no. 2 (1934): 467–70 [FP89a]; E. Amaldi, O. D'Agostino, E. Fermi, B. Pontecorvo, F. Rasetti, and E. Segrè, "Radioattività provocata da bombardamento di neutroni, VIII" (Radioactivity produced by neutron bombardment, VIII), *RS* 6, no. 1 (1935): 123–25 [FP90a]; E. Amaldi, O. D'Agostino, E. Fermi, B. Pontecorvo, and E. Segrè, "Radioattività indotta da bombardamento di neutroni, IX" (Radioactivity induced by neutron bombardment, IX), *RS* 6, no. 1 (1935): 435–37 [FP91a]; E. Amaldi, O. D'Agostino, E. Fermi, B. Pontecorvo, and E. Segrè, "Radioattività indotta da bombardamento di neutroni, X" (Radioactivity induced by neutron bombardment, X), *RS* 6, no. 1 (1935): 581–84 [FP92a]. And see Segrè, *Enrico Fermi, Physicist,* pp. 73 ff.

23. See E. Segrè, "A cinquant'anni dalla radioattività artificiale provocata da neutroni," *Rendiconti della Accademia nazionale delle scienze, detta dei XL, memorie fis.*, 5th ser., 8, pt. 2 (1984): 165.

24. Aristide von Grosse, "The Chemical Properties of Elements 93 and 94," *J. Am. Chem Soc.* 57 (1935): 440–41; E. Segrè, "An Unsuccessful Search for Transuranic Elements," *Phys. Rev.* 55 (1939): 1104–5; G. E. Villar, *Boletín de la Facultad de ingeniería* (Montevideo) 5 (1938): 231. These papers preceded the discovery of neptunium by Edwin McMillan and Philip Abelson. If the writings of Aristide von Grosse and of Ida Noddack (see n. 25 below) had been better appreciated, the whole history of fission and of the transuranics would have been different. The following papers appeared after the discovery of neptunium: M. Mayer, "Rare Earths and Transuranic Elements," *Phys. Rev.* 60 (1941): 184; G. T. Seaborg and E. Segrè, "The Trans-Uranium Elements," *Nature* 159 (1947): 863–65. Starting in 1944, Seaborg developed the concept of an actinide family; see, e.g., *The Transuranium Elements,* ed. G. T. Seaborg, J. J. Katz, and W. M. Manning (New York: McGraw-Hill, 1949): 1517–20. Seaborg's personal recollections may be found in *The Transuranium Elements* (New Haven: Yale University Press, 1957).

25. Ida Noddack, "Über das Element 93," *Angew. Chemie* 47 1934): 653–55.

26. Otto Hahn and Fritz Strassmann, "Über den Nachweis und das Verhalten der bei der Bestrahlung des Urans mittels Neutronen entstehenden Erdalkalimetalle," *Naturwiss.* 27 (1939): 39. The history of the discovery of fission is very complicated: important original sources are FP;

Frédéric Joliot and Irène Curie, *Oeuvres scientifiques complètes* (Paris: Presses universitaires de France, 1961); Otto Hahn, *Vom Radiothor zur Uranspaltung: Eine wissenschaftliche Selbstbiographie* (Braunschweig: Vieweg, 1962); and Fritz Krafft, *Im Schatten der Sensation: Leben und Wirken von Fritz Strassmann* (Weinheim: Verlag Chemie, 1981).

27. E. Fermi, E. Amaldi, O. D'Agostino, F. Rasetti, and E. Segrè, "Artificial Radioactivity Produced by Neutron Bombardment," *Proc. Roy. Soc.* (London) A146 (1934): 483–500. FP98.

28. Enrico Fermi, "Natural Beta Decay" (Int. Conf. on Physics, London, 1934), in *Nuclear Physics* (London: Physical Society, 1934), vol. 1. FP102.

29. E. Fermi, E. Amaldi, B. Pontecorvo, F. Rasetti, and E. Segrè, "Azione di sostanze idrogenate sulla radioattività provocata da neutroni" (The influence of hydrogenous substances on the radioactivity produced by neutrons), *Ricerca scientifica* 5 (1934): 282. FP105a.

30. Fritz Haber received the Nobel Prize for chemistry in 1919 for synthesizing ammonia from its elements.

31. Hermann Luedemann (1880–1959) was an engineer of liberal-socialist tendencies who turned to politics. From 1929 to 1932, he was *Oberpräsident* of Silesia—the highest civil official in the province—and Elfriede served as his secretary in that office. The Nazis put Luedemann in a concentration camp, but he survived.

32. On Pegram, see L. A. Embrey, "G. B. Pegram," in Nat. Ac. of Sciences, *Biographical Memoirs,* vol. 61 (Washington, D.C.: National Academy of Sciences, 1970), p. 357.

33. See J. R. Dunning, G. A. Fink, G. B. Pegram, and E. Segrè, "Experiments on Slow Neutrons with Velocity Selector," *Phys. Rev.* 49 (1936): 198–99; G. A. Fink, J. R. Dunning, G. B. Pegram, and E. Segrè, "Production and Absorption of Slow Neutrons in Hydrogenic Materials," *Phys. Rev.* 49 (1936): 199. Also see F. Rasetti, E. Segrè, G. A. Fink, J. R. Dunning, and G. B. Pegram, "Sulla legge di assorbimento dei neutroni lenti," *Rend. Lincei,* 6th ser., 23 (1936): 343–45, and "On the Absorption Law for Slow Neutrons," *Phys. Rev.* 49 (1936): 104.

Chapter 5

1. See U. Panichi; "Commemorazione del Corrispondente Carlo Perrier," *Rend. Lincei 6* (1949): 386.

2. There is still a copy of the lecture notes for this course in the library of the Palermo Physics Institute.

3. G. Bernardini, G. Gentile, Jr., and G. Polvani, *Questioni di fisica* (Physics topics) (Florence: Sansoni, 1947). Only the first volume was pub-

lished. The planned contents of the other two are given in it, but they never appeared.

4. On Lawrence, see Herbert Childs, *An American Genius: The Life of Ernest Orlando Lawrence, Father of the Cyclotron* (New York: Dutton, 1968), and J. L. Heilbron and Robert W. Seidel, *Lawrence and His Laboratory: A History of the Lawrence Berkeley Laboratory*, vol. 1 (Berkeley and Los Angeles: University of California Press, 1989). I thank the authors for access to the manuscript of the latter book.

5. See E. M. McMillan, "The Transuranium Elements: Early History," in *Les Prix Nobel in 1951* (Stockholm: Nobelstiftung, 1952), pp. 165–73.

6. Abelson later collaborated with McMillan in the discovery of neptunium. In time he became director of the Carnegie Institution of Washington and editor in chief of *Science*. His scientific work was mostly in geochemistry and isotope separation. See also L. W. Alvarez, *Adventures of a Physicist* (New York: Basic Books, 1987), ch. 4, and E. Segrè, "A cinquant'anni dalla radioattività artificiale provocata da neutroni," *Rendiconti della Accademia nazionale delle scienze, detta dei XL, memorie fis.*, 5th ser., 8, pt. 2 (1984): 165.

7. There is a vast literature, often fictional in character, on Oppenheimer. See esp. *Robert Oppenheimer: Letters and Recollections*, ed. A. K. Smith and C. Weiner (Cambridge, Mass.: Harvard University Press, 1980).

8. See C. Artom, G. Sarzana, C. Perrier, M. Santangelo, and E. Segrè, "Rate of Organification of Phosphorus in Animal Tissue," *Nature* 139 (1937): 836–38, and "Phospholipid Synthesis during Fat Absorption," *Nature* 139 (1937): 1105–6.

9. From my laboratory notebooks for 1937.

10. C. Perrier and E. Segrè, "Alcune proprietà chimiche dell'elemento 43," *Rend. Lincei*, 6th ser., 25 (1937): 723–30, and 27 (1937): 579–81. Also "Some Chemical Properties of Element 43," *Journ. of Chem. Phys.* 5 (1937): 712–16, and 7 (1939): 155–56.

11. We know today that the longest lived isotopes of technetium have a period of 4.2 million years, a time too short to permit survival from primordial material. Minute amounts of technetium produced in nature by the spontaneous fission of uranium were detected by P. K. Kuroda et al. in 1961.

12. Claimants to the discovery of Element 43 prematurely called it ilmenium, davyum, lucium, nipponium, and masurium, among other names, but their claims were not substantiated. See H. W. Kirby, "Technetium," *Gmelin Handbook of Inorganic Chemistry*, 8th ed. (Berlin: Springer, 1982).

13. See Hilde Levi, *George de Hevesy: Life and Work* (Copenhagen: Rhodos, 1985).

14. C. Perrier and E. Segrè, "Technetium: The Element of Atomic Number 43," *Nature* 159 (1947): 24.

15. See E. Segrè, "Italian Physics in Amaldi's Time," and Edoardo Amaldi, "The Years of Reconstruction," in *Perspectives of Fundamental Physics: Proceedings of the Conference Held at the University of Rome, 7–9 September 1978, Dedicated to Edoardo Amaldi*, ed. Carlo Schaerf (New York: Harwood Academic Publishers, 1979).

16. See also Edoardo Amaldi, "E. Majorana: Man and Scientist," in *Strong and Weak Interactions*, ed. A. Zichichi (New York: Academic Press, 1966), and E. Segrè, in *Storia contemporanea*, vol. 19 (Bologna, 1988), p. 107.

Chapter 6

1. For the text and the signers of the *Manifesto della razza*, and on the subsequent period, see R. De Felice, *Storia degli Ebrei italiani sotto il fascismo* (History of the Italian Jews under Fascism), 3d ed. (Turin: Einaudi, 1972).

2. On Jenkins, see *In Memoriam*, University of California, Berkeley, 1962, F. A. Jenkins, 1899–1960. On Brode, see ibid. 1986, R. B. Brode (1900–1986), p. 2.

3. On the history of the Radiation Laboratory, see the works cited in chapter 5, n. 4.

4. See works cited in chapter 5, nn. 5, 6.

5. See Martin D. Kamen, *Radiant Science, Dark Politics: A Memoir of the Nuclear Age* (Berkeley and Los Angeles: University of California Press, 1985), which vividly portrays the Radiation Laboratory and the scientific climate in the United States before the war.

6. See A. A. Noyes and W. C. Bray, *A System of Qualitative Analysis for the Rare Elements* (New York: Macmillan, 1927).

7. Tables of natural radioactive isotopes go back to Curie, Rutherford, and their colleagues. With the discovery of artificial radioactivity, they became much larger. A first one was drawn up in Rome by our group and by G. Fea. I drew a useful diagram, following Heisenberg, while still in Rome. Later keeping such tables current required several people, and these days isotope tables are as thick as telephone directories.

8. E. Segrè and G. T. Seaborg, "Nuclear Isomerism in Element 43," *Phys. Rev.* 54 (1938): 772; ibid. 55 (1939): 808.

9. See also the work cited in chapter 5, n. 7, as well as U.S. Atomic Energy Commission, *In the Matter of J. Robert Oppenheimer: Transcript of Hearing before Personnel Security Branch and Texts of Principal Documents and Letters* (1954; Cambridge, Mass.: MIT Press, 1971), which gives a vivid picture, not only of Oppenheimer, but also of many other persons who testified at those hearings, among them Edward Teller, L. W. Alvarez, W. M. Latimer, H. A. Bethe, General Leslie Groves, and I. I. Rabi.

10. L. I. Schiff, *Quantum Mechanics* (New York: McGraw-Hill, 1955).

11. On S. K. Allison, see *Bull. Atomic Scientists* 22 (1966): 2.

12. F. A. Jenkins and E. Segrè, "The Quadratic Zeeman Effect," *Phys. Rev.* 55 (1939): 52.

13. E. Segrè, R. S. Halford, and G. T. Seaborg, "Chemical Separation of Nuclear Isomers," *Phys. Rev.* 55 (1939): 55.

14. E. Segrè, "An Unsuccessful Search for Transuranic Elements," *Phys. Rev.* 55 (1939): 1104.

15. E. Segrè and C. S. Wu, "Some Fission Products of Uranium," *Phys. Rev.* 57 (1940): 552, and "Radioactive Xenons," ibid. 67 (1945): 142.

16. On Placzek, see also Edoardo Amaldi, "George Placzek (1905–1955)," *Ricerca scientifica* 26 (1956): 2038.

17. Cornog, in *Discovering Alvarez: Selected Works of Luis W. Alvarez, with Commentary by His Students and Colleagues,* ed. W. P. Trower (Chicago: University of Chicago Press, 1987), p. 26. And see D. R. Corson, K. R. MacKenzie, and E. Segrè, "Possible Production of Radioactive Isotopes of Element 85," *Phys. Rev.* 57 (1940): 459, and "Artificially Radioactive Element 85," ibid. 58 (1940): 672–78.

18. E. Fermi and E. Segrè, "Fission of Uranium by Alpha Particles," *Phys. Rev.* 59 (1941); 59. FP135.

19. E. Segrè, "Possibility of Altering the Decay Rate of a Radioactive Substance," *Phys. Rev.* 71 (1947): 274; R. F. Leininger, E. Segrè, and C. E. Wiegand, "Experiments on the Effect of Atomic Electrons on the Decay Constant of Be7," *Phys. Rev.* 76 (1949): 897, and ibid. 81 (1951): 284.

20. See V. A. Johnson, *Karl Lark-Horovitz: Pioneer in Solid State Physics* (New York: Pergamon, 1969).

21. Birge was well known for his studies on molecular spectra and on universal constants, and also an important administrator at the University of California, to which he was deeply devoted. One of the physics buildings at Berkeley bears his name, well-deserved recognition of his work. See E. McMillan, "R. T. Birge, 1887–1980," in Am. Phil. Soc. (Philadelphia), *Yearbook, 1981*, p. 430.

22. R. T. Birge, "History of the Physics Department" (University of California, Berkeley, 1966–?; 5 vols., mimeographed).

23. See the documentation in G. T. Seaborg, *Early History of Heavy Isotope Research at Berkeley*, Lawrence Berkeley Laboratory Publication No. 97 (Berkeley, 1976). The archives of the Bancroft Library at the University of California, Berkeley, contain much additional material, including a letter from Segrè to Fermi dated January 11th, 1941, signaling the beginning of the work.

24. Many of these documents are to be found in the Bancroft Library.

25. G. T. Seaborg and E. Segrè, "The Transuranium Elements," *Nature* 159 (1947): 159; E. Segrè, E. M. McMillan, J. W. Kennedy, and A. C. Wahl,

"An Account of the Discovery and Early Study of Element 94," UCRL report No. 2791, Dec. 23, 1942.

26. Letter of G. T. Seaborg to Fermi, January 11, 1941, Bancroft Library.

27. J. W. Kennedy, G. T. Seaborg, E. Segrè, and A. C. Wahl, "Properties of 94^{239}," *Phys. Rev.* 70 (1946): 555–56.

28. On A. L. Loomis, see L. W. Alvarez in Nat. Ac. of Sciences, *Biographical Memoirs,* vol. 51 (Washington, D.C.: National Academy of Sciences, 1980), p. 309.

29. See J. W. Kennedy and E. Segrè, "Component Analysis of Small Uranium Samples," Manhattan District Report MDDC-973, March 26, 1943.

30. E. Segrè, "Artificial Radioactivity and the Completion of the Periodic System of the Elements," *Scientific Monthly* 57 (1943): 57.

31. See, e.g., H. D. Smyth, *Atomic Energy for Military Purposes* (Princeton: Princeton University Press, 1945), the first and justly famous report on the atomic bomb. For details and documentation, consult R. G. Hewlett and O. E. Anderson, Jr., *A History of the United States Atomic Energy Commission,* vol. 1, *The New World, 1939/1946* (University Park, Pa.: Pennsylvania State University Press, 1962).

32. A letter to Fermi written in 1942, now in the Bancroft Library, gives my feelings about going to Los Alamos.

Chapter 7

1. The translation is from *Schiller's Historical Dramas* (New York: John D. Williams, n.d.).

2. Leslie R. Groves, *Now It Can Be Told* (New York: Harper & Row, 1962). See also K. D. Nichols, *The Road to Trinity: A Personal Account of How America's Nuclear Policies Were Made* (New York: Morrow, 1987).

3. On wartime Los Alamos, see *Reminiscences of Los Alamos, 1943–1945,* ed. Lawrence Badash, Joseph O. Hirschfelder, and Herbert P. Broida (Boston: Reidel, 1980), and Laura Fermi, *Atoms in the Family.* For a bibliography, see Richard Rhodes *The Making of the Atomic Bomb* (New York: Simon & Schuster, 1987), a well-written and accurately researched and documented work.

4. Winston S. Churchill, *The Second World War*, vol. 4, *The Hinge of Fate* (London: Cassel, 1951), p. 3.

5. Robert Serber, "The Los Alamos Primer" (Los Alamos, N. Mex.: University of California, Los Alamos Scientific Laboratory, photocopied typescript, 1973). Notes by E. U. Condon based on five introductory lectures given by Robert Serber in April 1943 in connection with the Manhattan Project. Declassified "secret limited," February 25, 1963. The titles

of the sections are: 1. Object; 2. Energy of fission process; 3. Fast neutron chain reaction; 4. Fission cross-sections; 5. Neutron spectrum; 6. Neutron number; 7. Neutron capture; 8. Why ordinary U is safe; 9. Material 49 (Pu^{239}); 10. Simplest estimate of minimum size of bomb; 11. Effect of tamper; 12. Damage; 13. Efficiency; 14. Effect of tamper on efficiency; 15. Detonation; 16. Probability of predetonation; 17. Fizzles; 18. Detonating source; 19. Neutron background; 20. Shooting; 21. Autocatalytic methods; 22. Conclusion.

6. See A. O. Nier, "J. H. Williams," in Nat. Ac. of Sciences, *Biographical Memoirs,* vol. 42 (Washington, D.C.: National Academy of Sciences, 1971), p. 339.

7. Josef Rotblat subsequently devoted himself to the Pugwash Movement and to the search for solutions to the immense problems created by atomic arms.

8. The first modest table of isotopes was published by a student in our group in Rome in the 1930s; see G. Fea, "Tabelle riassuntive e bibliografia delle trasmutazioni artificiali" (Comprehensive tables and bibliography of artificial transmutations), *Nuovo cimento* 12 (1935): 368–407. A similar compilation edited by C. M. Lederer and V. Shirley, *Table of Isotopes* (New York: Wiley, 1978), required 1,523 pages.

9. On G. I. Taylor, see *Biog. Mem. Fell. R. Soc.* vol. 22 (1976), p. 565.

10. E. Segrè, "Spontaneous Fission," *Phys. Rev.* 86 (1952): 21.

11. See Richard P. Feynman, *Surely You Are Joking, Mr. Feynman! Adventures of a Curious Character* (New York: Norton, 1985), pp. 120–21. A letter by Teller reported in Stanley A. Blumberg and Gwinn Owens, *Energy and Conflict: The Life and Times of Edward Teller* (New York: Putnam, 1976), p. 457, refers to the same episode. I do not remember having had exchanges with Teller on this subject at the time.

12. On the Trinity test, see W. L. Laurence, *Dawn over Zero: The Story of the Atomic Bomb* (New York: Knopf, 1946).

Chapter 8

1. R. T. Birge, "History of the Physics Department" (Berkeley: University of California, mimeographed, 1966–?), vol. 5, ch. 18.

2. Oppenheimer's letter is in the Bancroft Library.

3. The ideas of the theoreticians, in truth not all too brilliant, but based on ideas then current, are reflected in their programmatic documents for the Rad Lab.

4. The documentation relative to this chapter is contained in my private archive. Among the documents there are several texts of agreements; letters narrating previous events in preparation for my first return to Italy in 1947; hundreds of letters by my brother Angelo; accounts of different kinds; and

a file relative to the events of 1953, including my correspondence with Marco concerning our father's will and its legal ramifications, as well as legal opinions. For my visit to Italy in 1947, there is a series of letters to Elfriede, almost constituting a diary. Correspondence with Riccardo Rimini in Uruguay, before and after the war, often reveals our most confidential feelings and opinions.

5. See G. Bordignon Favero, *The Villa Emo at Fanzolo* (University Park, Pa.: Pennsylvania State University Press, 1972).

Chapter 9

1. For the general history of the Rad Lab in the postwar era, see J. L. Heilbron, Robert W. Seidel, and Bruce Wheaton, *Lawrence and His Laboratory: Nuclear Science at Berkeley, 1931–1961* (Berkeley: Office for the History of Science and Technology, University of California, 1981), and a communication by Seidel in the Proceedings of the 1985 International Symposium on Particle Physics: Pions and Quarks held at Chicago in 1985. I learned of many of the events of those times from these sources. Only a small circle, to which I did not belong, was privy to Lawrence's activities. See also H. F. York, *The Advisors: Oppenheimer, Teller and the Superbomb* (San Francisco: W. H. Freeman, 1976) and *Making Weapons, Talking Peace* (New York: Basic Books, 1987).

2. See, e.g., *Nucleon-Nucleon Scattering*, ed. K. Nishimura (Physical Society of Japan, Selected Papers in Physics, n.s., no. 26).

3. This series of investigations starts with J. Hadley, E. L. Kelly, C. E. Leith, E. Segrè, C. Wiegand, and H. F. York, "Angular Distribution in n-p Scattering with 90-Mev Neutrons," *Phys. Rev.* 73 (1948): 1114–15, and extends up to 1957. A partial review is given in E. Segrè, "High Energy Scattering and Polarization," *Physica* 22 (1956): 1079–90.

4. H. Stapp, T. Ypsilantis, and N. Metropolis, "Phase Shift Analysis of 310 Mev p-p Scattering Experiments," *Phys. Rev.* 105 (1957): 302.

5. *Experimental Nuclear Physics*, ed. E. Segrè, with contributions by H. Staub; H. Bethe and J. Ashkin; N. F. Ramsey; K. T. Bainbridge; P. Morrison; B. T. Feld; E. Segrè; G. C. Hanna; M. Deutsch and O. Kofoed-Hansen; and E. M. McMillan (New York: Wiley, 1953–59).

6. See E. Segrè, "Preface," *Ann. Rev. Nuclear Science* 26 (1976): vii–x.

7. See E. Segrè, "High Energy Scattering of Neutrons and Protons," *Helvetica Phys. Acta* 23, supp. 3 (1950): 197–205.

8. See G. R. Stewart, *The Year of the Oath* (Garden City, N.Y.: Doubleday, 1950), and D. P. Gardner, *The California Oath Controversy* (Berkeley and Los Angeles: University of California Press, 1967), as well as R. T.

Birge, "History of the Physics Department" (University of California, Berkeley, 1966–?; 5 vols., mimeographed), vol. 5, ch. 19.

9. Pius XI, *Per la azione cattolica* (encyclical, June 29th 1931). See, e.g., A. C. Jemolo, *Chiesa e stato in Italia negli ultimi 100 anni* (Turin: Einaudi, 1949), p. 664.

10. See York, *The Advisors*, and Rhodes, *Making of the Atomic Bomb.*

11. S. A. Blumberg and Gwinn Owens, *Energy and Conflict: The Life and Times of Edward Teller* (New York: Putnam, 1976) gives Teller's version of these events. E. Segrè, *Enrico Fermi, fisico,* 2d ed. (Bologna: Zanichelli, 1987) supplies some documents newly made available.

12. E. Segrè, *I nuovi elementi chimici: Chimica nucleare alle alte energie* (The new chemical elements: Nuclear chemistry at high energies) (Rome: Acc. naz. Lincei, Fondazione Donegani, 1953).

13. E. Segrè, "Über den Zeemaneffekt von Quadrupollinien" (On the Zeeman effect of quadrupole lines), *Zs. f. Physik* 66 (1930): 827–29; partial review in E. Segrè, "L'irradiamento dei quadrupoli" (Quadrupole radiation), *Nuovo cimento* 2 (1931): 28–37.

14. E. Amaldi and E. Segrè, "Einige spektroskopische Eigenschaften hochangeregter Atome" (Some spectroscopic properties of highly excited atoms), in *Zeeman Verhandelingen* (The Hague: Martinus Nijhoff, 1935), pp. 8–17.

15. E. Segrè, "Artificial Radioactivity and the Completion of the Periodic System of the Elements," *Scientific Monthly* 57 (1943): 12–16; J. W. Kennedy, G. T. Seaborg, E. Segrè, and A. C. Wahl, "Properties of 94^{239}," *Phys. Rev.* 70 (1946): 555–56; and G. T. Seaborg and E. Segrè, "The Trans-Uranium Elements," *Nature* 159 (1947): 863–65.

16. E. Segrè, R. S. Halford, and G. T. Seaborg, "Chemical Separation of Nuclear Isomers," *Phys. Rev.* 55 (1938): 321–22.

17. E. Segrè, "Possibility of Altering the Decay Rate of a Radioactive Substance," *Phys. Rev.* 71 (1946): 274 (abstract).

18. Partial summary of the former in E. Segrè, "High Energy Scattering and Polarization" (cited n. 3 above); partial summary of the latter in E. Segrè, "Antinucleons: Richtmyer Lecture 1957," *Am. Jour. of Physics* 25 (1957): 363–69.

19. E. Fermi, E. Amaldi, O. D'Agostino, F. Rasetti, and E. Segrè, "Artificial Radioactivity Produced by Neutron Bombardment," *Proc. Roy. Soc.* (London) 146 (1934): 483–500; E. Fermi, E. Amaldi, O. D'Agostino, B. Pontecorvo, F. Rasetti, and E. Segrè, "Artificial Radioactivity Produced by Neutron Bombardment, II," ibid. 149 (1935): 552–58.

20. Lewis Strauss, a well-known American investment banker and a protégé of President Hoover's, suffered a similar failure. He attributes this to the fact that the leaders of U.S. industry then thought that nuclear energy was science fiction; see Strauss, *Men and Decisions* (Garden City, N.Y.:

Doubleday, 1962). And see also Lawrence Badash, Elizabeth Hodes, and Adolph Tiddens, "Nuclear Fission: Reaction to the Discovery in 1939," *Proc. Am. Phil. Soc.* 130 (1986): 196–231.

21. J. W. Mihelich, A. Schardt, and E. Segrè, "Energy Levels in Po^{210}," *Phys. Rev.* 95 (1954): 1508–16.

22. C. L. Oxley, W. F. Cartwright, J. Rouvina, E. Baskir, D. Klein, J. Ring, and W. Skillman, "Double Scattering of High Energy Protons," *Phys. Rev.* 91 (1953): 419.

23. Enrico Fermi, "Polarization of High Energy Protons Scattered by Nuclei," *Nuovo cimento* 11 (1954): 417. FP267.

Chapter 10

1. The translation is from *Poems of Schiller* (New York: John D. Williams, n.d.).

2. See Stanley A. Blumberg and Gwinn Owens, *Energy and Conflict: The Life and Times of Edward Teller* (New York: Putnam, 1976), p. 374, for Teller's account of his visit to Fermi following my message.

3. Enrico Fermi, *Nuclear Physics,* notes compiled by J. Orear, A. H. Rosenfeld, and R. A. Schluter (Chicago: University of Chicago Press, 1949). See also C. N. Yang, FP239, for the impression received on a visit to Fermi at the hospital.

4. Enrico Fermi, *Collected Papers* (Accademia dei Lincei and University of Chicago Press, 1961, 1965).

5. I dream myself back to childhood
And shake my hoary head;
Why am I haunted by pictures
That I thought long ago to be dead?

High out of shady enclosures
A glistening castle looms large,
I know the embattled towers,
The stone bridge and also the gate.

From the escutcheon the lions are looking
Confidingly down at me
I greet these acquaintances gladly
And hurry across the court.

. . . .

Thus rises my ancestors' castle
Fixed loyally in my mind,
From the earth you have surely vanished,
The plow is traversing thy ground.

trans. Egon Schwarz

6. O. Chamberlain, E. Segrè, C. Wiegand, and T. Ypsilantis, "Observation of Antiprotons," *Phys. Rev.* 100 (1955): 947–50, and "Antiprotons," *Nature* 177 (1956): 11–12.

7. O. Chamberlain and C. E. Wiegand, CERN Symposium on High Energy Accelerators and Pion Phys., Geneva, *Proceedings*, vol. 2 (1956), p. 52; C. E. Wiegand, Inst. Radio Eng., 6th Scintillation Counters Symposium, *Proceedings* (Washington, D.C., 1958).

8. E. Amaldi, G. Baroni, C. Castagnoli, O. Chamberlain, W. W. Chupp, C. Franzinetti, G. Goldhaber, A. Manfredini, E. Segrè, and C. Wiegand, "Antiproton Star Observed in Emulsions," *Phys. Rev.* 101 (1956): 909–910, and E. Amaldi, G. Baroni, C. Castagnoli, O. Chamberlain, W. W. Chupp, A. G. Ekspong, C. Franzinetti, G. Goldhaber, E. J. Lofgren, A. Manfredini, E. Segrè, and C. Wiegand, "Example of an Antiproton-Nucleon Annihilation," *Phys. Rev.* 102 (1956): 921–23.

9. On November 15 and 16, 1985, there was a small conference at Berkeley to commemorate the thirtieth aniversary of the antiproton's discovery. The first morning was devoted to history; L. W. Alvarez presided. He started with an introduction explaining why he had not discovered the antiproton and why the discovery was not trivial and obvious. Next Lofgren spoke, giving an excellent presentation on the bevatron. Clyde Wiegand followed for about thirty minutes. He gave a detailed account of the work performed by himself and Chamberlain. Among other things, he said that he and Chamberlain had started planning the experiment secretly, outside of regular working hours. He mentioned Ypsilantis only peripherally, in connection with the addition of a counter to the apparatus. He never mentioned me, as though I had not existed. I was saddened by the performance. It must represent Wiegand's present state of mind; this must be his recollection of the discovery of the antiproton.

Others spoke following Wiegand. Finally, Piccioni spoke for a few minutes on the antineutron discovery in which he had participated; then, for about an hour, he renewed his accusations against Chamberlain and me, saying that we had stolen the plans of the apparatus used from him, and that I had by trickery excluded him from the execution of the experiment. Part of the public smiled; others did not seem to enjoy the performance. A historian commented to me: "See why historians put more trust in documents than in possibly distorted memories?" To avoid further unseemliness, I did not reply to Piccioni.

10. See B. Cork, G. Lambertson, O. Piccioni, and W. Wenzel, "Antineutrons Produced from Antiprotons in Charge-Exchange Collisions," *Phys Rev.* 104 (1956): 1193–96; they used a counter method.

11. J. Button, T. Elioff, E. Segrè, H. M. Steiner, R. Weingart, C. Wiegand, and T. Ypsilantis, "Antineutron Production by Charge Exchange," *Phys. Rev.* 108 (1957): 1557–61, and L. Agnew, T. Elioff, W. B. Fowler, L.

Gilly, R. Lander, L. Oswald, W. Powell, E. Segrè, H. Steiner, H. White, C. Wiegand, and T. Ypsilantis, "Antiproton-Proton Elastic and Charge Exchange Scattering at about 120 Mev," *Phys. Rev.* 110 (1958): 994–95.

12. See, e.g., L. W. Alvarez, in *Les Prix Nobel en 1968* (Stockholm: Nobelstiftung, 1969).

13. L. B. Auerbach, T. Elioff, W. B. Johnson, J. Lach, C. Wiegand, and T. Ypsilantis, "Study of Pion-Pion Interactions from Pion Production by Pions," *Phys. Rev. Letters* 9 (1962): 173–76. G. Kallen, *Elementary Particle Physics* (Reading, Mass.: Addison-Wesley, 1964, p. 185), reports these experiments to have established the resonance, but in the end they could not compete with the bubble chamber.

14. *Chemical and Engineering News* 48 (1947): 3572. The poll asked respondents to name the ten ablest chemists working in several specialized fields. In nucleonics, the list was P. H. Abelson, C. D. Coryell, F. Daniels, Gerhart Friedländer, J. W. Kennedy, W. F. Libby, G. T. Seaborg, E. Segrè, N. Sugarman, and A. C. Wahl, in alphabetical order.

15. E. O. Lawrence, letter to Harold C. Urey, of May 31, 1946 (Bancroft Library). See pp. 167–68 above for the relevant excerpt. The letter in which Lawrence joined with Seaborg in proposing me for the Nobel Prize in 1959, based on the discovery of antiprotons, is also among the Lawrence Papers at the Bancroft Library.

16. Many students, some of the greatest ability, participated in our work. From the postwar period at Berkeley I remember Herbert York, John Jungerman, S. C. Wright, S. N. Ghoshal, E. L. Kelly, G. Temmer, M. O. Stern, A. Bloom, G. Pettengill, R. L. Mather, F. N. Spiess, T. J. Thompson, W. John, J. E. Simmons, D. V. Keller, Joe Lach, J. Foote, H. Ruggs, Rein Silberberg, H. Stubbs, R. C. Weingart, L. E. Agnew, T. Elioff, R. R. Larsen, J. Button, M. Jakobson, P. Kijewsky, W. Lee, E. H. Rogers, R. E. Hill, D. A. Jenkins, S. R. Kunselman, and Gary Lum. Many of them have had distinguished careers that can be traced in *American Men of Science.* Herbert Steiner, W. Chinowsky, Gerson Goldhaber, Tom Ypsilantis, and R. Tripp joined the Berkeley faculty. Rae Stiening, who came at a later date, from MIT, where he had studied with Martin Deutsch, infused new vitality into our work. To this list one should add several postdoctoral fellows: Jonas Schultz, Paul Condon, and from outside the United States, von Dardel, Gilly, N. Lipman, Borghini, G. Ekspong, R. Mermod, N. Booth, B. Mashhoon, and much later Min Chen.

17. R. B. Bacastow, T. Elioff, R. R. Larsen, C. Wiegand, and T. Ypsilantis, "Measurement of the Branching Ratio for Pion Beta Decay," *Phys. Rev. Letters* 9 (1962): 400.

18. See, e.g., R. Seki and C. Wiegand, "Kaonic and Other Exotic Atoms," *Ann. Rev. Nucl. and Part. Science* 25 (1975): 241.

19. See, e.g., *Reminiscences about I. E. Tamm*, ed. E. L. Feinberg (Moscow: Nauka Publishers, 1987).

20. See R. Marshak, "The Rochester Conferences," *Bull. Atomic Scientists*, June 1970, p. 72.

21. See, e.g., *Les Prix Nobel en 1957* (Stockholm: Nobelstiftung, 1958). There is an extended literature on the discovery of the nonconservation of parity, including personal recollections by Lee, Wu, and Yang.

22. See E. Segrè, "Neue Atomarten und Antimaterie" (New atomic species and antimatter), *Angew. Chemie* 5 (1959): 173–76.

23. E. Segrè, *Enrico Fermi, fisico* (Bologna: Zanichelli, 1970, 1987).

24. Karl Manne Georg Siegbahn (1886–1978), winner of the 1924 Nobel Prize for physics, and his son Kai Manne Siegbahn (1918–), who won the physics prize in 1981.

25. See *Les Prix Nobel en 1959* (Stockholm: Nobelstiftung, 1960).

26. E. Segrè, "From Atoms to Antiprotons" (Faculty Research Lecture, University of California, 1960, unpublished).

27. See, e.g., *The Science of Materials Used in Advanced Technology*, ed. E. Parker and U. Colombo (New York: Wiley, 1973).

28. E. Segrè, *Nuclei and Particles* (New York: W. Benjamin, 1964, 1977); also translated into Italian (Bologna: Zanichelli, 1966, 1982) and Chinese.

29. E. Segrè, "The Consequences of the Discovery of the Neutron," 10th Int. Conf. of the History of Science, *Proceedings* (Paris: Hermann, 1965), vol. 1, p. 149–58.

Index

Abelson, Philip, 308n.6, 317nn.14,15; at Berkeley, 113, 135, 152, 153, 164, 165; and neptunium, 153, 306n.24, 308n.6
Accademia dei Lincei, 8, 61, 119, 239, 253, 279, 289
Accelerators, 283–84; and antiparticles, 255–59; Berkeley bevatron, 256, 316n.9; Berkeley cyclotron, 112–14, 133–36, 150, 155–64 passim, 171, 229; Berkeley synchrocyclotron, 229, 249; at Fermilab, 284; Los Alamos cyclotron, 184; and time assignments, 243–44; Washington University cyclotron, 209
Acque Albule, sulfur baths, 18, 215, 223, 291
Aebersold, Paul, 135
Aeby, Jack, 189
Africa, 275–77
Alikanian brothers, 263
Allegri, Natale, 16–17
Allison, Sam K., 142–43, 154, 187, 201, 251
Alvarez, L. W., 172, 241; at Berkeley, 135, 144, 155–56, 163, 207, 235, 237, 258–59, 267, 275, 316n.9; at Los Alamos, 185; memoirs, 297; Nobel Prize, 185; at Oppenheimer hearings, 309n.9; Soviet invitation to, 262
Alvaro, Almirante, 250
Amaldi, Edoardo, 28, 120, 149, 289; and antiprotons, 255, 258; in Cambridge, 92, 93; and CERN directorship, 275; and Fermi's papers, 268; German troops and, 195; and military service, 55, 60; and neutron work, 89, 92, 93, 94, 95, 245; in New York, 112; in Oslo, 68; in Rome, 40, 41, 47–52 passim, 61, 63, 83–95 passim, 105–6, 111, 119, 290; and University of Palermo competition, 103; in U.S., 144
Amaldi, Ginestra (Edoardo's wife), 90, 92, 93, 95, 120, 258
Amaldi, Ugo (Edoardo's father), 41, 288
Amaldi, Ugo (Edoardo's son), 93
American Chemical Society, 259
American College of Nuclear Physicians, 294
American Physical Society, 234, 299
Amsterdam, 100. *See also* Zeeman lab
Ancona, Giacomo, 22, 133, 222, 227, 237, 270
Ancona, Italy, 3, 4–5
Anderson, C. D., 197, 255, 268
Andrea Doria, 280
Angeli-Rimini reaction of aldehydes, 4
Annual Review of Nuclear Science, 231, 232
Antineutron, 258, 316n.9
Antiparticles, 255–59
Antiprotons, 243, 255–59, 260–61, 269–70, 273, 274, 316n.9
Anti-Semitism: finances for victims of, 100; Italian, 4, 55, 86, 111, 119, 132, 140, 217, 225; of Lo Surdo, 55; *Manifesto della razza*, 132, 140; Nazi, 79, 97, 100; Treves family and, 10; U.S., 102
Argentina, 250
Arizona, 294–95

Armenia, 263
Artom, Camillo, 115
Artom, Eugenio, 225
Artom, Giuliana (cousin), 225
Astatine (element 85), 155, 211, 243, 248
Aston, Francis, 117
Atombau und Spektrallinien (Sommerfeld), 68
Atomic bomb: Angelo on, 213; Fermi's comments on, 252; and Hiroshima, 209; Los Alamos and, 176–77, 179–206; Rotblat and, 189, 312n.7; Soviet, 214, 237, 238. *See also* Hydrogen bomb
Atomic Energy Act, U.S., 245
Atomic Energy Commission (AEC), U.S., 169, 246, 247, 284

Bacher, R. F., 182, 188
Back, E., 65, 67
Bainbridge, Kenneth, 201
Baker, Nicholas. *See* Bohr, Niels
Bakker, Cornelius J., 66, 90, 120, 275; at Philips, 99; at Zeeman's lab, 66, 67, 83, 136
Balbo, Italo, 149
Balewa, Abubakar Tafawa, 276–77
Barbella, Costantino, 17
Barresi, Ginetta, 110, 141, 149
Beck, Guido, 250
Benin, Oba of, 276
Berkeley, 206; move from (1955), 253; Piedmont Ave. house, 144; settling in, 114; Spruce Street house, 170, 253; visits, 112–14, 128, 131–78. *See also* Radiation Laboratory, Berkeley; University of California, Berkeley
Bernardini, Gilberto, 120, 239; book on elementary physics, 111; and CERN directorship, 275; at Columbia, 260; at Cosmic Ray Lab, 233; at University of Illinois, 239; and University of Palermo competition, 103
Beta decay, 152, 165, 261, 264; Fermi and, 86, 88, 157, 163, 165
Bethe, Hans A., 280; and accelerators, 135; in Berkeley, 176; at Cornell, 207; at General Electric, 213; and hydrogen bomb, 238; at Los Alamos, 185, 187; Nobel Prize, 185; at Oppenheimer hearings, 309n.9; in Rome, 63
Bhabha, Homi Jehangir, 282
Birge, Raymond T., 161–62, 211, 237, 310n.21; and "peace" physics, 172; and physics department staff, 208, 209, 210, 234
Bjerge, Dane T., 93, 94
Blackett, Patrick, 247
Bloch, Felix: in Berkeley, 144, 146; at Los Alamos, 185, 187, 193; and neptunium, 152; Nobel Prize, 185; in Rome, 63
Bogoliubov, Nikolai, 263
Bohr, Aage, 193–94
Bohr, Niels, 46, 305n.14; and Denmark, 122–23, 194, 268; Heisenberg and, 122–23; Hevesy and, 116; at Los Alamos, 193–95; and Nobel Prize, 151, 272
Bonfante, Giuliano, 18
Boni, Giacomo, 40
Boole, George, 192
Bordoni, U., 41
Bozzolo, Italy, 3
Bray, W. C., 136
Brazil, 249–50, 259
Bregman, Elizabeth, 285
Breit, Gregory, 169, 176
Bretscher, Egon, 163, 233
Briggs, Lyman J., 166, 170
British Mission, 192–93
Brobeck, William, 158
Brode, Bernice, 133, 142, 193
Brode, Robert B., 133, 142–43, 172, 187, 193, 233–34, 235, 237
Brookhaven National Laboratory, Long Island, 220, 240–41, 247, 284

Cacciapuoti, B. N., 110, 116
California, 63–64, 142–44; Lafayette, 253–54, 258, 286–87, 288. *See also* Berkeley
California Institute of Technology, Pasadena, 63–64
Calutrons, 197–98
Campetti, A., 103
Cannizzaro, Stanislao, 106
Carducci, Giosuè, 17, 25
Careri, Giorgio, 290
Carlandi, Onorato, 17

Carrelli, Antonio, 103, 107–8, 126
Castelnuovo, Guido, 38, 39, 55
Cavendish Laboratory, Cambridge, 92–93, 117
CERN, 290; directorship, 268, 275; international scientific conference, 265
Ceylon (Sri Lanka), 282
Chadwick, James, 87, 93, 190–91, 192, 247
Chalmers, T. A., 150
Chamberlain, Owen, 242, 260, 261; and antiprotons, 256, 257, 258, 260, 316n.9; at Berkeley, 172–73, 176, 227, 229–30, 231, 256, 258, 283; at Brookhaven, 247; at Chicago, 206, 230; at Harvard, 260, 270; at Los Alamos, 178, 185, 189; Nobel Prize, 185, 269–70; Soviet Union trip, 262
Chamisso, Adalbert von, "Schloss Boncourt," 254–55, 315n.5
Chemistry: Angeli-Rimini reaction of aldehydes, 4; at Berkeley, 211, 212; changes, 244; at Donegani summer school, 279–80; Hofmann medal, 266–67; Nobel Prize, 259–60. *See also* Elements
Cherenkov, Pavel, 263
Chew, Geoffrey, 235, 301n.4
Chicago: Fermi in, 47, 150, 175, 205–13 passim, 239, 248, 251–52; Fermilab, 284; Metallurgical Laboratory, 211. *See also* University of Chicago
Chromodynamics, 212
Churchill, W., 181
Cipolla, Michele, 106
Cockcroft, John, 93, 247
Coefficient of aquaticity, 96
Cohnen, H., 65–66
"Cold fusion," 299
Collegio Ghislieri fellowships, 4
Colombo, Umberto, 279–80
Columbia University, 101, 111–12, 132, 146, 167, 240, 264
Communist Party, U.S., 138, 139–40, 234
Competitions: Ferrara, 81; Palermo physics chairs, 101, 103, 125–26
Compton, A. H., 175, 208, 278–79
Condon, E. U., 182
Congress, U.S., 245
Congress of History of Science, Tenth International, 281
Consiglio nazionale delle ricerche (CNR), 89, 125
Conversi, Marcello, 17
Cooksey, Carleton, 135
Cooksey, Don, 113, 133, 135–36, 143–44, 167, 174, 241
Copenhagen: Bohr Institute, 268; conference, 122
Corbino, O. M., 38, 43–55 passim, 61, 90, 106, 231, 245; death, 118–19, 124; and La Rosa, 100; Rockefeller Foundation advisor, 63; and slow neutron patent, 96, 244
Cornog, Robert, 135, 155
Corson, Dale, 155
Coryell, C. D., 317n.14
Cosmic Ray Lab, Testa Grigia, 233
Coster, Dirk, 117
Crudeli, U., 103
Cuban missile crisis, 282
Curie, Marie, 88, 102, 115, 309n.7
Cyclotrons. *See* Accelerators

D'Agostino, Oscar, 89, 245
Dali, Gala, 274
Dali, Salvador, 273–75
Daniels, F., 317n.14
d'Annunzio, Gabriele, 17, 18, 25, 57, 76
Dante Alighieri, 1, 37, 104, 131, 293
Darrow, K. K., 45
DC amplifier, 133–34
De Bono, Emilio, 98
Debye, Peter, 65
Defoe, D., 143
de Franchis, Michele, 106
Denmark, 122–23, 194, 268
Deutsch, Martin, 189
Diffraction grating, 65, 66
Dirac, P. A. M., 82, 255, 273
Disarmament conference, U.S.-USSR, 265
Dodson, Richard, 198
Dolomites, 108–9, 120–21
Donegani, 279, 290
Donegani Lectures, 239, 279
Donegani school, 279, 286
Don Nello del Raso, 36
DuBridge, Lee, 133–34, 159
Dumas, J. B., 20

Dunning, J. R., 102
Duruy, Victor, 20

Education: in childhood, 14, 15, 18, 22, 31–32; in engineering, 37–38, 41, 43, 46, 48–49; high school, 24–26, 32, 33–34, 37; in Judaism, 35; in languages, 15, 25–26, 32, 33, 34; in mathematics, 22, 26, 38, 39–40, 41, 46, 52–53; in physics, 19–20, 26, 33–34, 37–56, 61–63; university, 37–56
Ehrenfest, Paul, 63
Einstein, Albert, 35–36, 46, 272, 273, 278, 292
Elements, 116–17; americium, 197; astatine, 155, 211, 243, 248; beryllium, 157; hafnium, 117; masurium, 90, 115, 117; neptunium, 152–53, 165, 306n.24; rhenium, 117, 152; rhodium, 138; thorium, 91; uranium, 91–92, 96, 113, 152, 163–67, 171, 197–98; xenon, 153. *See also* Plutonium; Technetium
Ellis, C. D., 87
Emo, Barbara, 226
Emo, Lorenzo, 113, 132–33, 271, 280, 292; in Berkeley, 118, 132–33, 135, 140, 142, 146; in Italy, 224, 225–26
Engineering: Brobeck and, 158; education in, 37–38, 41, 43, 46, 48–49
England: Cavendish Laboratory, 92–93, 117; *Proceedings of the Royal Society*, 62, 93; Royal Institution, 293; visits, 67–68, 78, 92–93, 293. *See also* London
English, S. G., 169
English language, 33, 82–83
Enriques, Federigo, 111
Enriques, Giovanni, 41, 42, 43, 44, 65
Euler, Hans von, 122
Everest, George, 193
Experimental Nuclear Physics (Segrè), 232, 262

Fano, Ugo, 71, 287
Fanzolo villa, 225–26
Faraday, M., 20, 88, 243, 293
Farwell, George, 178, 189, 206
Fascism, 103, 127, 139, 149, 214, 217; Avanguardisti youth organization, 28–29; Giuseppe Segrè and, 29, 84, 85; Italian consul in Germany and, 69; of Lo Surdo, 55–56; *Manifesto della razza*, 132, 140; Palermo people against, 105, 106, 108, 127; Rockefeller Foundation and, 63; Segrè uncles and, 8, 29. *See also* Mussolini, Benito
Fascist Party, 29, 139, 149
FBI, 170
Feenberg, Eugene, 63
Fermi, Enrico, 28, 79, 146, 194, 231, 261, 273; in Berkeley, 112, 156–57, 161, 232; and beta decay, 86, 88, 157, 163, 165; biography, 268; at Brookhaven, 247; in Chicago, 47, 150, 175, 205–13 passim, 239, 248, 251–52; at Columbia, 167; Compton and, 279; "crumbs," 61–62, 65; death (1954), 252–53; Donegani Lectures, 239; on Garwin, 240; and hydrogen bomb, 238, 251; international physics conferences, 45, 46, 88, 94; *Introduzione alla fisica atomica*, 52; in Italy, 41–69 passim, 81–95 passim, 118, 119, 251; judging at competitions, 80, 103; letters of recommendation by, 278; at Los Alamos, 126–27, 182, 183, 185, 187, 191, 197, 198, 202, 203–4, 232; in Michigan, 82, 101, 102; neutron work and patent, 88–95 passim, 111, 243, 244, 245–46, 247; in New Jersey, 162–63; and Nobel Prize, 151, 259–60; Oppenheimer contrasted with, 138, 139; and Palermo chair of theoretical physics, 125; papers published after death, 268, 280; and parity nonconservation, 265; Seaborg contrasted with, 168–69; and Segrè's Berkeley position, 150, 154–55; South American lecture tour, 92; stomach cancer, 227, 251–52; and Teller, 63, 251–52; and uranium, 152
Fermi, Laura, 59, 252, 260, 284, 287; *Atoms in the Family*, 252
Fermilab, 284
Fermi-Segrè formula, 81
Ferrari, Fabio, 301n.4
Ferro-Luzzi, Giovanni, 39, 40, 44, 85
Feynman, Richard P., 185
Finances: in Berkeley, 147–48, 153–54, 161, 170, 172, 210; for cyclotron, 135; for displaced Jews, 100; in Israel,

283; Italian professors, 223; Los Alamos earnings, 147; after marriage, 108; Nobel Prize, 272–73; patent, 247; Physics Institute, 90; Segrè family, 11, 14, 84–86, 101–2, 107, 108, 140–49 passim, 170, 216–22; after university, 57, 65, 69, 85
Fink, G. A., 102
Finzi family, 19
Fission: discovery of, 151–52, 245, 265, 299; spontaneous, 185, 195–97
Florence, 30. *See also* Treves villa
Forbidden lines. *See* Quadrupole radiation
Foscolo, Ugo, 30
Fraenkel, Stanley, 176
France: family education in, 4; Paris, 239. *See also* French language
Franchetti, Piero, 44
Francis Ferdinand, Archduke of Austria and Este, 2, 10
Franck, James, 154, 260
Frascati Laboratory, 290
French language, 18, 25–26, 299
Friedländer, Gerhart, 169, 176, 317n.14
Frisch, Otto, 69–70, 87, 192, 193, 305n.13
From Falling Bodies to Radio Waves (Segrè), 293, 298
From X-Rays to Quarks (Segrè), 281, 298
Frugoni, Prof., 22
Fuchs, Klaus, 193
Fulbright fellowship, 239

Ganot, Adophe, 20, 38
Garwin, Richard L., 240, 265, 278
General Electric, 213, 245
Gentile, Giovanni, Jr., 50, 111, 125–26
Gentile, Giovanni, Sr., 50, 98, 126
German Chemical Society, 266–67
German language, 15, 25–26, 32, 33, 299
Germany, 69, 123; Gino Segrè and, 8; Spiros, 96–97, 127, 149; surrender, 199. *See also* Hamburg; Hitler, A.; Nazis
Giannini, G. M., 82, 236–37, 245
Giordani, Francesco, 239, 289
Glashow, Sheldon, 234
Gofman, Jack, 169
Goldberger, Marvin, 235
Goldhaber, Gerson, 256, 260
Goldhaber, Maurice, 93
Gordon, William E., 70
Gordon Conference, 247–48
Goudsmit, S. A., 272
Grating, diffraction, 65, 66
Griggs, Helen, 169–70
Grimod, M., 25–26
Grosse, Aristide von, 306n.24
Groves, Leslie R., 177, 179–80, 182, 183, 201–2, 309n.9
Guggenheim Fellowship, 267–68

Haber, Fritz, 97
Hahn, Erwin, 240
Hahn, Otto, 91, 151, 152, 267
Haissinsky, Moise, 102
Halford, Ralph, 150
Hamburg, 123; experiments, 68–72, 87; romance in, 71–78, 79–80, 123, 269; visits, 122–24, 268–69
Heisenberg, Werner, 46, 122–23, 229, 273, 309n.7
Helmholz, Betty, 287
Helmholz, Carl, 287
Herskovits, M. J., 276
Hevesy, Georg von, 116–17, 250, 259, 281
Hilberry, Norman, 175
History of science, 280–81, 298
Hitchcock Lecturers, 156, 193
Hitler, A., 92, 132, 177, 200; Americans and, 139–40; coming to power (January 1933), 78; and immigration restrictions, 145; Mussolini and, 92, 127–28, 132; scholar victims, 63, 100; suicide, 199. *See also* Nazis
Hofmann medal, 266–67
Holland: Josephy in, 100; neutron conferences, 99–100; Philips Company, 99–100, 245; Zeeman lab, 65, 66–67, 78–79, 99, 128, 136
Hughes, A. L., 209
Hydrogen bomb, 237–38, 251. *See also* Atomic bomb
Hyperfine structure, 81–82, 87

Ibadan University, Nigeria, 275
IBM, 240
Illinois. *See* University of Chicago; University of Illinois, Urbana
India, 282

Inglis, D. R., 63, 82
Iran, 282–83
Isomerism, nuclear, 137–38, 150–51, 243
Israel, 282–83, 302n.5
Italy, 4, 63, 131, 208, 288–89, 292; Accademia dei Lincei, 8, 61, 119, 239, 253, 279, 289; Acque Albule sulfur baths, 18, 215, 223, 291; anti-Semitism/racial laws, 4, 55, 86, 111, 119, 132, 140, 217, 225; Civil Code, 7; Consiglio nazionale delle ricerche, 90, 125; consuls in Germany from, 69; military's diversity in, 60; physics in, 38, 46, 49, 119; postwar visits, 195, 213, 221, 222–27, 232–33; stay in 1970s, 288–92; touring, 41–45, 109, 124, 290–91, 292–93; and U.S. visas, 128–29; visit in 1960s, 282–83; and World War II, 79, 92, 103, 171, 173, 195. *See also* Fascism; Mussolini, Benito; *individual cities and universities*
Ivanenko, D. D., 263

Japan, World War II, 171, 203, 209
Japanese Americans, World War II, 173
Jaspers, Karl, 267
Jeffries, Carson D., 283
Jenkins, Francis A., 172, 275; in Berkeley, 133, 142–43, 150, 210, 233–34, 235, 249
Jenkins, Henriette, 133, 142, 143, 144, 249, 275
Jensen, Hans D., 70, 260
Jentschke, W., 268–69
Jews: brother Marco on, 28; education in Judaism, 35; finances for displaced, 100; military leave for Yom Kippur, 59; practicing/nonpracticing, 10–11, 35. *See also* Anti-Semitism
Johnson, Lyndon, 284
Joliot-Curies, Frédéric and Irène, 88–89, 91, 152, 239
Josephy, B., 69, 100
Jungerman, John, 189

Kahn, Milton, 176, 178, 189
Kalbfell, David, 135
Kamen, Martin D., 135, 144
Kammerlingh Onnes, Heike, 2
Kapitza, Peter, 93, 263
Keats, John, 228
Kennedy, John F., 281
Kennedy, Joseph W., 317nn.14,15; at Berkeley, 151, 164–76 passim; at Los Alamos, 178, 187, 195, 197; and plutonium patent, 246; stomach cancer, 247; at Washington University, 208, 209
Kerst, Donald, 187, 239, 259
King, Percival, 187
Kistiakowsky, George B., 197, 281
Klein, Felix, 111
Klein, Oscar, 268, 271–72
Knauer, F., 69
Konopinski, E. J., 176
Kristeller, P. O., 98
Kuhn, Thomas S., 280–81
Kurie, Franz, 113
Kuroda, P. K., 308n.11

Landau, E., 41
Landau, L. D., 263
Langsdorf, Alex, Jr., 137, 153
Languages: education in, 15, 18, 25–26, 32, 33, 34; English, 33, 82–83; French, 18, 25–26, 299; German, 15, 25–26, 32, 33, 299; Latin and Greek, 25, 34
Lansdale, John, Jr., 174–75
Lark-Horovitz, Karl, 160–61
La Rosa, Michele, 100–101, 105
Latimer, W. M., 168, 309n.9
Lattes, Dante, 35
Lattes, G. C., 249, 250
Lawrence, Ernest O., 70, 113, 134–35, 143–73 passim, 209–12 passim, 241–42, 266, 277, 313n.1; and antiprotons, 257, 258; and atomic weapons, 171, 237–38; and bevatron, 256; and calutrons, 197; colitis, 266; and Cooksey, 135–36, 241; and DC amplifier, 133; death, 266, 267; at disarmament conference, 265–66; and letters of recommendation, 278; and loyalty oath, 235; and Nobel Prize, 149, 151, 259, 266, 317n.15; and patents, 246; and *Physical Review* articles, 137–38, 170; and Piccioni accusations, 257; Seaborg contrasted with, 168; sending radioactive substances, 112,

115, 118; and Soviet Union trip, 261–62
Lawrence Berkeley Laboratory (LBL), 267, 283–84
League of Nations, 109, 182, 204
Lederman, Leon, 265
Lee, Tsung-Dao, 264, 265
Leininger, R., 211
Leitz, Fred, 169
Lenz, W., 70
Leone, Giovanni, 289
Leopardi, Giacomo, 24
Levi-Civita, Tullio, 38–39, 54, 63
Lewis, G. N., 136, 168
Leyden, John, 18
Libby, Willard F., 150–51, 317n.14
Lignola, Marquis, 58
Linenberger, G. A., 178, 189
Livermore Lawrence Laboratory (LLL), 267
Lofgren, Edward, 256, 316n.9
Lombroso, Cesare, 194
London, Fritz, 63
London: international physics conference, 94; parents in, 14; Royal Institution, 293; visits, 67–68, 78, 92, 293
Loomis, Alfred, 167
Loomis, W. F., 239
Los Alamos, 173, 177–78, 179–207, 230, 275; apartment, 186–87; Big House, 180, 182; earnings at, 147; Fuller Lodge, 180–81, 182, 185; Lansdale monitoring, 174–75; and loyalty oath, 236; Oppenheimer at, 139, 180; outings, 190–92; Pajarito, 186; physical surroundings, 180–81
"Los Alamos Primer," 182
Lo Surdo, Antonino, 52, 53–56, 64, 119, 173
Loyalty oath, 234–36, 238, 246
Luedemann, Hermann, 97, 307n.31

Macaluso, Damiano, 106
McCarthy, Joseph, 278–79
McMahon Act (1946), U.S., 246
McMillan, Edwin, 172, 241; at Berkeley, 113, 135, 144, 152, 153, 165, 207; at Los Alamos, 185, 187, 205; and neptunium, 153, 306n.24, 308n.6; and Nobel Prize, 185, 259, 270, 317n.15; phase stability discovery, 212; Rad Lab director, 267; Soviet invitation to, 262
Maggini, Signorina (teacher), 15, 18, 291
Maglic, Bogdan, 258–59
Majorana, Ettore, 229, 303n.3; disappearance, 126–27; and Feenberg, 63; and Palermo chair, 125–26; in Rome, 39–40, 46, 49–50, 51, 71
Majorana, Luciano, 126
Majorana, Quirino, 46, 80
Mandò, Manlio, 110
Manhattan District, 177, 179–206, 208, 246
Manley, John, 176, 178, 187
Marconi, G., 112, 126
Marshall, George, 203
Martinez, Maria, 187
Mathematics, education in, 22, 26, 38, 39–40, 41, 46, 52–53
Matteotti, Giacomo, 29
Maxwell, James Clerk, *Treatise on Electricity and Magnetism*, 70
Mayer, Maria, 260
Medicine, nuclear, 293–94
Meitner, Lise, 87, 91, 152, 271–72
Mendeleyev, D. I., 115, 264
Mesons, 283; K, 264; rho, 259, 261
Metallurgical Laboratory, Chicago, 211
Methyl bromide, 159
Metropolis, Nicholas, 230
Mexicali, Mexico, 145–46
Michigan. *See* University of Michigan
Military: and hydrogen bomb, 238; Italian service, 26, 28, 54–55, 57–61, 121; U.S., 177, 179–80, 182–84, 189, 200, 202–3, 232, 238
Millikan, Robert A., 46, 63
Millman, S., 102
Mines Segrè, Rosa, 284, 288, 294–95, 297–300
Minkowski, Hermann, 70
Miskel, John, 178, 189
Mitchell, D. P., 102, 187, 188
Moderation process, 95
Molecular beam work, 68–71, 83, 102, 243
Moll Flanders (Defoe), 143
Montanelli, I., 240
Montecatini chemical company, 239, 279, 290

Montemartini, 106, 114
Moon, P. B., 95, 192
Morandi, Luigi, 239, 279–80
Moyer, Burton J., 36
Mulford, Dean & Mrs., 170
Muons, 122, 264–65
Mussolini, Benito, 46, 92, 121, 127–28, 132, 171, 195

Naples, wedding trip, 107–8
National Bureau of Standards, 264
Nature, 61, 62, 64, 257
Nazis, 200, 214–15; Amelia Segrè captured by, 23, 195, 214; German romances and, 78, 79, 96–97, 269; and Luedemann, 307n.31; scientists and, 100, 154, 189, 194, 269. *See also* Hitler, A.
Neddermeyer, Seth, 197
Ne'eman, Yuval, 283
Nelson, Eldred, 176
Nepal, 282
Nernst, Walther, 38
Neutrons: antineutron, 258, 316n.9; discovery of, 88, 190–91, 281; patent, 96, 244–46, 247; Physics Institute work on, 88–96, 243; slow, 94–96, 244–46, 247. *See also* Nuclear physics
New England, 247–48
New Mexico, 187, 192, 200–202, 275. *See also* Los Alamos
New York: Angelo in, 146–47, 159, 213, 221; Columbia University, 102, 111–12, 132, 146, 167, 240, 264; Riccardo Rimini in, 248. *See also* Brookhaven National Laboratory
Neylan, J. F., 234, 235
Nigeria, 275–77
Nikotin, I. P., 263
Nobel, Alfred, 265
Nobel Prizes, 154, 165, 265, 269, 272, 281, 317n.15; Alvarez, 185; Bethe, 185; Bloch, 185; Chamberlain, 185, 269–70; Fermi, 151; Feynman, 185; Jensen, 260; Lawrence and, 149, 151, 259, 266, 317n.15; Lee, 265; McMillan and, 185, 259, 270, 317n.15; Mayer, 260; Onnes, 2; Rabi, 185; Raman, 60; Seaborg, 259; Segrè, 185, 221–22, 241, 259, 261, 265, 269–73, 317n.15; Siegbahns, 318n.24; Tamm, 263; Yang, 265; Zeeman, 66
Nobles, Bill, 189
Noddack, Walter K., 115, 116, 117–18, 124
Norway, 68, 268
Noyes, A. A., 136
Nuclear explosion, 199; in Japan, 203, 204; predetonation, 185; tests, 200–202, 203
Nuclear isomerism, 137–38, 150–51, 243
Nuclear medicine, 293–94
Nuclear physics, 162–63, 231–32, 245; at Berkeley, 134, 137–38, 195–96, 211–12, 228–31, 243, 280; at Chicago, 175–76; at Columbia, 102–3, 112; *Experimental Nuclear Physics*, 232, 262; Fermi's written lectures on, 252; in Holland, 99–100; in Italy, 86–96, 111; at Los Alamos, 179–206; patents, 96, 244–46; in Rome, 290; textbook on, 280. *See also* Accelerators; Elements; Fission; Neutrons; Radiation Laboratory, Berkeley
Nuclear weapons, 171, 237. *See also* Atomic bomb; Hydrogen bomb
Nuclei and Particles (Segrè), 280
Nucleon-nucleon interaction, 228–30, 233, 243, 248

Oak Ridge, Tennessee, 197–98
Occhialini, Giuseppe, 292–93
Oliphant, Marcus, 93
Oppenheimer, J. Robert, 138–39, 209, 263, 309n.9; at Berkeley, 113, 137–39, 156–57, 174–77 passim, 229; hearings, 251, 309n.9; and hydrogen bomb, 238; at Los Alamos, 139, 177–203 passim
Orvieto, Angiolo, 11
Ovid, 25

Palermo, 104, 109–10; apartment on Piazza Franceso Crispi, 119–20, 253; Elfriede leaving, 141; Hotel Excelsior, 108; Pensione Lincoln, 105, 108; Physics Institute, 105–6, 110, 268; postwar, 268; Rotary Club, 127; University of, 101, 103, 104–30, 208
Paneth, Fritz, 92, 116–17
Panofsky, Wolfgang, 235, 241, 260, 262, 265
Paper mill (family business), 9–10, 14,

84–85, 291; award ceremony for old employees, 269; father's estate and, 84–85, 215–22, 225; Marco and, 43, 84–85, 216–21, 225; Rome office, 22–23, 98
Parini, Giuseppe, 207
Paris, 239
Parity nonconservation, 264–65
Parravano, Nicola, 38, 119
Paschen, F., 65
Pasquali, Giorgio, 34
Patent: plutonium, 244, 246–47; on slow neutrons, 96, 244–46, 247
Pauli, Wolfgang, 46, 70, 88, 233
Peano, Giuseppe, 25
Pegram, G. B., 101, 112, 145, 163–64
Peierls, Rudolf, 63, 192, 247
Penney, William, 192, 247
Perlman, Isidor, 169
Perlman, Morris, 169, 176
Perosi, Don Lorenzo, 127
Perrier, Carlo, 105, 106, 116, 118, 127, 224
Persico, Enrico, 94, 125, 224, 268
Peru, 250–51
Philips Company, Eindhoven, 99–100, 245
Phipps, T. E., 69
Physical Review, 87, 137–38, 150–51, 166, 170, 257
Physics, 57, 231, 242–44; American Physical Society, 234, 299; "without apparatus," 242–43; "battleship experiments," 242–43, 258; at Berkeley, 154–55, 161–62, 172–73, 230–31, 233–36, 243, 280, 283, 288, 310n.21; book on nuclear and particle, 280; education, 19–20, 26, 33–34, 37–56; international conferences, 45–46, 88, 94, 233; lectures on, 281, 282; molecular beam work, 68–71, 83, 243; Moyer in, 36; Nobel Prize prestige, 272; Occhialini on, 293; particle, 261, 264, 280; at Purdue, 160–61; quadrupole radiation, 64–65, 67, 68, 83–84, 243; "swollen atoms"/ Rydberg states, 83–84, 150, 243; teaching, 110–11, 172, 280, 288; weak interactions, 88, 264–65. *See also* Antiparticles; Beta decay; Cyclotron; Nuclear physics; Quantum mechanics; Radiation Laboratory, Berkeley
Physics Institute: Palermo, 105–6, 110, 268; Rome (Via Panisperna), 46, 48, 55, 59–65 passim, 83–96 passim, 105–6, 119, 223, 309n.7
Piazzi, Giuseppe, 106
Piccioni, Oreste, 248, 258, 316n.9
Pittarelli, Professor, 38, 40
Pius V, Pope, 4
Pius XI, Pope, 236
Placzek, George, 63, 154, 213, 273
Planck, Max, 46, 272
Plutonium, 163–70, 175–76, 185, 196–97, 243; patent, 244, 246–47
Pochettino, Alfredo, 80
Poincaré, Jules-Henri, 5
Poles: and Nazis, 189; and U.S. visas, 128–29
Polo, Marco, 18
Pontecorvo, Bruno: at General Electric, 213; and neutron work, 94, 95, 245, 246; in Paris, 160; and rhodium, 138; in Soviet Union, 236–37, 246, 262; Tulsa job, 160, 237
Popovi, 187
Positron, 255
Powell, Wilson, 258
Prestwood, René J., 169, 189
Proceedings of the Royal Society, 62, 93
Pseudopotential, 84
Puccianti, Luigi, 80
Purdue University, 160–61
Pusterla, Emilia, 2, 16
Pusterla, Luigi, 16, 17
P waves, 212

Quadrupole radiation, 64–65, 67, 68, 78, 83–84, 243
Quakers, 143
Quantum mechanics, 52, 66, 138

Rabi, I. I.: at Columbia, 102; at Los Alamos, 180, 185; molecular beam work, 102, 243; Nobel Prize, 185; at Oppenheimer hearings, 309n.9; *Physical Review*, 87
Racah, Giulio, 86, 125, 126, 283
Racism. *See* Anti-Semitism
Radiation Laboratory, Berkeley, 210–11, 228–30, 241–43, 312n.3; administrative

work, 277–78; and antiparticles/antiprotons, 243, 255–59, 260–61; and Los Alamos project, 173, 177–78, 185, 195–96; and loyalty oath, 235; under McMillan, 267; and patents, 246; radioactive substances sent from, 113–18 passim; visit (1936), 112, 113–14; visit (1938–1943), 133–38, 147–78. *See also* Accelerators; Lawrence, Ernest O.; Lawrence Berkeley Laboratory
Radiations from Radioactive Substances (Rutherford, Chadwick, & Ellis), 87
Radiciotti, 17
Radioactivity, 309n.7; accident, 177; artificial, 88–89, 309n.7; measurement, 133–34; nuclear isomerism and, 137; Rad Lab sending substances, 113–18 passim; Seaborg and materials with, 211
Railroads, 4–5, 14
Raman, Chandrasekhara Venkata, 60
Raman effect, 60, 61, 63, 64
Ramsey, Norman F., 102
Rasetti, Franco, 28, 73, 119, 120, 149; and Fermi's papers, 268, 280; Germany trip, 79; and Guggenheim Fellowship, 267–68; in Morocco, 88; neutron work, 89, 94, 95, 245; in Oslo, 68; in Rome, 43–53 passim, 87, 89, 94, 95, 102; and U.S., 63–64, 81, 102–3, 114, 129, 133, 145, 159, 294
Ravenna, Renzo (cousin), 149, 224
Reiche, Fritz, 33, 38
Religion, 4, 10–11, 34–36, 187, 236
Rendiconti, 61
La ricerca scientifica, 90–91, 95
Righi, Augusto, 106
Rimini, Ada, 33, 99, 149, 214, 222, 223
Rimini, Bindo (second cousin), 4, 32–33, 85, 98–99, 108; and Giuseppe's illness, 121; and paper mill, 221–22; postwar visit with, 223; in Uruguay, 146, 149
Rimini, Enrico (cousin), 4, 32–33
Rimini, Riccardo (second cousin), 4, 32–33, 122–23; and Giuseppe's illness, 121; in New York, 247–48; and romantic difficulties, 79, 96, 123; and Rosa Mines, 284; in Uruguay, 146, 149, 174–75, 284
Rimini, Riccardo (uncle by marriage), 4
Rockefeller Foundation, 63, 67, 68, 69, 125, 275–76
Roman Forum, 8, 40
Roesler Franz, Ettore, 17, 302n.7
Rome: Appian Way, 29; childhood visits, 21; Donegani Lectures, 239; Guggenheim Fellowship, 267–68; move to Corso Vittorio (and high school), 5–6, 22–23, 24–36, 253; Physics Institute (Via Panisperna), 46, 48, 55, 59–65 passim, 83–96 passim, 105–6, 119, 223, 309n.7; postwar visits, 214–15, 222–27, 267–68; stay in 1970s, 288–92; Tivoli transport links, 14; Uncle Claudio in, 5–6, 21, 23, 28; wedding in, 107. *See also* University of Rome
Roosevelt, F. D., 140, 170, 198–99
Rose, Morris, 135
Rossi, Bruno, 80, 94; and antiprotons, 255; at Cornell, 207; at Los Alamos, 182, 187, 193, 194, 197, 198
Rossi, Nora, 194
Rostagni, Antonio, 103
Rotary Club, Palermo, 127
Rotblat, Joseph, 189, 312n.7
Royal Society, 92; *Proceedings of*, 62, 92
Russia: czarist, 5. *See also* Soviet Union
Rutherford, Ernest, 87, 92, 124; and Coulomb forces, 212; at international physics conference, 46; and natural radioactive isotopes, 309n.7; and neutron work, 92; and Nobel Prize, 272; *Radiations from Radioactive Substances*, 87; Taylor and, 192; and World War II, 100
Rydberg states/"Swollen atoms," 83–84, 150, 243

Salvini, Giorgio, 290
Santangelo, Mariano, 110
Sarton Lecturer, 281
Scaduto, G., 125
Scherbatskoy, Serge, 159, 160, 237
Schiff, Leonard, 138
Schiller, Friedrich von, 78, 179, 249, 305n.14
Schirbel, Peter, 97
Schnurmann, R., 69

Schrödinger, Erwin, 52
Science: history of, 280–81, 298. *See also* Chemistry; Mathematics; Physics
Scienza per tutti, 1–2, 301n.2
SCT. *See* Paper mill
S-D combinations, 64–65
Seaborg, Glenn T., 163–65, 168–69, 299, 317n.14; AEC chairman, 169, 284; and americium, 197; at Berkeley, 136, 137, 150–51, 163–70, 195, 211–12; at Chicago, 175–76, 211; and Nobel Prize, 259, 266, 317n.15; and patents, 246–47
Secrecy, 232, 262; Los Alamos, 183–84, 190, 193, 208; and patents, 245, 246; Rad Lab, 167, 170, 173–75
Segrè, Amelia Gertrude Allegra (daughter), 178, 186, 206, 270, 285, 286–87, 298
Segrè, Amelia Susanna Treves (mother), 3, 20, 22, 31, 141–42, 171; and automobile, 41–42; capture by Nazis, 23, 195, 214, 215; and Elfriede, 99, 124, 148; and Emilio's German romances, 79–80, 99; family background, 11–12; and Fermi, 62; and finances abroad, 86; and granddaughter's birth, 178; husband's illness, 121–22; last visit, 129–30; letters to and from U.S., 148–49; and Rome move, 23
Segrè, Angelo Marco (brother), 3, 26–27, 28, 31, 212, 218–19; and finances, 84, 146–47, 213–21 passim; in New York, 146–47, 159, 213, 221; parents' problems with, 5, 26, 27, 79–80, 99; science books left by, 33; and Simili, 216, 218–19, 221
Segrè, Angelo Miracolo (grandfather), 3, 4
Segre, Beniamino, 54, 289, 301–2n.4
Segrè, Bice (aunt), 4
Segrè, Bice (cousin), 11
Segrè, Claudio (son), 123, 124, 128, 129; in Berkeley, 144, 145; birth, 120; children, 285; in Indiana, 160–61; and Los Alamos, 186, 187, 206; marriage, 285; and mother's death, 286–87; Stockholm trip, 270; on trip to U.S., 142; and U.S. visa, 145–46
Segrè, Claudio (uncle), 3, 4, 14, 19, 45; and Angelo, 5, 26, 27; on engineering degree, 48; and Fascism, 29; in Italian intelligentsia, 8; legacy, 11; and Nobel Prize, 271; and physics, 20; in Rome, 5–6, 21, 23, 28; and sexuality, 30–31
Segre, Corrado, 5, 301–2n.4
Segrè, Egle Cases (grandmother), 3–4
Segrè, Egle (cousin), 11
Segrè, Elfriede Spiro (wife), 108, 111–12, 120–30 passim, 173; and Angelo's letters, 219; in Berkeley, 112–14, 140–49 passim, 160, 173, 177, 186, 227; birthday, 142; courtship and marriage, 96–99, 101–2, 103, 105, 107; death, 286–88; in Latin America, 249–51, 284; and letters from Emilio, 213, 222; and Los Alamos, 186, 190, 191, 192, 193, 200, 203, 206, 275; as Luedemann's secretary, 97, 307n.31; nickname, 144–45; in Palermo, 108, 109, 141; and paper mill sale, 222; and Rosa Mines, 284; Segrè Chart, 190; trip around globe, 282; U.S. citizenship, 205; and U.S. visa, 128–29, 141, 145
Segrè, Emilio Gino, 297–300, 317n.14; birth, 2, 12–14, 22; childhood, 1–3, 14–23; cigars, 67; death, 299–300; dress, 41; *Experimental Nuclear Physics*, 232, 262; family background, 1–12; Fermi's papers and biography, 268; fireworks in childhood, 18; *From Falling Bodies to Radio Waves*, 293, 298; *From X-Rays to Quarks*, 281, 298; Fulbright fellowship, 239; illnesses, 21–22, 25, 227, 286, 287, 299; marriage to Elfriede, 104, 105, 107, 124; marriage to Rosa, 288; military service, 54–55, 57–61; nicknames, 2–3, 51, 80; Nobel Prize, 185, 221–22, 241, 259, 261, 265, 269–73, 317n.15; *Nuclei and Particles*, 280; and religion, 10–11, 34–36; retirements, 283, 288, 290, 291–92, 298; seventy-seventh birthday, 295; sex and love before marriage, 30–31, 37, 42, 51, 71–80, 96–103 passim; sports/recreation, 31, 37, 40–45, 108–9, 143, 144, 190–91, 227, 285–86, 290–91; U.S. citizenship, 205. *See also*

Education; Finances; Physics; *individual locales*
Segrè, Fausta (cousin), 11, 19, 42–43, 224
Segrè, Fausta Irene (daughter), 206, 270, 285, 286–87
Segrè, Francesca (granddaughter), 285
Segrè, Gino (grandson), 285, 300
Segrè, Gino (uncle), 2, 4, 6–8, 11, 27, 29, 79
Segrè, Giuseppe Abramo (father), 3, 4, 8–10, 20, 22, 141–42, 171, 215–17; *cavaliere del lavoro*, 225; death (1944), 195, 217; eightieth birthday, 146; and Elfriede, 99, 108, 124; and Emilio's German romances, 79, 99; escape from Nazis, 214–15; and Fascism, 29, 84, 85; *fastidi grassi*, 240; on Fermi's "crumbs," 62; and financial gifts for Emilio, 11, 84–86, 108, 119–20, 127; and financial problems with Angelo, 147; illness, 121–22; last visit, 129–30; letters to and from U.S., 148–49; marriage, 12; and Palermo lodgings, 108; and paper mill, 9–10, 14, 22, 43, 84, 98, 215–16; and Rimini orphans, 33; and Rome move, 22–23; tomb, 215; training by, 31, 32; wills, 217, 218, 220–21
Segrè, Joel (grandson), 285
Segrè, Katja Schall (sister-in-law), 27
Segrè, Marco Claudio (brother), 3, 20, 26, 27–28, 30, 213, 222, 253; *cavaliere del lavoro*, 225; escape from Nazis, 214–15; father's illness, 121; father's will and, 217, 218, 220–21; and paper mill, 43, 84–85, 216–21, 225
Segrè, Rosa Mines (wife), 284, 287–88, 290, 294–95, 297–300
Segrè Chart, 190
Segrè family, 3–10, 26–28
Serber, Robert, 176, 182, 187, 235, 273
Sereni, Emilio, 32, 302n.2
Sereni, Enzo, 32, 302n.2
Severi, Francesco, 38, 39–40, 54
Siegbahn, Kai Manne, 268, 272, 318n.24
Siegbahn, Karl Manne George, 268, 318n.24
Simili, Silvestro, 216, 217–19, 221
Smithsonian Institution, 134
Smorodinsky, J. A., 263
Società cartiere tiburtine (SCT). *See* Paper mill
Solvay conference (1933), 88
Somigliana, Carlo, 80
Sommerfeld, Arnold, 63, 68, 99–100
Soviet Union: atomic bomb, 214, 237, 238; disarmament conference, 265; Hitler and, 139–40; Josephy in, 100; 1956 visit, 5, 261–63; Pontecorvo in, 236–37, 246, 262; scientists visiting Berkeley, 264
Specchia, O., 81
Special Engineering Detachment (SED), Los Alamos, 189, 202–3
Spectrography, 56, 64–65, 84, 87–88, 171, 197–98, 256
Spiess, Fred Noel, 278
Spiro, Elfriede. *See* Segrè, Elfriede Spiro
Spiro family, 97, 99, 127, 149
Sproul, R. G., 247
Stalin, J., 139–40
Stanford University, SLAC, 284
Stapp, Henry, 230
Stark effect, 55
Staub, Erika, 193
Staub, Hans H., 187, 193, 197, 207
Stearns, Joyce C., 208
Steiner, Herbert, 264
Stern, Otto, 68–72, 87, 243, 255, 269; Kuhn interviewing, 281; and Lawrence's cyclotron, 112; and Nobel Prize, 151
Stern Institute, 68–72, 243, 268–69
Stoughton, Ray, 169
Strassmann, Fritz, 91, 151
Strauss, Lewis, 314n.20
Sugarman, N., 317n.14
Superconductivity, 2, 265
Sweden: king of, 270; Stockholm, 268, 270–71
Switzerland, 44–45, 85; bank account, 86, 140, 142; Basel conference, 233; Breuil basin, 45, 233; Cosmic Ray Lab, 233
"Swollen atoms"/Rydberg states, 83–84, 150, 243
Szilard, Leo, 92, 132, 150, 180, 203

Tacke-Noddack, Ida W., 91, 115, 116, 117–18, 124, 306n.24

Tamm, Igor, 229, 262–63, 265–66
Tarchiani, Alberto, 208
Tasso, Torquato, 121
Tata (Austrian nanny), 14–15, 33
Taylor, Geoffrey, 192, 197
Technetium (element 43), 116–19, 124, 127, 243, 308n.12; on father's tomb, 215; isotopes, 116–19, 128, 293–94, 308n.11
Tel Aviv University, 283, 298
Telegdi, Valentin, 265
Teller, Edward: in Berkeley, 176; and Fermi, 63, 251–52; and hydrogen bomb, 237–38, 251; at Los Alamos, 187, 191, 193; at Oppenheimer hearings, 251, 309n.9
Teller, Mici, 191, 193
Terkel, Amir (grandson), 285
Terkel, Joseph (son-in-law), 285, 286
Terkel, Vivian (granddaughter), 285
Thomas, L. H., 135
Thomas-Fermi statistical method, 50
Thompson, Llewellyn, 214
Thompson, Stanley G., 169
Thomson, G. P., 250
Thomson, J. J., 68
Thornton, R. L., 135, 144, 209, 237, 241
Tieri, Laureto, 103
Tillman, J. R., 95
Tinkham, Michael, 234
Tissandier, Gaston, 20, 280
Tisserand, Eugene, 214
Tivoli, 1–2, 9, 12–23; Amelia Segrè street, 23; contemporary, 14, 253, 291; Elfriede and, 232–33; postwar visit, 223–24; sulfur baths near, 18, 215, 223, 291; vacations, 23, 25, 40, 128; Villa d'Este, 1–2, 10, 17–18, 31–32, 223, 291; Villini Arnaldi, Villino B Maria, 3, 12–14, 23, 253. *See also* Paper mill
Todesco, G., 103
Trabacchi, G. C., 87, 245
Tracer technique, 115
Transuranic elements, 91–92
Treves, Amelia. *See* Segrè, Amelia Susanna Treves (mother)
Treves, Elisa Orvieto (grandmother), 10
Treves, Emilia Finzi (aunt), 2, 11
Treves, Giuliana (cousin), 11
Treves, Guido (uncle), 11–12, 29, 224, 225
Treves, Jacopo (uncle), 11, 14
Treves, Marcella (cousin), 11, 19, 225
Treves, Marco (cousin), 11, 33
Treves, Marco (grandfather), 10, 302n.5
Treves, Silvia (cousin), 11, 19, 225, 302n.5
Treves family, 10–12, 302n.5
Treves villa (Marignolle), 10, 19, 29–30, 225, 253
Trilussa (Carlo Alberto Salustri), 127–28
Troccoli, Signor, 89
Truman, Harry S, 203
Tulsa, oil industry, 159–60, 237
Tumedei, Cesare, 290
Turner, L. A., 163
Tuscany, 292–93

Uhlenbeck, G. E., 82, 272
Union of Producers and Distributors of Electric Energy (UNIPEDE), 298
United Nations, 204
United States, 213, 247, 288, 292; AEC, 169, 246, 247, 284; Atomic Energy Act, 245; business and nuclear energy, 245, 314n.20; citizenship, 204–5, 237; disarmament conference, 265; foreign aid, 283; immigration law, 129, 145; Information Service, 282; McMahon Act (1946), 246; and patents in nuclear power, 245–47; State Department, 281; Supreme Court, 257; visas, 128–29, 141, 145–46; visit (1933), 82–83; visit (1935), 101–3; visit (1936), 111–14; visit (1938-1943), 128–29, 130, 131–78; White House, 281; in World War II, 149, 171, 172, 173, 177–78. *See also* Military; *individual states, cities, and universities*
Universities: Columbia, 102, 111–12, 132, 146, 167, 240, 264; Ibadan (Nigeria), 276; Purdue, 160–61; Tel Aviv, 283, 298; Washington (St. Louis, Missouri), 165, 208, 209. *See also* University of . . .
University of California, Berkeley, 133–78, 207–84 passim; Academic Senate, 275, 277; administrative work, 277–78; Birge tribute, 310n.21; committees, 275, 277, 280–81; Faculty Research

Lecture, 275; history of science, 280–81, 298; loyalty oath, 234–36, 238–39, 246; Nobel Prize winners' privileges, 270, 272; and patents, 246–47; physics, 154–55, 161–62, 172–73, 230–31, 233–36, 243, 280, 283, 288, 310n.21; Regents, 234–35, 236, 240, 246, 279; retirement age, 288; Russians visiting, 263–64. *See also* Nuclear Physics; Radiation Laboratory, Berkeley
University of Chicago, 208, 209; and anti-Semitism, 103; Fermi at, 47, 150, 175, 205–13 passim, 239, 248; Fermi lectures, 287; Fermi memorial service, 252; Fermi papers, 260; plutonium work, 175–76; Seaborg at, 175–76, 211; Telegdi at, 265
University of Chicago Press, 253
University of Ferrara, 81
University of Illinois, Urbana, 239, 242
University of Michigan, 82–83, 101, 112
University of Palermo, 100–101, 102, 104–30; chair of theoretical physics, 125–26; honorary degree, 268; politics, 105, 106; return offer, 208; written examinations, 110–11
University of Pavia, 4
University of Rome, 37–56, 118–19, 289–90. *See also* Physics Institute
Urey, Harold C., 161, 167
Uruguay, 250; Rimini cousins in, 146, 148, 174, 221, 250, 284; Rosa Mines in, 284

Vaccari, Lino, 17
Valle, C., 81
Vallery-Radot, René, 280
Veksler, Vladimir, 212, 263
Verdi, G., 228
Villa d'Este, Tivoli, 1–2, 10, 17–18, 31–32, 223, 291
Volterra, Vito, 11, 52–53, 54, 63, 115
von Neumann, John, 187, 193, 197

Wahl, Arthur C., 164–67, 175, 195, 197, 208, 246, 317nn.14,15
Warren, Earl, 234
Washington University, St. Louis, Missouri, 164, 208, 209
Watson, Thomas, Jr., 240
Weak interactions, 88, 264–65
Weinberg, Steven, 234
Weinrich, Gabriel, 265
Weisskopf, Victor, 61, 187, 193
Weizsäcker, C. F. von, 137
Westcott, H. C., 93
Wheatstone, Charles, 293
Wick, Gian Carlo, 71, 83, 111, 125, 126, 233, 235
Wiegand, Clyde E., 242, 260, 261, 270, 316n.9; and antiprotons, 169, 256, 257, 258, 260, 316n.9; at Berkeley, 157, 172, 176, 206, 211, 227, 230, 248, 256, 258, 259, 283; at Los Alamos, 178, 189
Wigner, Eugene, 61, 229
Williams, J. H., 188, 201
Williams, Wynn, 93
Wilson, Robert R., 135, 188, 284
Wolf Prize, 265
World War I, 2, 13, 16, 28
World War II, 180–81, 207, 222–23; atomic bomb, 176–77, 179–206, 209; Bohr and, 193–94; "enemy aliens," 172, 175, 179; events leading to, 79, 92, 103, 132, 139–42; Germany's surrender, 199–200; Japan's surrender, 203; Pearl Harbor, 170; start, 144, 160; U.S. in, 149, 171, 172, 173, 177–78. *See also* Hitler, A.
Wu, Chien-Shiung, 136–37, 153, 156, 157, 264, 265

Yang, Chen Ning, 264, 265
York, Herbert F., 230
Ypsilantis, Tom, 230, 248, 256–61 passim, 270, 316n.9
Yukawa, Hideki, 230

Zacharias, J., 102
Zeeman, Pieter, 65–70 passim, 84, 99, 293
Zeeman effect, 64–68 passim, 83–84, 150
Zeeman lab, 65, 66–67, 78–79, 99, 128, 136
Zeitschrift für Physik, 61, 62, 65

Compositor:	Impressions, a division of Edwards Bros.
Text:	10.5/14 Janson
Display:	Janson
Printer and Binder:	Edwards Bros.